Lecture Notes in Computer Science 737

Edited by G. Goos and J. Hartmanis

Advisory Board: W. Brauer   D. Gries   J. Stoer

Jacques Calmet   John A. Campbell   (Eds.)

Artificial Intelligence and Symbolic Mathematical Computing

International Conference AISMC-1
Karlsruhe, Germany, August 3-6, 1992
Proceedings

Springer-Verlag

Berlin Heidelberg New York
London Paris Tokyo
Hong Kong Barcelona
Budapest

Series Editors

Gerhard Goos
Universität Karlsruhe
Postfach 69 80
Vincenz-Priessnitz-Straße 1
D-76131 Karlsruhe, Germany

Juris Hartmanis
Cornell University
Department of Computer Science
4130 Upson Hall
Ithaca, NY 14853, USA

Volume Editors

Jacques Calmet
University of Karlsruhe
Postfach 69 80, D-76128 Karlsruhe, Germany

John A. Campbell
Dept. of Computer Science, University College London
Gower Street, London WC1E 6BT, United Kingdom

CR Subject Classification (1991): I.1-2, G.1-2

ISBN 3-540-57322-4 Springer-Verlag Berlin Heidelberg New York
ISBN 0-387-57322-4 Springer-Verlag New York Berlin Heidelberg

© Springer-Verlag Berlin Heidelberg 1993
Printed in Germany

Typesetting: Camera-ready by author
Printing and binding: Druckhaus Beltz, Hemsbach/Bergstr.
45/3140-543210 - Printed on acid-free paper

Foreword

The present volume reproduces the papers, updated in some cases, given at the first AISMC (Artificial Intelligence and Symbolic Mathematical Computations) conference which was held in Karlsruhe, August 3–6, 1992. This was the first conference to be devoted to such a topic after a long period when SMC made no appearance in AI conferences, though it used to be welcome in the early days of AI. Some conferences have been held recently on mathematics and AI or on applications of symbolic computing in AI; however, none covers similarly the topics of this conference.

Because of the novelty of the domain, authors were given longer allocations of time than usual in which to present their work. As a result, extended and fruitful discussions followed each paper. The introductory item in this book, which was not presented during the conference, reflects in many ways the flavour of these discussions and aims to set out the framework for future activities in this domain of research.

There was a unanimous demand from the attendees to hold conferences of this kind regularly. We plan to organize them biannually in even years.

This conference was organized mainly by the Institute of Algorithms and Cognitive Systems at the University of Karlsruhe. Sponsorships and financial supports from DFG, DEC Germany and IBM Germany are thankfully acknowledged.

We also express our gratitude to the members of the program committee and to the organizers, particularly to Karsten Homann and Ono Tjandra.

August 1993   Jacques Calmet and John A. Campbell

Conference Committee: Jacques Calmet (Germany – Conference Chairman)
John A. Campbell (UK – Program Committee Chairman)
David Y.Y. Yun (USA - Publicity Chairman)

Program Committee: L. Aiello (Italy)
J. Calmet (Germany)
J.A. Campbell (UK, Chairman)
E. Engeler (Switzerland)
J. Johnson (UK)
L. Joskowicz (USA)
M. Karpinski (Germany)
J.-L. Lassez (USA)
R. Lopez de Mantaras (Spain)
R. Loos (Germany)
A. Miola (Italy)
A.C. Norman (UK)
R. Parikh (USA)
M. Pohst (Germany)
Z.W. Ras (USA)
E. Sacks (USA)
E. Sandewall (Sweden)
S. Stifter (Austria)
D. Weld (USA)
D.Y.Y. Yun (USA)

Local Organizers: K. Homann, I.A. Tjandra, W. Fakler

Contents

Artificial Intelligence and Symbolic Mathematical Computations

J. CALMET
University of Karlsruhe

J.A. CAMPBELL
University College London

Abstract. This introductory paper summarizes the picture of the territory common to AI and SMC that has evolved from discussions following the presentation of papers given at the 1992 Karlsruhe conference. Its main objective is to highlight some patterns that can be used to guide both sketches of the state of the art in this territory and suggestions for future research activities.

The structure of the paper mirrors the emerging patterns of interaction between AI and symbolic mathematical computing (SMC). We begin with some historical considerations, to put the past relationship between the two subjects into perspective. 1971, the year in which the first major conference on computer algebra was held, can be regarded as the year in which the two subjects started to diverge significantly. The section that follows mentions the topics in AI that have never quite lost touch with SMC. We then consider the offerings that AI can make to SMC and vice versa, with a short intermediate section arguing for the idea of knowledge representation as a bridge between the two subjects. A concluding section looks at what we consider to be the most rewarding possibilities for future developments in research.

We make no claim that our treatment of the past, present or future is exhaustive. This introduction is best seen as a sketch of a survey paper. A more detailed version will appear elsewhere.

1 An Historical View

1.1 The first 15 years of modern AI

A watershed of AI was the summer workshop held at Dartmouth College in 1956, at which many of the founding fathers of the subject met to compare their work and to define an agenda for the near future [Mc]. During the 15 years that followed (until the time of the 1971 Los Angeles computer algebra conference), it was normal to find papers on SMC in AI sources and conferences. This is not surprising in retrospect, because of the mathematical backgrounds of the early AI researchers and because symbolic mathematics offered some problems of a manageable size where the criteria for successful solution were clear-cut.

Shortly after 1971, papers on symbolic computation disappeared almost completely from the places where AI material was published. Why was this? The answer, which we discuss below, has several parts, but a common strand of the explanation is that research on both sides of the fence was too successful.

The problems considered in early AI were basically search, computational logic (for theorem-proving and construction of plans, e.g. for carrying out searches, by steps similar to successive steps of a proof), pattern-matching, and symbolic mathematics. (Natural-language processing has also been a long-lived partner of AI, meeting up with AI at conferences but otherwise travelling mainly on a separate parallel track). What they had in common, and what distinguished them at the same time from early non-AI programming in Fortran and its predecessors, was that conventional programming dealt with algorithmic methods while AI relied on *heuristics*.

In the particular case of symbolic mathematics, it is not hard to find algorithmic applications (e.g. in differential calculus) that would involve novel uses of the computing technology of the 1950s. But the AI applications were heuristic. Good examples are Gelernter's geometrical theorem-prover [Ge] and Slagle's work on integral calculus [Sl]. These represented the state of the art in the early 1960s, when finding good heuristics and therefore making something new work for the first time was more important than thinking about regularities in what was found.

The best example of a change of emphasis towards looking at the regularities is Moses' [Mo] integral-calculus system SIN, which was publicized first in his PhD thesis at the beginning of 1967 and which was absorbed later into the development of the MACSYMA system for symbolic mathematics. There were basically two kinds of regularity. The one that was historically more important for the separation of AI and SMC was that the study of the heuristics collected for SIN and from other work on programs for integral calculus led to an *evolution away from heuristics and towards algorithms*. As integration in finite terms became more of a mathematical specialization and hence more algorithmic, it became less attractive as a topic for AI referees and hence AI conferences.

This story has been repeated for other areas of AI. It is clear from a reading of series of AI conference proceedings (the IJCAI series is the best source) from the late 1960s until today, that AI is a moving target: its contents change with time, and in particular some topics (e.g. symbolic mathematics) that were once popular disappear completely from later conferences. The paradox of AI is that "AI exports its successes". Success implies not only that something works for users but also that it works efficiently. Algorithms are usually more efficient than heuristics for the same job, so that there is a social pressure to turn good heuristics into good algorithms. And there seems to have been a tacit agreement, in many topics that were once focuses for AI research, that when something is algorithmic and no longer heuristic, it is not really suitable for reporting at AI conferences.

The second regularity that showed up in the heuristics collected for SIN, the system DENDRAL [Bc] for determination of molecular structure from experimental data, and several less well-known projects from the late 1960s onwards, was the *rule* structure, now standard in expert systems, for individual heuristics. In fact, it has been said that the SIN-MACSYMA system and DENDRAL were the first modern expert systems (through this is not an accurate picture of the former).

The general success of the rule-based (and therefore still heuristic) approach in AI forced attention in research towards the examination of rules, and towards a common labelling for rules and other schemes such as inheritance networks or semantic nets. Because they all held knowledge in some form, the label of "knowledge representation" emerged. The algorithmic direction of work on MACSYMA and similar software at the same time as this development led symbolic research. Hence the divorce proceedings with AI that started in 1971.

1.2   The first 20 years of SMC

It is convenient to date the origin of SMC to the two 1953 Master's theses of Kahrimanian [Ka] and Nolan [No], who wrote assembly-code programs to perform symbolic differentiation of simple mathematical expressions. These first examples were algorithmic and not heuristic.

After that piece of history, the first attempts to write software to solve serious problems were in celestial mechanics, and were associated with the group of A. Deprit [De]. These attempts, from the late 1950s, were motivated by the fact that accurate computations of trajectories of astronomical bodies required extremely long symbolic expressions, and that manual calculations could take a significant fraction of a mathematician's lifetime and could contain undetected errors. The computations typically involved substituting Poisson-series expressions in several symbolic variables into differential equations, and determination of solutions as series expansions.

The mainstream of symbolic computation during the 1960s was in application-oriented systems, either for general algebraic manipulations or specific extensions to high energy physics, celestial mechanics, general relativity or group theory. Many well-known modern systems such as REDUCE and CAYLEY started their development during this period.

As in the case of the 1951 work, there were no significant heuristic components in most of these systems. The SIN-MACSYMA activity was an isolated exception. Work on SMC could still be done and reported in circles close to AI because of the past honorary membership (dating from the choice of symbolic mathematics, mainly integral calculus, as a source of problems for AI after 1956) of the subject in AI. We have given part of the explanation for the post-1971 separation of AI and computer algebra in the sub-section above. The rest of the explanation is that the 1971 computer algebra conference and the later foundation of the ACM special-interest group SIGSAM and its Newsletter gave the work its own home; there was no further pressure to look for publication in outlets that were primarily devoted to AI.

1.3   MACSYMA as an expert system?

As we have said above, the SIN-MACSYMA work has been quoted as an early example of the development of an expert system (e.g. in [Ba]). To the extent that the heuristic knowledge in SIN could be read off as rules from the source code, this would be true. In practice it was not totally false, but one needed to be an expert excavator of LISP code to detect the traces of truth in the statement. Also, some of the heuristic information was more about pattern-matching than the use of rules. Finally, the architecture of the software was not set up to allow the rule-like material to be separated easily from the code that made use of it, i.e. there was no obvious way to distinguish a rule base in SIN from an inference engine.

DENDRAL has a much better claim to reflect the structure of an expert system, though this identification of the parts of the structure in the DENDRAL code would still not have been simple.

SMC did not invent expert systems, unfortunately. Nevertheless, it can claim to have come close. For example, in REDUCE and other early programs, the basic algorithm for symbolic differentiation must be supplemented by knowledge about the derivative $f'(x) = g(x)$ of any new function $f(x)$ that the programs have to differentiate. Each such pair is declared as a form of substitution, e.g. FOR ALL X LET DF(F(X),X)=G(X). The pairs are held in a format that allows the original meaning to be read off, and that distinguishes this knowledge from the code for the programs themselves.

This suggestion is a jeu d'esprit rather than a serious contribution to history - but it counters the belief in some quarters in AI that a particular symbolic computing system (MACSYMA) was almost the first expert system.

1.4 Why has a new convergence between AI and computer algebra been happening?

The original divorce between AI and computer algebra was driven by the fact that AI continued to focus on and develop its understanding of the notion of heuristics while symbolic computation became more and more algorithmic. The algorithmic emphasis was increased by a migration of professional mathematicians into research on computer algebra.

We have now reached a point where concentration on algorithmics rarely leads to qualitatively new capabilities in symbolic computing systems. On the other hand, users of the systems can still be seen doing pen-and-paper calculations, or trying alternative ways of expressing instructions to the systems, because some of their needs have not been met. Generally, the reason why the systems do not service these calculations (or service them well) is that the needs require heuristic treatment. For example, the control of intermediate expression swell is in general still a heuristic exercise.

Heuristic knowledge about symbolic mathematics exists in many places. The number of attempts to capture it systematically and put it to work via AI-influenced methods is not yet large, but there is an increasing trend towards doing such research. The contribution of H. Hong to this book is one excellent example of the trend. There are potential payoffs for AI, in the evaluation of its methods in a new area of application, and for users of symbolic computing systems, who can expect qualitatively new kinds of support for their calculations when the systems become "intelligent assistants".

From the AI side, some of its topics for research and applications have a mathematical flavor, after quite a long period since 1971 when there was not much contact between mathematics and AI. The AI topic that has had the greatest effect in this direction is qualitative reasoning, which typically requires the use of a mixture of formal operations and heuristics on mathematical entities such as equations. We pay more attention to qualitative reasoning in section 2.3.

Qualitative reasoning is simply the most visible instance of the more general effect that formal (mathematical) systems of description of phenomena in different corners of AI lead to non-trivial calculations, which symbolic computing systems can perform, either immediately or after relatively simple changes or additions. Another such instance [Wa], in which the systems can be used immediately, emerges from a method for proving geometrical theorems (a classical AI exercise) constructively by the manipulation of simultaneous linear equations. A further area in which future applications of symbolic computation in AI can be expected is in robotic path planning and varieties of the general "piano movers" problem.

In such topics, at present, the knowledge expressed in AI systems and software can be used to produce statements of problems that are then (most often after some manual translation or massaging) suitable as inputs to symbolic computing systems. A challenge for the near future is to integrate the two types of systems and cut out as much as practicable of the present human intervention.

There is a still further extension of the ideas expressed immediately above. It is increasingly common for conceptual structures in AI to be expressed in mathematical language, usually algebraic and quite often in terms of definitions that specify some unusual algebra. Here, one is interested not just in doing symbolic calculations, e.g. in the sense of path planning or qualitative reasoning, but in exploring the internal features of the algebra and its consequences for the AI problem that gave rise to it. In this sense, symbolic computing systems are possible components of an "AI algebraist's laboratory". Integrating them with other relevant software produced from AI research is basically the same kind of challenge that we have mentioned in the previous paragraph.

The final reason for the existence of converging paths between AI and SMC is that specialists in knowledge representation are becoming interested in capturing and using the considerable amount of heuristic knowledge that mathematicians possess about suitable symbolic structures (e.g. differential equations). Apart from the technical interest of doing this, there is the attraction that success will make it possible to build practical systems that combine operational mathematical knowledge with the calculational capabilities of present systems like MACSYMA and REDUCE. Any substantial progress in this area will give such systems their first significant boost in functionality since the early 1970s.

2 Long-lived AI areas of special interest

2.1 Theorem-proving

Theorem-proving is probably the first topic that comes to mind when one is looking for examples of the interaction of AI and mathematics (see for instance [Bu]). This opinion can only be reinforced if one reads the proceedings of the first relevant topical conference [Do]. Two of the main concepts involved in theorem-proving are logic and deduction. So many conferences and similar published activities have been devoted to these topics that it was decided not to give them any special emphasis within the scope of the 1992 Karlsruhe meeting. Even so, it is difficult (and not desirable) to avoid them completely. An example is the paper of Hähnle, which deals with the constraint tableaux upon which one recent theorem-prover has been designed. Another is the contribution of Cioni et al., on deduction, using the sequent-calculus method which has recently attracted attention in several areas of computer science.

2.2 Expert systems

We have discussed MACSYMA and the claims that have been made [Ba, Ha] for its expert system-like properties, in section 1.3. In summary: while it is not an expert system, it contains a significant amount of mathematical expertise that may in principle be extracted by inspection and reverse-engineered into some future mathematical expert system.

In the historical account in section 1.2, we have pointed out that the SIN package for symbolic integration has motivations that were clearly in the style of 1960s AI. Today, however, its successors use purely algorithmic methods of solution, which are practicable and efficient even for the most difficult definite integrals. This situation is nevertheless somewhat unusual even when one considers just the typical applied mathematicans' set of problem-solving techniques. An example is the algorithmic construction of the closed-form solution of (linear) ordinary differential equations (ODEs). Following the work of Ulmer (see [Ul] and the references therein to the previous work of Singer), it is known that algorithmic solutions are available for ODEs of any order, n. However, even for the lowest orders ($n = 2$ and 3) which are being implemented at present, the algorithms are very inefficient timewise. Therefore, for any practical method embodied in software, one may predict safely that any solution will involve an expert-system scheme mixing algorithmic and heuristic methods. The DELIA system [Bo] for symbolic solution of ODEs already displays some of this structure, and one can predict that a similar approach will be required for future developments in the ODE area.

Although symbolic integration can now be treated algorithmically, there is a case for returning to the study of integration heuristics if one wants to model human (e.g. students in calculus sources) understanding of the process, help human learners by representing this understanding in slow motion (by displaying traces of program execution) during teaching, and/or experiment on a well-defined problem area with AI techniques (e.g. for machine learning) that are strongly heuristic. Good examples of work of this kind are the LEX system [Mi], where learning methods are used to solve integral problems, and CAMELIA [Vi], where plans are generated within the integration component of a computer algebra system.

If we make reference to the situation for numerical computation, for which numerous expert-level packages area available (a well-known example is ELL-PACK [Ho]), and take into account the present trend to integrate symbolic and numerical methods, one may conclude that expert systems will become more and more present in SMC. Nevertheless, there is still the 1971-199? cultural gap between AI and computer algebra, which will have to be bridged before such activities become common.

Another link between AI and expert systems can be found in the design of systems that enable users without deep mathematical knowledge to compute in highly non-trivial domains [Ca].

2.3   Qualitative reasoning

There are several names for this subject, indicating differences of emphasis in one basic idea. Other significant names are qualitative physics, naive physics, and commonsense reasoning about processes that are (broadly speaking) non-social. The origin of the subject, within AI, was cognitive and philosophical. It took up the observation that humans can make consistent and realistic models of phenomena (behavior of liquids, everyday economic systems, etc.) without having any access to the terms or the modelling schemes used by technical experts on those phenomena, and asked the question "What must we do in order to capture this knowledge in a form in which computation can reproduce human reasoning?" The "Naive Physics Manifesto" of P.J. Hayes [Hn] is still be best description of this outlook.

Although the philosophical flavor of the original emphasis suggested that we would need a wide variety of representations, including logics, for this job, the subsequent developments have become more technically mathematical - but therefore closer and more accessible to SMC. For example, given a mathematical description (usually a system of equations) of a problem where an exact analytic solution is not possible, it is sometimes possible to learn all that one needs to know in a specific situation (e.g. for a complicated physical system with multiple coupled components, which way does component Z move when component A is moved downwards?) by assuming that the variables have values drawn from the set $(+, 0, -)$ rather than being real or integer. Several large AI programs exist to perform such computations, and they are usually underpinned by formal theories whose contents one could never have guessed from a "naive physics" standpoint. "Qualitative process theory" is one such example [Fo]. A good concentrated source for results of this kind is a book [Hb] that reprints a collection of papers that have defined the structure of the field.

When such computations become complex, the AI programs use methods that are standard in SMC, and in some cases they have reinvented SMC wheels. There is certainly scope in future for this kind of effort to be replaced by efforts devoted to integration of symbolic qualitative reasoning and SMC packages into larger software systems - which should also increase the flexibility of what the qualitative reasoners can do.

Lately, the emphasis in qualitative reasoning has moved away from using a calculus of qualitative variables $(+, 0, -)$, i.e. a primarily mathematical emphasis, to the understanding of qualitative choices and modelling in design, particularly engineering design. This overlaps longer-lived lines of research into the qualitative understanding of kinematics (e.g. in mechanisms such as clocks) and into diagnosis of faulty operation (which sometimes follows from faulty design) of mechanisms, circuits etc. Good examples of work of these kinds are given by [Fa] and [Ki] respectively. As they indicate, there is still a substantial mathematical content, but it is governed by the problems being studied rather that driving the study itself. The survey by P. Struss in this book illustrates the trend further.

This change of emphasis has broadened the range of types of knowledge that an automated qualitative reasoner should be able to handle. For example, in addition to whatever mathematical framework is contained in a design problem, it must be routine for expert system-like knowledge to be included and used. Knowledge that a designer normally draws in diagrams or other forms that express kinematic information, also, must often be treated - which suggests that the resulting computing system should have access to simple logics. (It is no accident that many groups working on knowledge-intensive problems in engineering have chosen to use the language Prolog, because it fits this last need very well.)

The conclusion that one can draw, from the SMC perspective, is that symbolic mathematical methods can extend the functionality of the real or projected systems of present-day researchers on qualitative reasoning, and/but that the first challenge to be met here is (as in several other areas where AI and SMC overlap) the integration of traditional SMC modules with the software that supports expert systems, simple logic (first-order predicate calculus or the Horn-clause subset of it that Prolog uses), and other methods for formal manipulation (for example, Qualitative Process Theory [Fo]).

3 Modern offerings from SMC to AI

Mainly because of their relevance to advanced robotics, two methods of SMC are at present playing an important part in AI. They are Gröbner bases and cylindrical algebraic decomposition.

An early use of symbolic methods was reported in [Hu], which examined the problem of deciding whether the trajectory of a mechanical system in a dynamically-changing situation should be totally recomputed or merely updated by correction. More routinely at present, computer algebra systems are used to complete many "service" computations for robotics or machine vision. To integrate such capabilities directly into robots appears rather utopian for the near future, because of the size of the systems. However, the situation may improve if more attention is given to a track that has been very little investigated until now: design of the basic algorithms directly for the chip or VLSI level.

3.1 Gröbner bases

Gröbner bases can be thought of as a canonical rewriting system for equational theories. They thus provide a powerful technique to solve systems of polynomial and algebraic equations. Their applications range from economics, where they solve systems of equations that express given models of world economic growth, for example, to robotics. The papers of Pfalzgraf and of Monfroy in this book illustrate the robotic applications. Gröbner bases are particularly useful to detect the singular configurations of a robot system. A more generic characterization is to state that they are very useful to solve many problems in geometric reasoning. Their use in such domains is at present well established.

3.2 Cylindrical algebraic decomposition

This domain is a very good illustration of the fact that there is a significant interface area between SMC and AI. The method of cylindrical algebraic decomposition (see the contribution of Hong) was first designed by Collins to solve the quantifier-elimination problem of Tarski, a problem in logic. It was then realized that the same technique could help to solve some problems in robotics. The method is algorithmic but very inefficient (its complexity is doubly exponential). However, in practice AI can come to the rescue, because suitable heuristics can be used to speed up the running time, as Hong shows. This is therefore an example where each of SMC and AI has been able to make some contribution to the usefulness of the other.

3.3 Differential and Integral expressions

These are expressions that often arise when one is dealing with engineering problems that occur in some domains of AI, for instance spatial geometrical or kinematical reasoning. Such problems are often translated into mathematical form as integral or differential equations. Except for partial differential equations, for which a general theoretical background for the construction of symbolic solutions is still lacking, they can be solved by well-understood and efficient constructive algorithms. Hence, as we have pointed out already, these problems can be tackled by routine use of standard SMC facilities when they arise in AI contexts.

3.4 Computational group theory

This is a domain for the future, but its outlines are evident in the present. It may well become at least relevant, and even important. The underlying idea is that group theory may be a good framework for looking at certain problems in modelling and design, because of its potential for expressing the inner structure and symmetries of objects in a consistent and generic manner. The same line of reasoning may well be extended to other domains of AI where the properties of entities allow natural descriptions that have some correspondence to the language of group theory. This suggests that, in addition to questions of integration of mainstram SMC systems with AI software, it is worth considering how to achieve a similar integration for specialized group-theory symbolic systems such as CAYLEY [Cn] (the next version will be named MAGMA) for instance that are equally powerful in their own area of SMC.

4 Knowledge representation: a possible bridge between SMC and AI

When AI and SMC took different directions after 1971, the main reason for the divergence could be summed up as "heuristics vs. algorithms". But there is another and probably more fundamental way of saying the same thing. This is that, although both subjects were concerned with knowledge and how to represent it, the representations that were natural for their needs became progressively more different in the years after 1971. In SMC the question of "representation" was not often mentioned at all: researchers simply got on with the job of writing the relevant (algorithmic) knowledge into their standard (procedural) kinds of programs. The relevant representational problems were basically problems of efficient use of standard data structures. Also, there was no particular reason to make an explicit separation between "knowledge" and programs.

The contrast with the story of AI in the 1970s is instructive. First, because of the success of DENDRAL and other experiments, the concept of a rule-base (not necessarily an efficient way to support computation, but a way of making it possible to display the information clearly and edit it) to hold heuristic knowledge evolved, while SMC was in the business of converting heuristics to algorithms and burying them deeply (efficiently) inside programs. Shortly after rule-bases and rule interpreters started to grow into "expert systems", the value of frames and inheritance networks for holding knowledge that did not fit neatly into the rule format became recognized. It was appreciated that all these structures belonged to a common abstraction because of their common purpose. In fact, the name that it acquired simply expressed this purpose: *knowledge representation*.

AI and SMC are coming together again at present, for a variety of particular and application-oriented reasons. But all of these reasons have a single rationale: extending the range of SMC's capabilities by making use of heuristic knowledge.

Because the expert-system paradigm is well known, SMC researchers have used it first in trying to find frameworks for their heuristics. There have certainly been consequent successes, but in other situations progress has been slow because rules are not always sufficient to express mathematical knowledge. At this point, AI can offer a way forward for SMC. Several different types of knowledge representation are relevant for symbolic mathematical knowledge. A first step for SMC specialists seeking to take advantage of what AI is offering is to become familiar with the AI picture of what "knowledge representation" actually covers. Two useful sources for general background are a collection of important early papers [Br] and a more recent book [Ri] that aims to be a survey of representations. But the number of interesting representations is still increasing, so it is helpful to follow up these references by looking at the papers in the knowledge-representation sections of proceedings of the most important series of AI conferences (IJCAI and AAAI).

The bridge that we are suggesting here can carry traffic in both directions. From SMC towards AI, the most promising traffic is in ideas for applications. In and around kowledge representation in AI, one can easily get the impression of solutions looking for problems, or suggestions looking for tests of feasibility. SMC is a good potential source of both problems and tests. More generally, a way of connecting SMC and AI through knowledge representation is to ask (from the SMC side) "What is our mathematical knowledge, and how can we go about representing it, for practical use, in computing systems?" The question was asked before 1971, but in retrospect the answer given at that time was too limited because of the focus on algorithmics. The time is now right for the question to be used again, as a guide to setting a research agenda common to AI and SMC.

5   Modern offerings from AI to SMC

5.1   User interfaces

Today it is regarded as normal good practice never to propose a system to users without paying attention to the design and provision of a good user interface. It is also not unusual to find that the providers describe these interfaces as "intelligent", which is partly an acknowledgement of the impact of AI techniques and ideas on the design of such modules. (Systematic research on human-computer interaction occurred first within AI, though HCI now has a life of its own: another example of the "AI exports its successes" proverb that we have mentioned in section 1.1)

In the field of SMC one may distinguish between two approaches to interfaces. The first one is found in systems whose main goal is to be successful commercial product. The interface then emphasizes graphical and iconic capabilities, probably because this is seen as a good marketing feature. In addition, recent commercial programs use a "notebook" metaphor in their user interfaces. The second approach is to provide systems for intended use by people who are highly skilled in other particular domains. The contribution of Clarkson in this book consider the use of MACSYMA by physicists as a prototype of the situation. The role of the interface is then mainly to help a new or recent user to become acquainted quickly with what the computing system can offer to amplify his or her special skills. The motivation is that such users are not often willing to invest time in learning all the details of how to use a new system. They are typically familiar with some existing package or language and will use it with brute-force methods rather than sacrifice the time that is needed to master another one, even if the advantage of that one is that it contains new features that are highly relevant to the users' current problems.

Another promising line of research is to design an interface that allows a user to access the vast collection of mathematical and statistical software that is now available to scientists and engineers [Bi], e.g. through the expert-system approach that has been pioneered in REX [Ga]. Although [Bi] is intended mainly for scientific numerical computing, there is no reason why similar projects cannot be tackled for SMC resources also.

5.2 Transforming logic into algebra

We have already mentioned one prototype of this topic in section 3.2 in connection with the quantifier-elimination problem, and in section 1.4 where the relevant issue was the proving of geometrical theorems. There are many further possible instances of the topic, as there are many problems of logic that can be transormed into algebraic problems that can then be solved by algorithmic methods.

Probably a better-known example than either of those that we have mentioned is the PRESS system [St], which was built to solve transcendental non-differential equations symbolically. The main motivation was to explore some AI ideas about controlling search in mathematical reasoning by using meta-level descriptions and strategies. The main tool of PRESS is in fact meta-level inference.

The transformation from logic to algebraic problems is found in still further different settings such as rewriting techniques for equational theories, and type inference mechanisms for computer algebra systems, to quote just two that receive attention later in this book.

Two of the advantages of such transformations are obviously the potentiality of algebraic algorithms to compute efficiently (unlike the typical state of play in logic-based problems) and the simplification capabilities associated with these algorithms. SMC is used routinely to perform huge computations whose statements or descriptions are very often straightforward. Analysis and observation of the internal details of such computations have shown that most of the operations that are performed consist of substitutions and simplifications. These parts of the best SMC systems have therefore had considerable effort at optimization spent on them by their developers. This is an asset which can be expected to pay off strongly whenever one can manage to transform a logical problem into an algebraic one.

5.3 Constraint-based programming

SMC is at home with computations on suitable systems of equations, e.g. linear equations. Considerable time has been devoted to building efficient parts of SMC software for manipulating and solving such systems. In essence, constraint-based programming is the same kind of activity with the important difference that inequalities can occur, in place of or in addition to equations, in the systems.

Many examples of constraint-based programming deal with inequalities and their solution, especially where alternative solutions can be rated via a payoff function and where the solution with the largest payoff is required. Algorithms for such searches exist in the literature on optimization, and a selection of good and provably sound ones has been incorporated in the extended Prolog system Prolog III [Ru] built by A. Colmerauer and his collaborators. In terms of performance and functionality in handling inequalities, no SMC software can compete with Prolog III, but Prolog III is not intended as a package for SMC. The challenge of integrating the capabilities of Prolog III and current SMC is well worth considering.

There are many possible types of constraint-based computation apart from work that is narrowly related to optimization. A more general meaning of "constraint" is: any condition that restricts the set of possible solutions of a problem because some candidate solutions may fail to respect it. The paper by Ladkin and Reinefeld below fits this general interpretation. Further, a well-known example of the use of constraints (constraint propagation) to solve an AI problem is in the Waltz filtering procedure for reducing (ideally to 1) the number of 3D figures that are consistent interpretations of a 2D image or drawing of a 3D scene [Wi]. Many of these constraint problems have a SMC character, or contain symbolic mathematical sub-problems. Because the size of the resulting computations can be large even for problems that look small, it is important to consider how SMC methods can improve the state of the AI art in this area.

A further development in constraint-based programming that is worth mentioning is that languages or systems intended primarily for such programming exist [Le, Gu]. This is in addition to the fact that one can regard Prolog and its near relatives as languages for constraint programming, and that some of them (e.g. CHIP) are described explicitly as "constraint logic languages" by their builders.

5.4 Subsymbolic computing

This heading expresses an idea that has become particularly important in AI since about 1986. It is that the symbolic paradigm that was taken for granted since the early days of AI (with a few isolated disturbances, such as the brief flurry of interest in Perceptrons around 1960) is not the only one that should be considered in a fully-rounded view of what AI is. In addition, it is widely believed that there are some activities of intelligence (e.g. recognition of multidimensional patterns) where an approach operating at some lower level than a level of description in symbols is more appropriate than the traditional logical-symbolic approach.

Neural nets and genetic algorithms are the most popular subsymbolic approaches at present. It is well known that both can be fitted into the wider framework of optimization, description of interactions among large numbers of agents, etc., that is the province of statistical mechanics. Therefore, for example, statistical mechanics has provided some of the most effective features of neural nets. But statistical mechanics is also a field in which it is easy to find worthwhile applications of methods and techniques drawn from SMC. This is an obvious communication channel between neural nets and SMC that deserves consideration for future research.

In the opposite direction, neural nets have already been used to design devices that allow handwritten inputs into a piece of software [Ma].

The contribution of Garigliano and Nettleton in this book shows how the mathematical modelling of qualitative properties of genetic algorithms can help in the selection of optimal parameters for such algorithms.

5.5   Learning

Machine learning includes the inference of possible regularities (rules, classfications etc.) from volumes of data or examples. In the past it has concentrated on symbolic methods, but now the fastest-growing parts of the subject belong under the subsymbolic heading of section 5.4. Nevertheless, there are still topics in which the symbolic approach has some value for SMC.

The best source for information on symbolic techniques and results from AI is the "Machine Learning" series of edited books, which has 3 volumes [Mh]. A book that is more narrowly focused and that presents its results in a useful didactic way is by Forsyth and Rada [Fr]. Current research receives significant attention in papers in the sections on Learning in IJCAI and AAAI proceedings, where this is always an active area.

The regularities that are of interest in SMC are regularities in the behaviour or the outputs of SMC systems.

In connection with the effective heuristic control of size and structure of intermediate expressions in symbolic computations, it may be possible to infer automatically what settings of simplification-controlling flags in SMC systems and what sequences of steps in a user's program are likely to lead to optimal management of storage. Also, observations of curious patterns in trace information may allow the tentative inference that a user's program is sub-optimal or contains faults. There is an example of human identification of this kind of behaviour in an old paper by Fitch and Garnett [Fi], and a general discussion of the knowledge-based use of trace information by Gardin and Campbell [Gr], but no subsequent work connecting the examples presented there with machine learning. There is certainly a case for the connection to be examined further now.

A further cue for automated learning of a rather simple but still useful kind is available in outputs from SMC systems. In the history of AI the BACON system and several others produced by the group of P. Langley [La], for automated inference of symbolic relationships between variables so that these relationships are satisfied in given stocks of numerical data, are well known. One can reproduce this type of computing for more practically useful SMC problems (e.g. see [Cm]), or even extend it to situations where there are stocks of mixed numerical and symbolic data. Here, several techniques that are already available from AI await exploration in SMC.

A final example of the idea of learning in SMC is given in the paper by Roque below.

6 Areas with potential to generate future research

6.1 Spatio-temporal reasoning

There are many lines of activity in AI research on reasoning about spatial and temporal information. Apart from the kinematic reasoning mentioned in section 2.3, these are basically expressed in terms of logics. For temporal reasoning, most of the logics are non-standard in some way, and raise consequent problems of logical soundness, completeness etc., as well as computational efficiency.

There are two possible ways for SMC to relieve the situation. One has already been mentioned in section 5.2: conversion of logical problems to alternative algebraic forms. This involves showing the existence of a suitable formal equivalence, as in the examples quoted in that section. The second way is analogous, but rather less formal. It relies on the fact that the purpose of many instances of spatio-temporal reasoning is just to predict trajectories in terms of spatial and/or temporal variables. Where these trajectories are required in exact symbolic form or as symbolic approximations (e.g. as trajectories interpolating or extrapolating given sets of points), all the apparatus of classical mechanics and SMC is available to assist the process. Constraint-based computing (see section 5.3) is also relevant, because many trajectory problems are stated at least in part through constraints. Traditional AI sticks to methods of logic for spatio-temporal reasoning, with occasional recent gestures towards constraint logic programming, but the substantial potential contribution of SMC has been neglected in AI so far.

We should also mention that the idea of transforming the contents of problems of spatio-temporal reasoning into trajectory-prediction problems is not the only possible approach to reducing the computational difficulties that always seem to come with non-standard logics. For example, it may also be possible to find alternative representations of such problems in terms of graphs, as the contribution by Golumbic below indicates.

6.2 Knowledge relevant to robotics

The longest-lived AI view of robotics is that planning is the relevant topic, because robot actions should be governed by plans derived from information about a robot's goals and the environment by methods very similar to theorem-proving methods. However, realistic problems of modern robotics are just as likely to need SMC treatment as AI. Kinematics of compound or multi-jointed robot motion requires intensive symbolic computation, e.g. using constraints plus standard methods of classical mechanics. The same kinds of mathematics are relevant also for novel methods of finding trajectories for robot arms, or mobile robots as a whole, in environments containing obstacles. These methods treat target positions for given points on the robot as sources of attractive potentials, and obstacles as producing repulsive potentials. A trajectory is then determined by the use of a variational principle, e.g. through methods one can find in standard texts on classical mechanics. It is also possible, at the price of more complexity, to use the principle to find trajectories simultaneously in coordinate and velocity spaces, so that a good path also moves a robot arm smoothly (without sudden changes in accelerations and higher derivatives, which cause undue wear on components) into contact with a target at a local speed close to zero – rather desirable, for example, if a robot has the job of picking up a delicate object such as an egg.

The ideas expressed above now have quite a wide circulation in robotics. What has been lacking so far, but which it is practicable to achieve, is the solution of realistically large robot trajectory problems by the explicit use of SMC to do the mathematical manipulations.

There are several further connections between SMC and robotics, as the contributions by Hong, Monfroy and Pfalzgraf to the present book show. It is clear from what these authors write that the connections can be strengthened by future research that also involves AI.

6.3 Case-based reasoning

Case-based reasoning (CBR) is a fairly recent development in AI, but it has already led to much interest and activity because it is well adapted to the ways in which many types of experts express their knowledge. (In this sense, cases are yet another significant knowledge representation to add to the list given in section 4). The October 1992 issue of the journal IEEE Expert contains several articles on the subject and its properties and achievements. The only book on the subject so far [Sc] does not take modern developments fully into account, but others are apparently in an advanced state of preparation in 1993.

We all know informally what a case is, but there is no standard answer in AI to the technical question of "What is a case?". The question is likely to remain unsettled for some time yet, so it can be left on one side here. However, it is clear that the question of finding the case that is "most similar" to a given case will always occupy a central place in CBR. There are several facets to the concept of similarity, among which semantic and pragmatic similarity may be able to draw on help (in suitable problem-areas) from SMC. For example, the meaning or semantics attached to a case may have a quite precise interpretation according to the typing of a case or its parts, wherever the imposition of a typing scheme is meaningful, and distances between cases can then receive contributions from distances between their types in pre-stated type hierarchies. In this respect we can hope to adapt ideas of typing that are discussed elsewhere in this book so that they become usable for CBR, and to automate searches and other manipulations of type hierarchies. Also, SMC may be useful in the future through allowing determination of pragmatic similarities between cases (x and y are pragmatically similar if outcomes, components etc. are usable for similar purposes or have similar consequences, and this idea can include outputs that are symbolic answers of a similar kind, such as having the same asymptotic behaviour) to be speeded up.

6.4 Intelligent tutoring or assistance

We have mentioned above that a trend in designing user interfaces for computer algebra systems is clearly linked to the very idea of an "intelligent assistant" system. This line of research is in fact already well defined, as shown by the history of work by several groups that G. Butler reports in his survey below. A further interesting idea consists of setting up intelligent tutoring schemes which can make reasonable automated attempts at constructing models of a user's level of competence from this user's behaviour during a session of computation. In connection with SMC, such ideas were presented during the ISSAC'91 conference by Roque but did not appear in the proceedings. Such an assistant is likely to rely heavily on methods of machine learning.

In dynamic user-modelling, the achievements available in AI are well ahead of those reported in the specific application area of SMC. There are thus many opportunities for SMC researchers to follow up leads which may improve the quality of the assistance that our computing systems can give to users, in addition to the interesting work on automated mathematical tutoring that is in progress in the group of L. Aiello (e.g. see [Ai]). A book by Wenger [We] is a good example of the more general AI perspective.

6.5 Specification Issues

Specification issues – and more precisely, issues of algebraic specification – are becoming progressively more important throughout computer science. In terms of our interests, this supports the expansion of activity in design of specification languages for SMC. A well-known illustrative example can be found in the KADS approach for knowledge acquisition [Hi].

This approach now extends to specification even of knowledge bases. The contribution of Calmet and Tjandra belongs in a similar setting for symbolic computing in general and computer algebra in particular. Their approach enables (in addition to the activity of specification itself) one to define horizontal and vertical polymorphism [Tj], and to transform program specifications into "knowledge" that is processed by a hybrid knowledge representation system.

While this is not possible in general, careful selection of the underlying type system (as in their work) can make it possible in particular situations.

References

[Ai] L. Aiello and A. Micarelli, *Applied Artificial Intelligence*, 4, 15, 1990.

[Ba] A. Barr and E.A. Feigenbaum (eds.), *Handbook of Artificial Intelligence*, Morgan Kaufmann, Los Altos, 1985.

[Bc] B.G. Buchanan and E.A. Feigenbaum, Artificial Intelligence, 11, 5 1978.

[Bo] A.V. Bocharov, *Will DELIA grow into an expert system?* In A. Miola (ed.), Design and Implementation of Symbolic Computation Systems. Proc. of DISCO'90. LNCS 492, Springer-Verlag, 1990.

[Bi] R.F. Boisvert, *Toward an intelligent system for mathematical software selection.* In P.W. Gaffney and E.N. Houstis (eds.), Programming Environments for High-Level Scientific Problem Solving. IFIP Transactions A-2, North-Holland, p. 79, 1992.

[Br] R. Brachman and H. Levesque (eds.), *Readings in Knowledge Representation*, Morgan Kaufmann, Los Altos, 1985.

[Bu] A. Bundy, *The computer modelling of Mathematical Reasoning.* Academic Press, London, 1983.

[Ca] J. Calmet and I.A. Tjandra, *An Expert System for Correctness of Symbolic Computation.* In J. Liebowitz (ed.), Expert Systems World Congress Proceedings, Vol. 3, Pergamon Press, 1991.

[Cm] J.A. Campbell, J. Phys. A, 12, 1149, 1979.

[Cn] J.J. Cannon, *A draft description of the group theory language CAYLEY*, SYM-SAC'76, ACM, 1976.

[De] A. Deprit, J. Henrard and A. Rom, *Lunar Ephemeris: Delaunay's Theory Revisited*, Science 168, 1569–1570, 1970.

[Do] B. Donald, D. Kapur and J. Mundy (eds.), *Symbolic and Numerical Computation for Artificial Intelligence*, Academic Press, London, 1992.

[Fa] B. Faltings, *Artificial Intelligence*, 56, 139, 1992.

[Fi] J.P. Fitch and Garnett, 1971 conference paper, citation in [Gr].

[Fo] K.D. Forbus, *Artificial Intelligence*, 24, 85, 1984.

[Fr] R. Forsyth and R. Rada, *Machine Learning*, Ellis Horwood, Chichester, England, 1984.

[Ga] W.A. Gale (ed.), *Artificial Intelligence and Statistics*, Addison Wesley, 1986.

[Ge] H.L. Gelernter, *Realization of a geometry theorem-proving machine*, in Computer and Thought, E.A. Feigenbaum and J. Feldman (eds.), McGraw Hill, New York, 1963.

[Gr] F. Gardin and J.A. Campbell, *Tracing occurences of patterns in symbolic computations,* in Proc. 1981 ACM Symposium on Symbolic and Algebraic Manipulation, P.S. Wang (ed.), p. 233, ACM, New York, 1981.

[Gu] H.-W. Güsgen, *CONSAT: a System for Constraint Satisfaction,* London, Pitman, 1989.

[Ha] F. Hayes-Roth, D.A. Waterman and D.B.Lenat, *Building Expert Systems*, Addison-Wesley, Readind, Mass., 1983.

[Hb] J. Hobbs and R. Moore (eds.), *Formal Theories of the Commonsense World* American Elsevier, New York, 1985.

[Hi] F.R. Hickman et al., *Analysis for Knowledge-Based Systems – A Practical Guide to the KADS Methodology*, Ellis Horwood Ltd., Chichester, England, 1989.

[Hn] P.J. Hayes, *The Naive Physics Manifesto,* in Expert Systems the Microelectronic Age, ed. D. Michie, pp. 139 – 147, Edinburgh Univ. Press, Edinburgh, 1979.

[Ho] E.N. Houstis and J.R. Rice, *Parallel ELLPACK*. In P.W. Gaffney and E.N. Houstis (eds.), Programming Environments for High-Level Scientific Problem Solving. IFIP Transactions A-2, North-Holland, p. 229, 1992.

[Hu] M.A. Hussain and B. Noble, *Application of Symbolic Computation to the Analysis of Mechanical Systems, Including Robot Arms*. In E.J. Haug (ed.), Springer Verlag, Berlin, 1984.

[Ka] H.G. Kahrimanian, *Analytical differentiation by a digital computer*, M.S. Thesis, Temple Univ., Philadelphia, 1953.

[Kl] J. de Kleer, A.K. Mackworth and R. Reiter, *Artificial Intelligence*, 56, 197 1992.

[La] P. Langley, *Strategy acquisition governed by experimentation*, in Progress in Artificial Intelligence, L. Steels and J.A. Campbell (eds.), p. 52 (Ellis Horwood Ltd., Chichester, England, 1985.

[Le] W. Leler, *Constraint Programming*, Addison Wesley, Reading, MA, 1989.

[Ma] R. Marzinkewitsch, *Operating Computer Algebra Systems by Handprinted Input*. In S.M. Watt (ed.), ISSAC'91, p. 411, ACM Press, New York, 1991.

[Mc] P. McCorduck, *Machines Who Think*, Freeman, San Fransisco, 1979.

[Mh] R. Michalski, J. Carbonell and T.M. Mitchell (eds.), *Machine Learning*, 3 vols, Tioga Press, Palo Alto, CA, and Springer-Verlag, Berlin, 1983 and subsequent dates.

[Mi] T.M. Mitchell, P.E. Utgoff and R.B. Banerji, *Learning by experimentation: Acquiring and refining problem-solving heuristics*. In R.S. Michalski et al. (eds.), Machine Learning, Tioga Publ. Co., Palo Alto, CA, 1983.

[Mo] J. Moses, *Symbolic Integration, the Stormy Decade*, Communications of the ACM, 14, pp 548–560, 1971.

[No] J. Nolan, *Analytical differentiation on a digital computer*, S.M. Thesis, MIT Cambridge, 1953.

[Ri] G. Ringland and D.D. Duce (eds.), *Approaches to Knowledge Representation*, Science Research Associates/ Wiley, Chichester, England.

[Ru] R. Rueher, *Software - Practice & Experience*, 23, 1993.

[Sc] R. Schank, *Inside Case-Based Reasoning*, L. Erlbaum Associates, NY, 1989.

[Sl] R. Slagle, A heuristic program that solves symbolic integration problems in freshman calculus. In Computer and Thought, E.A. Feigenbaum and J. Feldman (eds.), McGraw Hill, NY, 1963.

[St] L. Sterling, A. Bundy and L. Byrd, *Solving symbolic equations with PRESS*. In J. Calmet (ed.), Proc. of EUROCAM'82, LNCS 144, p. 108, Springer Verlag, Heidelberg, 1982.

[Tj] I.A. Tjandra, *Spezifikation mathematischer Rechenstrukturen und Typenpolymorphismen in der Computeralgebra*, PhD. Thesis, Univ. Karlsruhe, July 1993.

[Ul] F. Ulmer, *On Liouvillian Solutions of Linear Differential Equations*, AAECC Journal, Vol.2, No. 3, pp. 171-194, 1992.

[Vi] M. Vivet, *Expertise mathématique: CAMELIA, un logiciel pour raisonner et calculer*. Thèse d'état, Univ. Paris IV, 1984.

[Wa] D. Wang, *On Wu's Method for proving constructive geometrical theorems*, Proc, IJCAI-89, 1, 419, 1989.

[We] E. Wenger, *Artificial Intelligence and Tutoring Systems*, Morgan Kaufmann, Los Altos, CA, 1987

[Wi] P.H. Winston, *Artificial Intelligence*, 2nd edition, Addition-Wesley, Reading, MA, 1984.

Qualitative Modeling of Physical Systems in AI Research

Peter Struss

Computer Science Dept., University of Technology
Orleansstr. 34, W-8000 Munich 80
struss@informatik.tu-muenchen.de

Abstract. Much of human reasoning about the physical world, in every-day life as well as in science and engineering, is performed at a conceptual and non-quantitative level. Providing formalisms, languages, and systems for the acquisition and use of qualitative models is the goal of a very active research area in Artificial Intelligence. In this introductory survey, we discuss motivations for this research, illustrate different approaches by presenting some "classical" systems, point out some issues in relation to mathematics, and provide some references to current work in research and applications.

1   Introduction

1.1   Motivation

If we are intrested in reasoning about and predicting the behavior of physical systems, then why do we consider this as a task that sets new problems for AI research and AI programming? After all, there is physics that describes the *laws* "controlling" physical systems with mathematical formalisms (e.g. differential equations). So, why look for new theories? And there exist systems, such as numerical simulation, that allow us to actually *compute* solutions to the equations. Do we really have to develop new computational formalisms for our purposes?

Let us discuss these questions briefly to motivate our efforts and to clarify their goal. As an example, consider a ball being thrown or hit upward, say, in a squash hall. If we determine the initial location of the ball to be (x_0, y_0) (ignoring one dimension) and its initial velocity to be (v_{x_0}, v_{y_0}), if we further assume friction to be negligible and gravity to apply a constant acceleration $(0, g)$ with $g < 0$, then Newtonian mechanics predicts a movement of the ball along the trajectory

$$(x_0 + v_{x_0}*t, y_0 + v_{y_0}*t + 1/2*m*g*t^2),$$

and a numerical solution would approximate it arbitrarily close (Fig. 1a). Isn't this the desired prediction of the ball's behavior over time? What's wrong with it? What's wrong is that this is *not reasoning about the physical world*. It is a description of a curve in an abstract mathematical space. A space that contains the (abstraction of the) ball and nothing else.

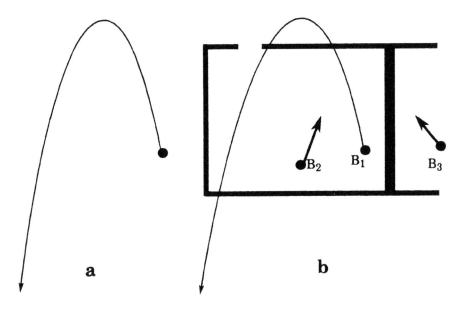

Figure 1 Throwing a ball ... a) in physics b) in the real world

The physical world is different. The squash hall has a ceiling, walls, and a floor. And the ceiling has a hole through which the ball might exit (perhaps after breaking the respective window pane). And there is another person next to us also throwing a ball. And so on (Fig. 1b). So, what about the ball following the trajectory? Oh well, you may say, that's simple. Of course, the trajectory describes the movement of the ball only until it hits another object, such as the ceiling or the second ball, B_2. That's not a problem, we can handle this by adding restrictions, such as $y < y_c$ for expressing the constraint imposed by the ceiling.

The second ball is a bit trickier, but we can compute both trajectories and check whether the two balls move to the same location at the same time. If such boundary conditions are met, we know that a different law applies (e.g. elastic collision), and we can simulate it. Immediately after this, we can apply Newton's laws again.

No problem? The problem is, that *we* have to analyze and state the conditions for the applicability of the different laws. They are not part of the differential equations. $y(t) = 1/2*m*a_y*t^2$ does not tell you whether and how t (or y) is bounded. We can make the applicability conditions (e.g. no other forces acting on the ball) explicit. But there remains the task of analyzing in a particular scenario, such as the squash hall in Fig. 1b, whether or not, when and how long these conditions hold. This includes noticing that the movement of the second ball, the height of the ceiling, etc. may affect B_1. It includes also to determine what is of *no importance*. When choosing the appropriate laws, we decided that friction is negligible for the class of predictions we are interested in, and that gravity can be treated as constant in the hall etc. And while analyzing the scenario, we knew that we could ignore the third ball, B_3, which is being hit in the hall next door, because the wall inbetween will prevent it from interacting with B_1.

Note that all this reasoning about a physical system happens *prior to* the use of any differential equation, even *before establishing* any differential equation. Still stronger, it forms a *prerequisite* for the derivation and computational use of a mathematical model of the system. It can and has to be preformed at a structural and qualitative level, independent of a quantitative or numerical model or, at most, interleaved with its tentative and qualitative use. Reasoning at a similar level is necessary when controlling the *application* of the mathematical model (e .g. when switching from Newton's law to elastic collission) and, in particular, when *interpreting* its results.

Obviously, this is, almost by definition, beyond the scope of traditional simulation systems (forming a "front-end" for them). While this answers one part of our initial question, we have to ask whether we need to invent a new physics. There is no doubt that the task of conceptual modeling we discussed above is a part of a physicist's (or engineer's) skills. Even more, being able to analyze a physical phenomenon in order to derive a mathematical model, to hypothesize laws, to interpret observations and numerical data, and to design useful experiments is probably what distinguishes a good and experienced physicist from a bad or novice one, rather than applying a mathematical machinery to an established model. However, although being part of the physics as a system of research and application activities, this task of conceptual modeling and reasoning about physical systems has not been formalized in traditional physics, so far. As a result, it is hardly taught in a systematic way; it is acquired mainly through experience (if at all), and remains an art.

If research in AI is dedicated to "qualitative physics", this is not to say that new or different laws of physics should be detected and replace those derived and turned into an effective formal system by traditional physics. It is also not confined to the development of abstractions of the known physical laws for qualitative simulation (in contrast to numerical simulation). Rather, its primary goal should be seen in the task of formalizing what is not part of the formal system of physics, the modeling step itself, the conceptual layer of reasoning about physical systems, the intuitive explanation of system behavior, etc. This provides the foundation for languages that allow us to express our conceptual knowledge about a physical domain and store it in a knowledge base, and for inference mechanisms that automate (part of) reasoning about physical systems.

But, perhaps, we have to justify this objective, too - beyond the principled argument that it helps to provide insight into human reasoning capabilities. A more practical reason for building such automated reasoning systems is a tremendous increase of the generality, robustness and application scope of computer systems, which even is a necessity for some areas, such as autonomous systems. A simulation system could be taylored to make correct predictions for the scenario in Fig. 1b. It requires substantial extensions to enable it to cope with arbitrary, but standard "squash hall scenarios". We cannot simply enumerate all possible situations (n balls, different shapes of walls etc.), but rather have the system "reprogram itself" when it encounters a new scenario, e.g. generating a new model. And yet, this system gets stuck, or makes wrong predictions, if some action or unanticipated event introduces a new object or phenomenon (a strong wind blowing through the hole in the ceiling, somebody digging a hole in the ground etc.), even though it could, in principle, be handled within the scope of the laws that are represented in the system. A system should be able to recognize this, change its model, and behave appropriately.

Now, letting artificial persons, robots, play a real squash game is probably as odd and worthless as letting real humans play squash in a virtual reality. But a robot operating in a warehouse or a factory would establish similar requirements. Or, leaving the world of autonomous systems, a powerful diagnosis system that can handle new types of faults and a system supporting conceptual design need the same flexibility and power of reasoning about new and unanticipated scenarios.

Thus, research on qualitative reasoning (QR) addresses an important part of problem solving activities in physics and engineeering: modeling. Analyzing a particular physical scenario, identifying the important objects, interactions, and influences is an important prerequisite for interpreting or even getting equations and differential equations. Not only, rather: not so much, *solving* the (differential) equations is the concern of QR, but understanding ways of *deriving* them. QR aims at formalizing this modeling step and, based on this, building systems that provide support for it. This is the ultimate reason why QR is not really competing with traditional (and useful!) systems that simulate and control physical systems, but rather complementing them by allowing for a cognitively adequate representation and control of such systems. And this also explains why QR has been and can only be performed as an area of Artificial Intelligence research.

1.2   Two Origins of Qualitative Reasoning Research

Qualitative reasoning is involved both in reasoning about natural phenomena and every-day physics and in the performance of scientific or engineering problem solving. These two domains, although not completely distinct, can be identified as origins of QR research. Due to the nature of AI, the commonsense aspect was strong in the beginning. The *Common Sense Algorithms* of [Rieger-Grinberg 77], and the *Naive Physics Manifesto* of [Hayes 78] are results of this interest.

Another origin can be identified as the *Engineering Problem Solving Project* at M.I.T. which established an environment leading to many concepts and systems that still have an important impact on QR and other fields of AI (causal reasoning, contraint systems, truth maintenance etc.). In particular, the EL system, actually a family of systems for circuit analysis ([Sussmann-Stallmann 75], [Stallman-Sussman 77]), incorporated a number of concepts and techniques that form the basis for a whole branch of QR. Based on the observation of methods applied by experienced engineers, one important step was to *localize* the analysis of a circuit as opposed to the manipulation of large systems of equations. This was done by basing the analysis on *component-centered* laws and *local one-step deductions*, the propagation of constraints. The approach of EL, deriving the overall state or behavior of a system from a description of its structure and of its components, characterizes a mainstream of QR research (*structure-to-function*). This branch recently gained importance under the influence of commercial interests in *Expert Systems* and the paradigm of model-based reasoning.

In the early eighties, limitations of the existing methods for building expert systems became obvious. In an invited lecture at IJCAI 1981, R. Davis presented an analysis of achievements and shortcomings of the "Accepted Wisdom" of systems based on "empirical associations" and formulated goals for new types of representations and inference mechanisms ([Davis 82]). Analyzing an example from the domain of technical diagnosis, he demonstrated that an *"understanding of behavior derives from reasoning about structure and function"* of the device and that, besides empirical associations, this was crucial for solving the diagnosis task. Hence, R. Davis argued for the construction and use of *causal models* for a new generation of expert systems, and he and his group tackled the task in the domain of troubleshooting of circuits ([Davis 84]).

1.3   Overview

The general goal of this paper is, besides motivating this line of research, to survey some achievements and problems of the fundamental work started about ten years ago and to highlight and reference some of the more recent research topics and applications. These are quite diverse, as illustrated by recently published collections of papers ([de Kleer-Williams 91], [Faltings-Struss 92], [Weld-de Kleer 90]), and the debate about goals and directions of QR in [CompIntell 92], and cannot be covered comprehensively in this survey. We completely disregard the special, though important problem of qualitative reasoning about time intervals which is covered by a separate contribution to this volume.

In the following section, we briefly characterize three "classical" approaches to QR. We highlight their distinctions as far as the underlying ontologies are concerned, and their commonalities with respect to the incorporated reasoning about quantities based on intervals. Then some technical details are given about the respective systems, ENVISION (section 3), QPT (section 4), and QSIM (section 5). Section 6 summarizes results concerning properties and problems of the interval-based qualitative calculus. Order-of-magnitude reasoning, as a different way of expressing and exploiting qualitative information is presented in section 7. Finally, we provide some references to current research topics and to steps towards applications. A bibliography is appended.

2 Three "Classical" Approaches to Qualitative Reasoning

The three sections following this one will describe approaches which, until recently, together covered most of QR research, each represented by one system, *ENVISION* [de Kleer-Brown 84], *QPT* [Forbus 84], and *QSIM* [Kuipers 86]. Before giving details about these systems, it appears to be useful to clearly state their principal differences and commonalities. It turns out that besides technical issues their essential *difference* lies in the *ontological primitives* they use for describing a physical system, whereas their mechanisms for inferring behavior based on *reasoning about quantities* share a *common* abstract basis.

2.1 Ontologies

In the *Qualitative Process Theory (QPT)*, the overall behavior of a system is established by processes and their interaction via conditions and results. *ENVISION* uses *components* as behavioral primitives which interact via connections, thus constructing the behavior of the composed device. *QSIM* includes *no ontology* about the physical situation. It uses mathematical abstractions, (differential) equations or relations combined via variables, to represent a system.

Processes versus components - is there a real difference behind the names? Are not both referring to parts whose combined activity causes the global behavior? Beyond this abstraction, the two paradigms reflect quite different perspectives on changes physical objects are involved in. In QPT, objects are passive entities, subject to changes imposed by the acting processes. A lit candle and an ice cube do not act, hence not interact - it is a process called *heatflow* that acts on both of them, or: it is the interaction of the objects. And not the ice cube reacts to the heatflow, but a process *melting* affects it, which means destroys the ice cube object and creates a new object, liquid water.

In ENVISION, activity and changes originate in the objects. Everything that happens is caused by some component. If the pedal of a bike is pushed, then not a process rotation is created and acting, but the pedal transports the force to the toothed wheel which transmits it to the chain, from there it reaches via another toothed wheel the bike's wheel which reacts by rotating, etc.

To a certain extend, these different perspectives reflect differences between the subjects and our reasoning about them: On the one hand, there are natural phenomena. From observation and experience we know "what happens". Different objects may be involved and changed, and their configuration may also change. On the other hand there are artifacts. They contain a number of building blocks in a fixed structure. What happens, is a result of combining the specific functions of the components.

Processes appear to be more suitable for expressing primitive elements of a causal chain. They are *causally* (and hence temporally) *directed*. There is no heatflow from the destination to the source. Though being spatially bounded, processes may *act on several objects. Components* may work in *several directions* dependent on their context. A valve permits a flow of water in either direction. Components are *completely local*, they know nothing about their actual context.

Although these are only relative statements and arguable in specific cases, the ontology and the stated goals of QPT are very much inspired by Hayes' Naive Physics, whereas ENVISION's roots are methodologically and personally originating in the EPSP.

2.2   The Essence of Qualitative Reasoning about Quantities

Though ENVISION, QPT, and QSIM are rather different in the nature of the behavioral primitives, their methods for describing and inferring the composite behavior can be reduced to the following scheme for dealing with quantities, or at least they contain it as an essential part:

Represent the system's

- *behavioral constituents* (processes, components, ...) by relations between their local variables and derivatives

- *interaction of these constituents* by identifying variables in these relations

- *qualitative values* by certain *landmarks* (real numbers) or intervals between landmarks

- *states* by qualitative values of characteristic variables (and potentially of their derivatives)

- *behaviors* by sequences of states.

The process of *inferring* the possible behaviors of a system contains three tasks (which are not necessarily distinguished, subsequent steps): determine

- *possible states* of the system, i. e. sets of qualitative values consistent with the relations specifying the behavioral constituents. This is a problem of constraint satisfaction.

- *possible state transitions*, i.e. steps from one state to another that are in accordance with the derivative relations and continuity conditions. Fig. 2. shows an example of this transition analysis, where ∂x denotes the qualitative value of the derivate of x (the sign) and crossed arcs are impossible transitions.

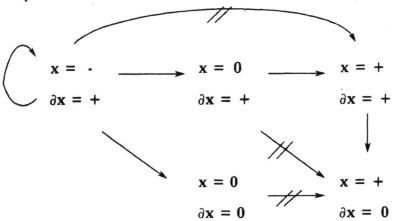

Figure 2 Transition analysis

- *possible sequences of state transitions.* This is the weak point in many current QR methods, because they do *not offer any citeria* for checking the global correctness of such sequences, and, hence they, implicitly or explicitly, assume that any sequence of possible state transitions is admissible, and establishes a possible behavior.

Note that the process of describing the behavior by solving differential equations in the real domain has no correspondence in the qualitative (interval) domain. It is substituted by solving equations over the qualitative domain and checking continuity conditions .

In the following presentation of the three approaches, we will characterize them with respect to the main aspects stated in the introduction:

- *Representation of structure*: what are the primitives used for composing the model of a physical system, and how are they linked?

- *Representation of behavior*: how is the functionality of these primitives described? In particular, what are the qualitative values of quantities?

- *Inference of behavior*: how are the structural and behavioral descriptions used to derive the global behavior of the system?

3 Structure-to-Function - ENVISION

3.1 Component-Oriented Ontology

ENVISION ([de Kleer-Brown 84]) treats systems that are composed of *components* and *conduits* which are connected via *terminals*.

- *Components* are physically disjoint objects of different types that exhibit a predictable behavior characteristic for this type, e.g. a resistor, a valve, or a toothed wheel.

- *Conduits* are passive channels for transporting material or information from one component to another, e.g. a wire without resistance or a pipe. They do not store or modify anything.

- *Terminals* connect components with conduits. Components cannot communicate with their environment except through their terminals.

Hence, the components are the "behavioral constituents" of the device. Its *structure* is considered to be *fixed*. Although plausible for large classes of artifacts, such as circuits, this is an important restriction that makes the treatment of mechanical systems difficult, because moving parts may continuously create and destroy connections. Problems arise also in diagnostic tasks when the cause of a malfunction lies in a violation of the designed structure (e.g. a bridge fault). In ENVISION, a device has a flat structure. This appears to be inappropriate for reasonably practical examples. In [Struss 87], a system is presented that allows hierarchical and aspect-oriented splitting of the structural description.

3.2 Qualitative Behavior Models

Variables in ENVISION take on values of the **quantity space** $Q = \{ -, 0, + \}$ which are derived from their real values by

$$[.] \quad : \mathbb{R} \to Q$$

$$[x] := \begin{cases} - & \text{if } x < 0 \\ 0 & \text{if } x = 0 \\ + & \text{if } x > 0 \end{cases}$$

i. e. their signs are taken. Their qualitative derivatives are the qualitative values of the real derivatives:

$$\partial x := [dx/dt]$$

Higher order qualitative derivatives may be used. However, they can only be obtained from the respective real derivatives:

$$\partial^n x = [d^n x/dt^n],$$

there is no "qualitative differentiation". Qualitative values can be added and subtracted according to the tables in Fig. 3. Of course, the sum of - and + is undefined.

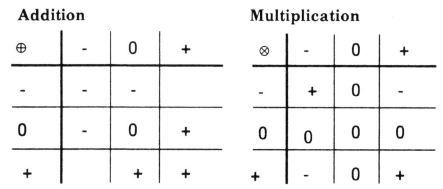

Addition

$\oplus$	-	0	+
-	-	-	
0	-	0	+
+		+	+

Multiplication

$\otimes$	-	0	+
-	+	0	-
0	0	0	0
+	-	0	+

Figure 3   Addition and multiplication of signs

Component descriptions are given in terms of relations that hold on the qualitative values of their variables and parameters. These so-called *confluences* are derivable from linearized quantitative equations and differential equations. For instance, the confluence corresponding to Ohm's Law $\Delta u = R \cdot i$ is

$$[\Delta u] = [i]$$

since $[R] = +$. In principle, confluences can also be used to express naive knowledge which has no quantitative refinement. Note that e.g.

$$\partial x = \partial y$$

states only that E and h are monotonically dependent, but does not specify, how.

Although involving the equation sign, confluences are not equations as becomes obvious when their solutions are defined:

A *solution of a confluence* is a tuple of qualitative values for which either

- the equation sign holds, or

- the confluence contains an operation with an undefined result.

For example, $[x] = +$ and $[y] = +$ is a solution to

$$[x] \oplus [y] = +,$$

and $[x] = +$ $[y] = -$ solves

$$[x] \oplus [y] \oplus [z] = 0$$

regardless of the value of [z]. The deeper reasons and the consequences are analyzed in [Struss 88, 90].

3.3 The No-Function-in-Structure Principle

As an example, a valve shall be modelled by confluences. Our first attempt could be to state that the flow, Q, has the direction of the pressure drop, P, and that the change of the flow, ∂Q, is positively influenced by a change in the pressure drop, ∂P, and by a change in the area, ∂A:

$$[P] = [Q]$$
(3.1) $\partial P \oplus \partial A = \partial Q.$

However, this is not correct for a *negative* pressure drop, since (3.1) implies, for instance,

$$\partial A = + \wedge \partial P = 0 \quad \Rightarrow \quad \partial Q = +$$

but opening the value increases the *amount* of the flow, which has a *negative direction*:

$$\partial A = + \wedge \partial P = 0 \wedge [Q] = [P] = - \quad \Rightarrow \quad \partial Q = -.$$

(3.1) implicitly assumes pressure drop and flow only in one direction. Perhaps the valve was designed in order to regulate the flow only in one direction. But who guarantees that no faulted situations occurs in which the fluid is forced to flow in the opposite direction? Our model would suggest the wrong counter action for stopping this.

This leads to the formulation of the *no-function-in-structure principle*. "The laws of the parts may not presume the functioning of the whole". Obeying this requires that the component models should be context-free and should not refer to any other part but only to its internal parameters and the information reaching it via its terminals. Otherwise, the aim of constructing the device behavior from purely *local* actions and interactions would be violated.

The ideal formulation of the no-function-in structure principle would require to construct models that exhaustively describe all possible behaviors of a component in arbitrary contexts. This is infeasible, but should not lead to an absolute negation of the principle which is essential for analyzing of unexpected or faulty behavior. Hence, the goal should be to construct local models that are valid for a wide class of systems, not presupposing their (intended) function, and to make the assumptions explicit.

An improved confluence would be

(3.1') $\partial P \oplus [P] \otimes \partial A = \partial Q$

But still, it is not the ultimate solution, since it implies

$$\partial A = 0 \wedge \partial Q = 0 \quad \Rightarrow \quad \partial P = 0$$

which is not true for a closed valve, since, in this case, the pressure drop and its change may be arbitrary. This leads to the notion of qualitative states of a component, which are governed by different confluences. A model for the valve has to account for its three states:

- OPEN $\quad\quad$ $(A = A_{MAX})$ $\quad$ $[P] = 0, \partial P = 0$
- WORKING $\quad$ $(0 < A < A_{MAX})$ $\quad$ $[P] = [Q], \partial P \oplus [P] \otimes \partial A = \partial Q$
- CLOSED $\quad\quad$ $(A = 0)[Q] = 0, \partial Q = 0$

3.4 The State Diagram

The (consistent) combinations of the qualitative states of the individual components define qualitative states of the whole device. For each of these *device states* a certain set of confluences must be satisfied. ENVISION determines the set of possible solutions for each state by a combination of constraint propagation and generate-and-test. This set is called the *intrastate behavior* and contains all combinations of qualitative values that are consistent with the description of the given state.

The second kind of behavior occurs when one or more variables reach a state defining threshold, e. g. A reaches A_{MAX} of the valve. Such transitions between device states are part of the *interstate behavior* of the device. Determining it requires to analyze whether the qualitative derivatives and continuity conditions allow variables to reach thresholds and, if there are several, which one is the first. This *transition analysis* applies a number of rules to check whether a device state can be a successor of another one.

The result of this analysis can be represented in the *state diagram* containing the consistent device states and the possible transitions between them. Fig. 4 shows a possible diagram for a mass on a spring, where X is the deviation of the mass from the rest-length of the spring (the transitions from and to the equilibrium state, $[X] = 0 \ \wedge \ \partial X = 0$, could be ruled out by considerations about analytic functions).

4 Causal Knowledge - QPT

4.1 Process Vocabularies

Qualitative Process Theory (QPT) ([Forbus 84]) has been developed in order to provide a representational framework for naive physics. *"To understand commonsense physical reasoning we must understand how to reason qualitatively about processes, when they will occur, their effects, and when they will stop"* [Forbus 84]. Processes are the crucial entities in QPT, the origins of changes. A process acts on objects, and its occurrence depends on the existence of these objects in a certain configuration.

Descriptions of such portions of possible worlds are called *individual views* in QPT. Changes are introduced by *processes*. According to Forbus, a process is *"something that acts in time to change the parameters of objects in a situation"*. If one wants to avoid this spirit-like "something", one could perhaps regard a process as a mental image of the interaction of some

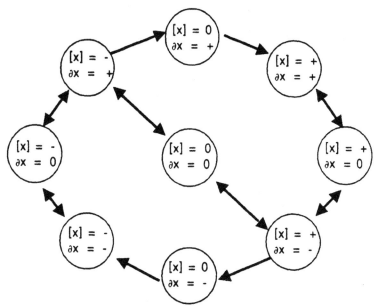

Figure 4   State diagram for a mass on the spring

objects and its overall result without a further structuring or explanation, how this happens.

Processes are described by five slots:

- *Individuals* are the objects involved

- *Preconditions* are externally influenced conditions (not deducible by the QPT system) whereas

- *Quantity conditions* are predictable by means of QPT's inference procedures about quantities

- *Relations* are statements that hold for the process, in particular about interrelated parameters.

- *Influences* which specify the effects of the process in terms of changes of parameters.

Process instances are

- *created* for each collection of objects that satisfy the *individuals* description and

- *active* whenever the *preconditions* and the *quantity conditions* hold. Then also the relations hold, and influences occur.

An example of a process is *heat-flow* (Fig. 5), which will be further discussed in the following.

```
PROCESS HEAT - FLOW

INDIVIDUALS
    src an object, Has - Quantity (src, heat)
    dst an object, Has - Quantity (dst, heat)
    path a Heat - Path, HeatConncetion (path, src, dst)

PRECONDITIONS
    Heat - Aligned (path)

QUANTITY - CONDITIONS
    A [temperature (src)] > A [temperature (dst)]

RELATIONS
    Let flow - rate be a quantity
    A [flow - rate] > ZERO
    flow-rate ∝_Q + (temperature (src) - temperature (dst))

INFLUENCES
    I - (heat (src), A [ flow-rate ] )
    I + (heat (dst), A [ flow-rate ] )
```

Figure 5   The process "heat-flow"

The interaction of processes is mediated by the objects they act on. Since processes can create or destroy objects, and change their parameters, they may affect the individuals or quantity conditions of other processes and, hence, trigger the creation, activation, or deactivation of process instances. For instance, a process *boiling* could destroy the *dst* of the *heat-flow* process by complete evaporation of the water.

Hence, there is no fixed "system structure" in QPT that would correspond to the device topology in the component-oriented approach. Rather, there is collection of the potential processes occurring in a certain domain or task, a *process vocabulary*, which serves as a source for process instances which are activated and deactivated according to the actual conditions, thus establishing a permanently changing *dynamic structure*.

4.2   Description of Changes

Quantities are representations of physical properties of objects, called *parameters*. They are described by their *amount*, A, and their *derivative*, D, which are reals and in turn characterized by two terms: their *magnitudes*, A_m and D_m, respectively, are non-negative real numbers, and their *signs*, A_s and D_s, take on values out of {-1, 0, 1}. Each quantity has a *quantity space* associated, a finite set of numbers with a partial order. Every quantity space contains the element ZERO. The value of a number or magnitude is decribed by ordering relations w.r.t. the elements of this quantity space. Arithmetic on signs is extended in comparison to ENVISION: when a positive and a negative sign lead to ambiguities, additional information may be obtained from the ordering relations involving the magnitudes.

As one source of restrictions imposed on values of parameters, we have encountered the *relations* in individual views and processes. They express functional dependencies of parameters that hold whenever the respective individual view or process is active. There are various ways of expressing such dependencies. Of course, there is *equality*. The most important relation, however, is

- *qualitative proportionality*, which expresses a monotonic dependency. For example,
  $$Q_1 \quad \propto_{Q+} \quad Q_2 \qquad \text{means that} \qquad Q_1 = f(..., Q_2, ...)$$
  where f is some monotonically increasing function w.r.t. Q_2. An example is
  $$\text{flow-rate} \propto_{Q+} (\text{temperature (src)} - \text{temperature(dst)})$$
  in the process *heat-flow*. Analogously, $\propto_{Q-}$ is defined with a monotonically decreasing function.

The ultimate causes for changes of parameters are the
- *influences.* Considering *heat-flow* again,
  $$\text{I- (heat(src), A[flow-rate])}$$
  indicates that the flow-rate will decrease the heat of the source, provided there is no other influence acting on it. In the same process, $I+$ expresses an increase of heat(dst). An influence $I_+(f,g)$ is equivalent to stating $\partial f = [g]$ in ENVISION.

It is important to note that both qualitative proportionality and influences are *directed*. This establishes a difference to constraints, in particular to confluences in ENVISION'S device models. The motivation is to reflect *causality* in QPT's models. *"Naive physics attempts to uncover the ideas of physical reality that people actually use in daily life. Thus the notions that physics throws away (objects, processes, causality) ... are precisely what we must keep"* [Forbus 84].

4.3 Histories

The change of a parameter's value over time is described by a *parameter history*. In QPT, a history is composed of *episodes*, which occur over an *interval* of time, and *events*, which are *instantaneous*. An episode has a start and an end which are events, and may be an end and a start, respectively, for the adjacent intervals. The structure of a parameter history is induced by changes of the parameter value or its parts w.r.t. its quantity space. Fig.

6a presents an example, the history for *temperature* when water is heated, boiling, and completely turned into gas.

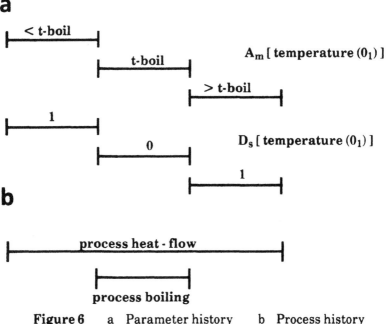

Figure 6   a   Parameter history   b   Process history

The set of all parameter histories of the involved objects can be considered as a counterpart to the state diagram used as a behavior representation in ENVISION. However, because of the crucial role of processes in QPT, state changes imposed on an object include not only its parameter histories, but also its *process histories*. In process histories, the maximal periods of activity and inactivity of a process is recorded. Hence, in the water-heating example, we have to include the process histories shown in Fig. 6b.

The task of deriving the parameter histories and process histories includes several types of deductions. The system has to determine in a loop

- *possible processes* by checking whether their *individuals* specification is met

- *activity of process instances* by checking preconditions and by searching in quantity spaces in order to find fulfilled quantity conditions

- *changes in quantities* by analyzing the combined (either direct or indirect) influences on a quantity and checking whether a limit in the quantity space may be reached. Then possible changes in ordering relations are determined on the basis of derivative signs.

- *quantity hypotheses,* i.e. consistent conjunctions of possible changes. For instance, it has to be checked whether one change blocks another, e.g. by destroying an object. The ordering of changes (changes from equality occur at once) has also to be considered.

- *limit hypotheses,* i.e. changes in the set of active processes.

Thus the task being performed is, starting from a description of some initial situation, to construct the parameter histories and process histories as a representation of the subsequent "behavior", hence a *partial envisionment.*

Among the current goals pursued by the QPT group in Urbana are the construction and usage of large-scale models [Falkenhainer-Forbus 91] and the introduction of actions [Forbus 89].

5   "Qualitative Simulation" - QSIM

5.1   Qualitative States and Constraints

The QSIM-system ([Kuipers 86]) is not dedicated to a specific ontology for representing the physical structure of a given system. QSIM starts with a structural description that is abstracted from a mathematical description in terms of differential equations.

Hence, a structural description of a system in QSIM consists of

- *a set of functions* representing the interesting physical parameters, and

- *a set of constraints* relating these functions, including a "qualitative version" of differentiation.

Each function has a set of *landmarks,* $L = \{L_i\} \subset \mathbb{R} \cup \{-\infty, \infty\}$ associated with it representing the crucial distinctions of values for the respectivce physical parameter. It always includes the landmarks $-\infty$, 0, ∞ (which are the ones that ENVISION uses in its $\{+, 0, -\}$-quantity space. In contrast to QPT, these landmarks are totally ordered: $l_i < l_j$ for $i < j$. Another difference is established by the fact that the landmark set is not considered to be completely predefined; new landmarks that differentiate between parts of the function may be discovered by the inference process, and are added to the set. The idea is that the landmarks capture (at least) the critical points of the function.

Since the landmarks indicate the essential distinctions, they define the *qualitative values* of f being either a landmark or the interval between two adjacent landmarks.

$$\text{qval}\,(f, t) \quad := \quad \left\{ \begin{array}{ll} l_j & \text{if } f(t) = l_j \in L \\[2ex] (l_j, l_{j+1}) & \text{if } f(t) \in (l_j, l_{j+1}) \end{array} \right.$$

Accordingly, the *qualitative state* of a function, f, is given by the pair

$$QS\,(f, t) := \; < qval\,(f, t), qdir\,(f, t) > \,,$$

where qdir, the *qualitative direction*, indicates whether f is increasing, decreasing, or steady

$$qdir\,(f, t) \quad := \left\{ \begin{array}{lll} inc & if & f'(t) > 0 \\ std & if & f'(t) = 0 \\ dec & if & f'(t) < 0 \,. \end{array} \right.$$

The set of functions for the physical parameters together with their (initial) landmark sets form one part of a system description. The structural aspect is given by a set of constraints on these functions. QSIM operates with the following set of primitive constraints, which, besides the *arithmetic constraints*, contain

- *the derivative constraint*
  $DERIV\,(f, g) \quad :\Leftrightarrow \quad f'(t) = g\,(t)$

- *monotonic functional dependencies*
  $M^+\,(f,g) \qquad :\Leftrightarrow \qquad f(t) = H\,(g(t)) \;\; \wedge H'(x) > 0$
  $M_0^+\,(f,g) \qquad :\Leftrightarrow \qquad M^+\,(f,g) \qquad \wedge \quad H\,(0) = 0$
  $M^-\,(f,g) \qquad :\Leftrightarrow \qquad f(t) = H\,(g(t)) \;\; \wedge H'(x) < 0$
  $M_0^-\,(f,g) \qquad :\Leftrightarrow \qquad M^-\,(f,g) \qquad \wedge \quad H\,(0) = 0\,.$

Hence, $M^+(f,g)$ corresponds to $\partial f = \partial g$ in ENVISION.

5.2 Qualitative Simulation

A behavior will be described as changes in qualitative states over time. Formally, for a function, f, a (finite) set of *distinguished time-points*, $\{t_1, ..., t_n\}$, is introduced capturing all time points where f takes on landmark values:

$$\{t' \,|\, f\,(t') \in L\} \subset \{t_1, ..., t_n\}$$

However, in QSIM the time-points will not receive real values, only their order is important.

For an interval, (t_i, t_{i+1}), the qualitative state

$QS\,(f, t_i, t_{i+1}) : = QS\,(f, t)$ for some arbitrary $t \in (t_i, t_{i+1})$

is well-defined. Thus, the *qualitative behavior of a function*, f, can be defined as a sequence of qualitative states at distinguished time-points and the intervals between them:

$$(QS\,(f, t_0), QS\,(f, t_0, t_1), QS\,(f, t_1), ..., QS\,(f, t_{n-1}, t_n), QS\,(f, t_n)).$$

If $F = \{f_1, ..., f_m\}$ is the set of functions defining a system, and its distinguished time points are the union of those of the f_i, a *qualitative state of a system* can be defined by the tuple of the qualitative states of the f_i:

$QS\,(F, t_i) : = (QS\,(f_1, t_i), ..., QS\,(f_m, t_i))$

$QS\,(F, t_i, t_{i+1}) : = (QS\,(f_1, t_i, t_{i+1}), ..., QS\,(f_m, t_i, t_{i+1}))$

and, finally, we have the *qualitative behavior of a system* given by the sequence of its qualitative states:

$(QS(F, t_0), QS(F, t_0, t_1), QS(F, t_1), ..., QS(F, t_{n-1}, t_n), QS(F, t_n))$.
It is obviously a correspondent to a combination of parameter histories in QPT.

The goal of the qualitative simulation performed by QSIM is to determine the possible qualitative behaviors, starting with a given initial state. Due to ambiguities of the qualitative analysis, this is, in general, a tree of qualitative states.

The algorithm that perfoms this analysis can be described as follows:

Initialize ACTIVE-STATES to contain the initial state

For STATE in *active-states* do

1. For each f_i, determine the possible set of *transitions* in STATE

2. For each *constraint*, check *consistency* of these changes

3. For any two *adjacent constraints*, check *consistency* of the changes of the shared parameters (Waltz filtering)

4. Generate all possible *successor states* of *STATE*: the consistent sets of transitions

5. *Filter* out cycles, final and unchanged states, and append the rest to ACTIVE-STATES.

The *transition analysis* for the functions is based on 16 possible types of transitions which are in accordance with continuity conditions w.r.t. the f_i and with their derivatives. Since there may be several consistent sets of transitions, there may be more than one successor state (and, unfortunately, there are often many of them). The result of QSIM is a tree of states, and for each of these states, there exists a system behavior leading to this state. The latter can be displayed by QSIM as an assembly of diagrams for the behaviors of the involved functions.

QSIM has been extended by integrating temporal abstractions [Kuipers 87], the use of quantitative information [Berleant-Kuipers 92], and a process-oriented modeling layer ([Crawford-Farquhar-Kuipers 92]).

6 Mathematical Aspects of Qualitative Reasoning

Originally, QR pursued the goal of envisioning the qualitative behavior of a system with very strong expectations about the correctness of the prediction. The hope was to derive all and only the behaviors exhibited by "real systems". And this term was often identified with the real-valued description of the system's function. In [Struss 88, 90], a formal framework for QR methods was developed which allows to define and analyze their properties such as

- *completeness*,

- *soundness*, and

- *stability* (i.e. insensitivity to formal transformations)

of QR methods, taking "exact" mathematical system descriptions in terms of algebraic or differential equations as the "gold standard".

The inference mechanisms used by the standard approaches for reasoning about parameters and qualitative states of a system, can be formalized by some kind of interval arithmetic. Based on this formalization, properties of the inference schemes can be analyzed. The main result is that *interval-based QR methods* are

- *complete*, i.e. do not miss real-valued solutions, but

- *not sound*, i.e. potentially predict states that do not correspond to any real-valued solution ([Struss 88, 90]).

The cause for these spurious solutions is identified: the multiple occurrence of variables may prevent a globally consistent real-valued solution of a constraint network (or a set of equations) although it is satisfied by intervals. The same reason is responsible for the

- *non-stability* of QR methods w.r.t. elimination of terms in sets of equation.

There is a tension between the goal of operating with a finite number of qualitative values and the wish to have natural properties of operations:

- Except for the sign algebra, a QR method enforcing a *finite number of qualitative values* (i.e. intervals) *conflicts with associativity* of addition and multiplication.

Since these results refer to the inference of qualitative *states*, they reveal problems more fundamental than those underlying the prediction of spurious *behaviors* based on ordinary differential equations which are analyzed in [Kuipers 86] and [Struss 88a]. Major sources of weaknesses in behavior prediction resulting in huge envisioning spaces, are

- the *loss of temporal information* and

- the *merging of behaviors* belonging to different instances of the described class of systems.

The last problem is due to the lack of any means for expressing a global view on behaviors. In order to overcome these limitations,

- *global* criteria for *filtering behaviors*

for autonomous differential equations of second order are developed in [Lee-Kuipers 88], [Struss 88a].

These criteria are based on the uniqueness of the solutions for fixed initial conditions, and the algorithms applying them are inspired by a *phase portrait representation* of the solution space. They are shown to eliminate some spurious predictions of current QR methods for the simple mass-spring system. But, being confined to second order differential equations, they are only a first step towards expressing global constraints on behaviors.

Beyond this, the analysis of dynamic systems based on mathematical techniques, in particular phase space analysis, has become a specialized branch in QR (see e.g. [Dordan 92], [Sacks 90], [Yip 91]).

Also, ways to cope with spurious states have been developed. A principles approach to eliminating spurious states can be based on the relation between algebras at the quantitative and the qualitative level. If qualitative values are intervals (or, more generally, sets) of the original values (e.g. real numbers), then, for instance,

$[x+y] \subseteq [x] \oplus [y]$

holds. As a consequence, stepping down in the lattice shown in Fig. 7 potentially leads to more precise results. In [Williams 88, 92], *hybrid algebras* are introduced which combine algebraic operations at different levels and enable a system to eliminate qualitative ambiguities and spurious solutions by changing the order of qualitative abstraction and algebraic operations, if necessary.

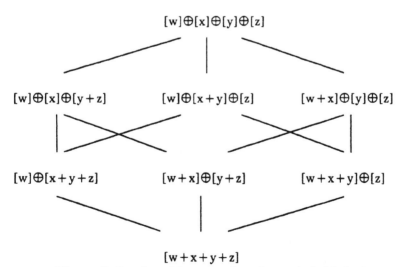

$[w]\oplus[x]\oplus[y]\oplus[z]$

$[w]\oplus[x]\oplus[y+z]$ $[w]\oplus[x+y]\oplus[z]$ $[w+x]\oplus[y]\oplus[z]$

$[w]\oplus[x+y+z]$ $[w+x]\oplus[y+z]$ $[w+x+y]\oplus[z]$

$[w+x+y+z]$

Figure 7   Lattice with expressions from a hybrid algebra

7 Order-of-Magnitude Reasoning - FOG

The theories and systems described so far, have a limited expressive power in dealing with quantities: it is mainly confined to characterizing values by

- their signs,

- intervals of real numbers, and

- ordering relations to other values ("greater than").

In some cases, there is more, but still qualitative, information available about the relationship of values. For instance, we may know, and want to express, that one value is negligible w.r.t. another one. Or, that values are of the same order of magnitude. *Order of magnitude reasoning*, introduced by [Raiman 86, 91], attempts to formalize and use this kind of knowledge in order to resolve ambiguities produced by the classical approaches. For instance, consider two masses of very different weights approaching each other with almost the same amount of velocity from opposite directions. The arithmetic of signs fails to derive any information about the direction of their movement after the impact.

[Raiman 86] suggests the introduction of three relations:

- is-negligible-to   (*Ne*)

- is-close-to   (*Vo*),   where $A\ Vo\ B$ means that   (A-B) *Ne* B

- is-comparable-to   (*Co*),   i.e. has the same sign and order of magnitude,

   allowing to deduce
   $$A\ Ne\ B\ \wedge\ B\ Co\ C \quad \Rightarrow \quad A\ Ne\ C.$$

A *Co* B is weaker than A *Vo* B, and does *not* imply that (A-B) *Ne* B.

The formal system for order of magnitude reasoning, FOG, then consists of a number of rules capturing

- the link to *sign arithmetic*, e.g. by
   $$A\ Vo\ B \quad \Rightarrow \quad [A] = [B]$$

- properties of the relations, such as *symmetry*, e.g.
   $$A\ Vo\ B \quad \Rightarrow \quad B\ Vo\ A,$$
   or *transitivity*, e.g.
   $$A\ Co\ B \quad \wedge \quad B\ Co\ C \quad \Rightarrow \quad A\ Co\ C.$$

- links between the relations, such as
   $$B\ Vo\ A \quad \Rightarrow \quad B\ Co\ A,$$
   $$A\ Ne\ B \quad \wedge \quad B\ Co\ C \quad \Rightarrow \quad A\ Ne\ C,$$
   and

- relationships to *algebraic operations*, e.g.

$$A \; Ne \; B \quad \wedge \quad C \; Vo \; D \quad \Rightarrow \quad (A \otimes C) \; Ne \; (B \otimes D) \, ,$$
$$(A \oplus B) \; Vo \; C \quad \wedge \quad B \; Ne \; A \quad \Rightarrow \quad A \; Vo \; C.$$

This formal system, which has an interpretation in non-standard analysis ([Raiman 86]), solves the collision problem stated above, i.e. from

$$V_i \; Co \; \text{-} \; v_i$$

(the initial velocities of the masses, M and m, being close), and

$$m \; Ne \; M$$

(one mass being negligible w.r.t. the other one),
it deduces

$$[v_f] = [V_f] \, ,$$
$$V_f \; Co \; V_i \, , \text{and}$$
$$v_f \; Co \; V_i$$

for the velocities, v_f and V_f, respectively, after the collision.

This approach to order of magnitude reasoning was used in DEDALE, a system for trouble shooting analog circuits ([Dague-Devès-Raiman 86]) where the deviation of values by orders of magnitude are considered as symptoms, i.e. indications of a malfunction.

It has two features that may cause problems in some cases:

● Transitivity of the Co and Vo relations raises the problem that, for instance, a chain of *many* close-to relations could connect two values that are no longer close to each other.

● There is no "smooth" transition between adjacent orders of magnitude due to the fact that there is no upper bound for infinitesimals. This is sometimes appropriate since we cannot really define a boundary where e.g. "small" turns into "medium". However, we know that a "small" value might grow "smoothly" to become medium.

The second problem was addressed by [Mavrovouniotis-Stephanopoulos 87] and [Dague 88]. The latter presents a solution based on the introduction of a fourth relation "is-distant-from" (the negation of is-close-to). Order-of-magnitude reasoning has also been applied to qualitative simulation based on differential equations ([Davis 87]). The exaggeration technique of [Weld 88a] shares much of the spirit with order-of-magnitude reasoning.

8   Ongoing Work

There exists a great variety of attempts to extend the scope of existing methods and to develop new formalisms to cover other features of qualitative reasoning. To mention just a few topics:

● *analytic solutions* to qualitative differential equations ([Schaefer 91]),

● *qualitative vector algebra* ([Weinberg-Uckun-Biswas 92]),

● axiomatization of *topological relations* and reasoning ([Randell-Cohn 92]),

- *employing fuzzy logic* in qualitative simulation to cope with imprecise values ([Shen-Leitch 92]),

- *dimensional analysis* ([Nigam-Bhaskar 92]).

A major theme in current research addresses the fundamental modeling issue again. It is based on the insight that, normally, not a unique model of a system is required, but a set of different models which reflect different contexts, physical aspects, tasks, levels of detail etc. *Multiple modeling* involves a number of issues, such as

- *using multiple ontologies* ([Zheng-Yang 92]),

- determining different *relations between models,* such as abstraction, simplification, and approximation, and constructing *graphs of models* ([Addanki-Cremonini-Pernberthy 91], [Iwasaki 92], [Struss 92], [Weld 92]),

- *selecting* the appropriate model for a given task and *switching models* during the problem solving process (see e.g. [Falkenhainer-Forbus 91], [Weld-Addanki 92] for query answering, [Hamscher 91], [Struss 92] for diagnosis).

Emphasizing the modeling task again, which we introduced as the core problem in QR, is not a retreat into pure research. On the contrary, the use of different abstractions and simplifications of models appropriate for the respective task and case turns to be a prerequisite for building effective and efficient problem solvers based on qualitative modeling.

Recently, significant progress has been made in applying QR techniques to real problems, although the number of actually deployed systems is still small. Among the tasks being tackled, we find

- *diagnosis,* the most advanced application area (e.g. [Dague-Devès-Raiman 87], [Dvorak-Kuipers 89], [Hamscher 91], [Struss 92]),

- *measurement interpretation* ([deCoste 91]),

- generation of *numerical simulation* systems ([Forbus-Falkenhainer 92]),

- *sensor placement* ([Doyle et al. 92]),

- *mechanisms* ([Forbus-Nielsen-Faltings 91]),

- *traffic modeling* ([Nypelseer 92]),

- *innovative design* ([Williams 92]).

Thus, the research field of qualitative reasoning has reached a state of maturity that produces useful techniques for application systems. One has to keep in mind that it is concerned with one of the fundamental problems in understanding human intelligence, namely reasoning about and interacting with the physical world, which still prompts for more basic research. Even though the field has not yet developed a closed set of formalisms and a uniform methodology, it is probably one of the most successful and promising area of Artificial Intelligence.

9   Bibliography

[Addanki-Cremonini-Penberthy 91]
  Addanki, S., Cremonini, R., Penberthy, J.S.: Graphs of models. Artificial Intelligence 51 (1-3), October1991.
[Berleant-Kuipers 92]
  Berleant, D., Kuipers, B.: Combined Qualitative and Numerical Simulation with Q3. In: [Faltings-Struss 92].
[Bobrow 84]
  Bobrow, D.: Qualitative Reasoning about Physical Systems. Artificial Intelligence 24 (1-3), 1984.
[Chiu-Kuipers 92]
  Chiu, C., Kuipers, B.: Comparative analysis and qualitative integral representations. In: [Faltings-Struss 92].
[CompIntell 92]
  Computational Intelligence 8(2), 1992.
[Crawford-Farquhar-Kuipers 92]
  Crawford, J., Farquhar, A., Kuipers, B.: QPC: A Compiler from Physical Models into Qualitative Differential Equations. In: [Faltings-Struss 92].
[Dague 88]
  Dague, P.: Order of Magnitude Revisited. 2nd Workshop on Qualitative Physics, Paris, August 1988.
[Dague-Deves-Raiman 87]
  Dague, P., Deves, P., Raiman, O.: Troubleshooting: When Modeling is the Trouble. AAAI 1987.
[Davis 82]
  Davis, R.: Expert Systems: Where Are We? And Where Do We Go From Here? The AI Magazine, Spring 1982.
[Davis 84]
  Davis, R.: Diagnostic Reasoning Based on Structure and Behavior. Artificial Intelligence 24 (1-3), 1984.
[Davis 87]
  Davis, E.: Order of Magnitude Reasoning in Qualitative Differential Equations. New York Univ. Tech. Report #312, 1987.
[DeCoste 91]
  DeCoste, D.: Dynamic Across-Time Measurement Interpretation. Artificial Intelligence 51 (1-3), October 1991.

[de Kleer-Brown 84]
    de Kleer, J., Brown, J.S.: A Qualitative Physics Based on Confluences.
    Artificial Intelligence 24 (1-3), 1984.
[de Kleer-Williams 91]
    de Kleer, J.,Williams, B.C. (eds.): Artificial Intelligence (Special
    Volume on Qualitative Reasoning About Physical Systems II), 51(1-3),
    1991.
[Dordan 92]
    Doordan, O.: Mathematical Problems Arising in Qualitative
    Simulation of a Differential Equation. Artificial Intelligence 55 (1),
    1992.
[Downing 92]
    Downing, K.: The Qualitative Criticism of Circulatory Models Via
    Bipartite Teleological Analysis. In: [Faltings-Struss 92].
[Doyle et al. 89]
    Doyle, R.J., et al.: A Focused, Context-Sensitive Approach to
    Monitoring. In: Proceedings of IJCAI 1989.
[Doyle et al. 92]
    Doyle, R.J., et al.: Sensor Selection in Complex System Monitoring
    Using Information Quantification and Causal Reasoning. In: [Faltings-
    Struss 92].
[Dvorak-Kuipers 89]
    Dvorak, D., Kuipers, B.: Model-Based Monitoring of Dynamic Systems.
    IJCAI 1989.
[Falkenhainer-Forbus 91]
    Falkenhainer, B., Forbus, K.D., Compositional Modeling of of Physical
    Systems. Artificial Intelligence 51 (1-3), October 1991.Also in:
    [Faltings-Struss 92].
[Faltings-Struss 92]
    Faltings, B., Struss, P. (eds.): Recent Advances in Qualitative Physics.
    MIT Press, 1992.
[Forbus 84]
    Forbus, K. D.: Qualitative Process Theory. Artificial Intelligence 24 (1-
    3), 1984.
[Forbus 86]
    Forbus, K. D.: Interpreting Measurements of Physical Systems. In:
    Proc. 5th National Conf. on Artificial Intelligence, Philadelphia, PA,
    1986.
[Forbus 89]
    Forbus, K.D.: Introducing Actions into Qualitative Simulation. In:
    Proceedings of IJCAI 1989.
[Forbus 90]
    Forbus, K.D.: The Qualitative Process Engine. In: [Weld-de Kleer 90].

[Forbus 92]
Forbus, K.D.: Pushing the Edge of the (QP) Envelope. In: [Faltings-Struss 92].

[Forbus-Falkenhainer 92]
Forbus, K.D., FalkenhainerW B.: Self-explanatory Simulations: Integrating Qualitative and Quantitative Knowledge. In: [Faltings-Struss 92].

[Forbus-Nielsen-Faltings 91]
Forbus, K.D., Nielsen, P., Faltings, B.: Qualitative Spatial Reasoning: the CLOCK Project. Artificial Intelligence 51 (1-3), October 1991.

[Fouche-Kuipers 92]
Fouche, R., Kuipers, B.: An Assessment of Current Qualitative Simulation Techniques. In: [Faltings-Struss 92].

[Hamscher 91]
Hamscher, W.: Modeling Digital Circuits for Troubleshooting. Artificial Intelligence, 51 (1-3), 1991.

[Hamscher-Console-de Kleer 92]
Hamscher, W., de Kleer, J., Console, L. (eds.): Readings in Model-based Diagnosis: Diagnosis of Designed Artifacts Based on Descriptions of their Structure and Function. Morgan Kaufmann, San Mateo, 1992.

[Hayes 78]
Hayes, P.: The Naive Physics Manifesto. In: Michie, D. (ed.): Expert Systems in the Micro-Electronic Age. Edinburgh, 1978.

[Hayes 85]
Hayes, P.: The Second Naive Physics Manifesto. In: Hobbs, J., Moore, R. (eds.): Formal Theories of the Commonsense World. Norwood, 1985.

[Hellerstein 92]
Hellertstein, J.: Obtaining Quantitative Predictions from Montone Relationships. In: [Faltings-Struss 92].

[Hibler-Biswas 92]
Hibler, D.L., Biswas, G.: TEPS: the Thought Experiment Approach to Qualitative Physics. In: [Faltings-Struss 92].

[Hobbs-Moore 85]
Hobbs, J., Moore, R. (eds.): Formal Theories of the Commonsense World. Norwood, 1985.

[Hyun-Kyung 92]
Hyun-Kyung, K.: Qualitative Kinematics of Linkages. In: [Faltings-Struss 92].

[Iwasaki 88]
Iwasaki, Y.:Causal Ordering in a Mixed Structure. AAAI 1988.

[Iwasaki 92]
Iwasaki, Y.: Reasoning with Multiple Abstraction Models. In: [Faltings-Struss 92].

[Kuipers 86]
   Kuipers, B.: Qualitative Simulation. Artificial Intelligence 29 (3), 1986.
[Kuipers 87]
   Kuipers, B.: Abstraction by Time-Scale in Qualitative Simulation. AAAI 1987.
[Kuipers-Chiu 87]
   Kuipers, B., Chiu, C.: Taming Intractable Branchung in Qualitative Simulation. IJCAI 1987.
[Kuipers et al. 91]
   Kuipers, B., Chiu, C., Dalle Molle, D.T., Throop, D.R.: Higher-order Derivative Constraints in Qualitative Simulation. Artificial Intelligence 51 (1-3), October 1991.
[Kuipers-Berleant 88]
   Kuipers, B., Berleant, D.: Using Incomplete Quantitative Knowledge in Qualitative Reasoning. In: 2nd Workshop on Qualitative Physics, Paris, 1988.
[Lee-Kuipers 88]
   Lee, W.W., Kuipers, B.: Non-Intersection of Trajectories in Qualitative Phase Space: A Global Constraint for Qualitative Simulation. AAAI 1988.
[Mavrovouniotis-Stephanopoulos 87]
   Mavrovouniotis, M.L, Stephanopoulos, G.: Reasoning with Orders of Magnitude and Approximate Relations. AAAI 1987.
[Murthy 88]
   Murthy, S.: Qualitative Reasoning at Multiple Resolutions. AAAI 1988.
[Nigam-Bhaskar 92]
   Nigam, A., Bhaskar, R.: Qualitative Reasoning about a Large System Using Dimensional Analysis. In: [Faltings-Struss 92].
[Nypelseer 92]
   van Nypelseer, P.: Qualitative Change Waves - the Automatic Detection of Traffic Accidents. In: [Faltings-Struss 92].
[Raiman 86]
   Raiman, O.: Order of Magnitude Reasoning. AAAI 1986.
[Raiman 91]
   Raiman, O.: Order of Magnitude Reasoning. Artificial Intelligence 51 (1-3), October 1991.
[Randell-Cohn 92]
   Randell, B., Cohn, G.: Exploring Naive Topology: Modeling the Force Pump. In: [Faltings-Struss 92]
[Rieger-Grinberg 77]
   Rieger, C., Grinberg, M.: The Declarative Representation and Procedural Simulation of Causality in Physical Mechanisms. IJCAI 1977.

[Rose-Kramer 91]
Rose, P., Kramer, M.A.: Qualitative Analysis of Causal Feedback. AAAI 1991.

[Sacks 85]
Sacks, E. Qualitative Mathematical Reasoning. IJCAI 1985.

[Sacks 90]
Sacks, E.: Automatic Qualitative Analysis of Dynamic Systems Using Piecewise Linear Approximation. Artificial Intelligence 41 (1990).

[Schaefer 91]
Schaefer, P.: Analytic Solution of Qualitative Differential Equations. AAAI 1991.

[Shen-Leitch 92]
Shen, Q., Leitch, R.: Integrating Common-Sense and Qualitative Simulation by the Use of Fuzzy Sets. In: [Faltings-Struss 92].

[Simmons 86]
Simmons, R.: "Commonsense" Arithmetic Reeasoning. AAAI 1986.

[Stallman-Sussman 77]
Stallman, R. M., Sussman, G. J.: Forward Reasoning and Dependency - Directed Backtracking in a System for Computer-Aided Circuit Analysis. Artificial Intelligence 9 (2), 1977.

[Struss 87]
Struss, P.: Multiple Representation of Structure and Function. In: J.Gero (ed.): Expert Systems in Computer-Aided Design. Amsterdam, 1987.

[Struss 88]
Struss, P.: Mathematical Aspects of Qualitative Reasoning. In: International Journal for Artificial Intelligence in Engineering 3 (3)1988.

[Struss 88a]
Struss, P.: Global Filters for Qualitative Behaviors. AAAI 1988.

[Struss 90]
Struss, P.: Problems of Interval-Based Qualitative Reasoning - Revised Version. In: [Weld-de Kleer 90].

[Struss 92]
Struss, P.: What's in SD? Towards a Theory of Modeling for Diagnosis. To appear in: [Hamscher-Console-de Kleer 92].

[Sussman-Stallman 75]
Sussman, G. J., Stallman, R. M.: Heuristic Techniques in Computer-Aided Circuit Analysis. IEEE Transactions on Circuits and Systems CAS-22 (11), 1975.

[Weinberg-Uckun-Biswas 92]
Weinberg, J., Uckun, S., Biswas, G.: Qualitative Vector Algebra. In: [Faltings-Struss 92].

[Weld 88]
Weld, D.S.: Comparative Analysis. Artificial Intelligence 36, 1988.

[Weld 88a]

Weld, D.S.: Exaggeration. Proceedings of AAAI 1988.

[Weld-Addanki 92]

Weld, D.S.,.: Approximation Reformulations. In: [Faltings-Struss 92].

[Weld-de Kleer 90]

Weld, D.S., de Kleer,J. (eds.): Readings in Qualitative Reasoning about Physical Systems,Morgan Kaufmann, San Mateo, CA, 1990.

[Weld-Addanki 92]

Weld, D.S., Addanki, S.: Task-Driven Model Abstraction. In: [Faltings-Struss 92].

[Williams 86]

Williams, B.: Doing Time: Putting Qualitative Reasoning on Firmer Ground. AAAI 1986.

[Williams 88]

Williams, B.C.: MINIMA: A Symbolic Approach to Qualitative Reasoning, AAAI 1988, pages 105-112.

[Williams 92]

Williams, B.C.: Interaction-based Invention: Designing Devices from First Principles. In: [Faltings-Struss 92].

[Williams 92a]

Williams, C.P.: Analytic Abduction from Qualitative Simulation. In: [Faltings-Struss 92].

[Yip 91]

Yip, K.M.-P.: Understanding Complex Dynamics by Visual and Symbolic Reasoning. Artificial Intelligence 51 (1-3), October 1991.

[Zheng-Yang 92]

Zheng-Yang, L.: A Charge-carrier Ontology for Reasoning about Electronics. In: [Faltings-Struss 92].

On the Topological Structure of Configuration Spaces

Dr. Jürgen Sellen

Fachbereich 14 – Informatik
Universität des Saarlandes, Germany

Abstract. The presented work investigates the topological structure of the configuration spaces of mechanisms. We will demonstrate the practical importance of the considered questions by giving some motivations, especially from the viewpoint of qualitative reasoning.

Although topological invariants such as the fundamental group are computable, they cannot provide us with a useful tool for classifying arbitrary mechanisms. We will show that for any finitely represented group one can construct a simple mechanism in the plane such that the fundamental group of its configuration space is isomorphic to the given group. The construction shows the undecidability of some interesting problems, e.g. the problems of homeomorpy and homotopy equivalence of configuration spaces.

Key words: qualitative reasoning, qualitative kinematics, mechanisms, configuration spaces, topology, homeomorphy, homotopy, fundamental group.

1 Introduction

In this paper we consider mechanisms consisting of rigid objects in a 2- or 3-dimensional euclidian space, the physical world. We denote a collection of objects as a *kinematic scene* since we will not consider any dynamic problems.

The notions of *configuration space* and *free space* are central for the wide field of robotics. The configuration space of a kinematic scene is the space which is spanned by all the parameters that determine the positions of the movable objects bijectively. At a fixed time point the position of all parts of the mechanism can be described by a point of the configuration space, which we call a *configuration*.

If we consider as an example two rotary gearwheels in the plane, then the configuration space of this scene is a torus: The rotation angles ϕ and ψ of both gearwheels span a rectangle. By identifying opposite sides the torus is generated.

As the parts of a mechanism generally constrain their positions, the configuration space can be divided into parts consisting of physically legal and illegal configurations, the so called *free space* and *blocked space*. The boundary of these parts consists of all those configurations, in which two or more objects are in contact.

If in our example the teeth of the gears affect such that the rotation of one gear forces a rotation of the other, then the free space consists of several strips around the torus. If we are only interested in one connected component, we can regard the free space as nothing else than a circle line, corresponding to a simultaneous rotation of both gears.

In this work we investigate the topological structure of the free space of a given kinematic scene. In order to demonstrate that understanding and characterizing properties of free spaces is both a relevant and interesting subject of research, we give a motivation from the viewpoint of qualitative reasoning.

When reasoning about the physical world, people normally use qualitative arguments but not strategies provided by classical physics such as solving differential equations over real-valued functions.

Qualitative reasoning is a growing field in Artificial Intelligence that is concerned with a description of physical processes using qualitative terms, aiming to provide the next generation of expert systems with more "human skills".

Pursued goals are e.g. to explain or to predict the behaviour of a given physical system, or even to give an envisionment of all possible behaviours. Existing programs try to achieve this by detecting qualitatively different states and constructing a diagram which represents possible state transitions (cf. e.g. [1]).

Qualitative kinematics adresses the problem of reasoning about mechanical systems, e.g. to allow engineers to understand or evaluate mechanisms. This goal can not be achieved by classical numerical simulation techniques: the behaviour of a complex mechanism may depend on several parameters, but simulation only traces a path in the free space starting at a given initial configuration.

Generating a representation of all possible behaviours can be achieved by decomposing the free space in regions representing qualitative states, thus characterizing the working of mechanisms by identifying operating modes and mode transitions.

Despite the high computational complexity, there are several pragmatic approaches to compute such decompositions in regions satisfying adequate criteria, e.g. exploiting the fact that most mechanisms are built modularly of simple subassemblies ([2]) or taking into account acting forces ([3]). However, these methods can only deal with special classes of "feasible" mechanisms (shown to be fairly large in [2]), and in order to get a manageable output, further abstraction is necessary to reduce complexity.

To illustrate our considerations, we study as a small example two gearwheels A, B in the plane that can be coupled by moving a third one C between them, this motion being restricted to linear translation. This situation can e.g. be found as simple coupling mechanism in a tape recorder, one gear attached to the motor by a kinematic chain, and the other driving the tape (in old tape recorders such couplings are established mechanically by pressing the operating mode buttons $REW/FF/PLAY$).

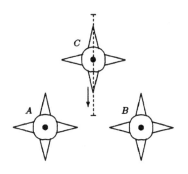

The topological structure of the free space for fixed positions of the gear C gives us a distinction of these positions in 3 intervals:

If the teeth cannot collide, the free space is a 3-dimensional torus. If the motions of the gears A,C and B are fully coupled, the free space corresponds to a circle line. Between both extremes there is a region of interplay where the free space consists of a torus with holes corresponding to collisions of the teeth tips.

A decomposition of the free space in e.g. convex regions would have a complexity $O(\#\text{teeth}^3)$, which would be further increased when taking into account kinematic chains attached to the gears. As each tooth affects the decomposition, the resulting description may get too detailed for many problems, e.g. classification. We observe that the change of the topological structure of the free space describes the function of the coupling mechanism. Thus topological considerations could be useful when dealing with the question of mechanism understanding, e.g. as a tool to distinguish qualitatively between parameter values or to perform further abstraction in a region decomposition.

While connectivity of the free space is too weak to reason about mechanisms, taking into account all knowledge about the geometric shape may be too restrictive. Topology as mathematical discipline can be looked upon as an effort to perform qualitative reasoning about geometric structures, providing properties of the free space lying between both extremes. Therefore it is natural to ask if it is feasible to characterize or even to *compare* mechanisms according to such properties. Is it possible to construct *normal forms* to arbitrary mechanisms with respect to topological invariants of their free space ?

We first give a brief mathematical intoduction to topological invariants of arbitrary spaces and discuss their computability with respect to applications in robotics. Then we will turn to the question if these invariants can be used to classify mechanisms. The answer will be provided by a construction which shows that even simple mechanisms have very arbitrary free spaces whose structure according to the invariants in consideration cannot be interpreted in the sense of Turing-Decidability.

2 Topological Preliminaries

To investigate computability we first have to specify the notion of kinematic scenes. As objects we allow arbitrary semi-algebraic, connected subsets of $\mathbf{R}^d$ $(d = 2, 3)$, i.e. an object is described by a boolean combination of inequalities of the form $p(x) > 0$ or $p(x) \geq 0$, with p a multivariate polynomial with integer coefficients. One should mind that, for the sake of simplification, objects may have "open" and "closed" sides corresponding to the two kinds of inequalities. The possible motions of an object are given by the group of rigid transformations of $\mathbf{R}^d$. Some objects may have no freedom of motion and hence are called obstacles.

Under these general suppositions the configuration space equals $(\mathbf{R}^d \times SO(d))^m$, m being the number of movable objects and $SO(d)$ denoting the orthogonal group of $d \times d$-matrices, and hence can be embedded in $\mathbf{R}^{m \cdot (d+d^2)}$ as semi-algebraic subset. Consequently the free space, consisting of all configurations in which the objects do not overlap, is a semi-algebraic subset of the euclidian space $\mathbf{R}^n$. Schwartz and Sharir have shown that it is possible to compute the *homology groups* of the free space using Collin's algorithm in order to produce a decomposition as regular cell complex with semi-algebraic cells (cf. [4],[5]). This provides us with a general solution to the motion planning problem (remark that the 0-th homology group yields the number of connected components of the free space).

The homology groups, especially the Betti numbers, describe the geometrical complexity of the free space: figuratively speaking, the k-th Betti number is the number of k-dimensional holes. But in order to compare mechanisms we are interested in topological properties that allow to transform configurations and motions of one mechanism to those of another one and vice versa. So it is natural to ask for the two most common characterizations of topological spaces, namely *homeomorphy* and *homotopy equivalence*, for which we will give intuitive definitions now.

As we are only interested in subspaces of $\mathbf{R}^n$ we do not have to take care of how the topology may be defined: the significant topologies, i.e. those based on any norm, are equivalent and so we can speak of continuity in the common sense. Two subspaces of $\mathbf{R}^n$ are called homeomorphic iff there is a bijection f between them such that both f and f^{-1} are continuous. Examples are scalings, rotations or affine transformations. Homeomorphisms between the free spaces of mechanisms allow to map motions, i.e. continuous paths in the free space, bijectively between mechanisms and thus to translate motion planning problems. The notion of homotopy equivalence provides us with a weaker subdivision of topological spaces into classes: two subspaces of $\mathbf{R}^n$ are homotopy equivalent iff they can be transformed to each other by a process of "blowing up" and "shrinking" (more formally: iff there exists another space which contains both as deformation retracts). This allows us to compare paths in the free spaces of mechanisms up to continuous deformations and seems to be a more adequate criterion of classifying mechanisms than homeomorphy.

A useful invariant of homotopy equivalence is the *fundamental group*. We consider paths in a subspace $X \subseteq \mathbf{R}^n$ to be equivalent iff they are homotopic, i.e. iff they can be transferred to each other by a continuous deformation process in X. The fundamental group $\pi_1(X, x)$ consists of the homotopy classes of all loops at a fixed base point x. The group multiplication is induced by the connection of loops. If X is connected, the base point x is redundant. By the fundamental group of a kinematic scene we denote the fundamental group of the connected component of its free space we are interested in.

If we ask for an overview of the possible motions of a complex mechanism, a representation of the fundamental group by generators and relations could help: the generators give an overview of the essential different cyclic motions, the relations describe the degree of interaction of the parts. This may be illustrated by two gearwheels, each having exactly one tooth: if the teeth of both gearwheels can collide such that the free space corresponds to a torus with one hole, then the fundamental group of this scene is the free group with two generators a, b, each one corresponding to a rotation of a gear. If no collision is possible, the fundamental group turns to the free abelian group with two generators, i.e. a group with generators a, b and relation $ab = ba$ stating that a and b are "independant" such that their order may be exchanged.

Natural questions arising from this observation concern the computability and the possible complexity of the fundamental group. The cell decomposition of the free space which can be obtained by a variant of Collin's algorithm can serve as a basis to compute a finite representation of the fundamental group. The underlying algorithm is often denoted as Poincaré's procedure:

- Consider the 1–dimensional skeleton X^1 of the cell complex X to be a graph and compute a spanning tree B of X^1.

- Direct the edges of $X^1 \setminus B$ arbitrarily and mark them with symbols $a_1, \ldots, a_k$, the *generators* of the fundamental group.

- For each face F in the 2–dimensional skeleton X^2 choose a closed path in X^1 around this face. If this path enters edges marked with $a_{i_1}, \ldots, a_{i_l}$ in this order let $\gamma_F := a_{i_1}^{\varepsilon_1} \ldots a_{i_l}^{\varepsilon_l}$; $\varepsilon_j = \pm 1$ denoting the direction in which the path runs through a_{i_j} according to its orientation. The relations $\gamma_F = 1$ form the *relations* of the fundamental group.

The running time of the entire algorithm is determined by the running time of Collin's algorithm, which is doubly exponential in the dimension of the free space. Examples like the planar "game" consisting of $n^2 - 1$ sliding boxes of size 1×1 arranged in a box of size $n \times n$ show that the smallest number of generators needed to represent the fundamental group may grow at the rate faculty of the number of movable objects (respectively the dimension of free space).

Though we cannot expect efficient algorithms for the general case, we nevertheless have seen that it is possible to compute a representation of the fundamental group of an arbitrary kinematic scene. But another question is if we can interpret this representation, i.e. if we are for instance able to detect algebraic properties or to use the representations to compare mechanisms. These problems are closely related to the problems of homeomorphy and homotopy equivalence and we will attack them in the next sections.

3   The Main Theorem

Starting from the fact that there is no algorithmic solution to the word problem for finite group representations, a huge variety of group theoretic problems have

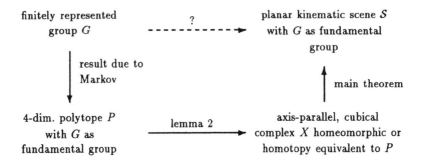

Figure 1: what we prove

been proved to be undecidable. By a construction showing that each finite group representation can occur as fundamental group of a 4–dimensional polytope, Markov could reduce the problem of homeomorphy to the isomorphism problem of groups (cf. [6]). With similar constructions, Boone, Haken and Poénaru could show the undecidability of further topological classification problems such as the problem of homotopy equivalence (cf. [7]).

So if we want to classify kinematic scenes according to the stated topological invariants of their free space, we first have to answer the question, how "arbitrary" the free spaces of the considered scenes can be:

Is it possible to construct for any given 4–dimensional polytope a "real" mechanism whose free space is homeomorphic / homotopy equivalent to the given polytope ?

(Remark that this also would imply that for any given finite group representation one can construct a mechanism which has the represented group as its fundamental group.)

Our main theorem answers this question with yes - thus giving a negative answer to the classification problem.

In order to prove this result, we take a roundabout way via simple subsets of the euclidian space built up by axis-parallel boxes, i.e. boxes for which each boundary edge is parallel to one of the coordinate axes (cf. diagram 1). Furthermore, we require the considered sets to be compact and to be embedded in the unit cube $[0, 1]^n$. Analogous to simplicial complexes, we define axis-parallel, cubical complexes:

Definition 1 *An* axis-parallel, cubical complex $X \subseteq \mathbf{R}^n$ *is a set of disjoint, axis-parallel, open boxes of the form $B = I_1 \times \ldots \times I_n$, I_j being an open interval or a point interval $\forall\ 1 \leq j \leq n$, such that the sides of each box are again contained in X.*

For simplicity, we also denote with X the union of all contained boxes.

The following lemma states that the confinement to axis-parallel, cubical complexes is no essential restriction with respect to the topological invariants we consider:

Lemma 1 *It is possible to construct for each finite polytope $P \subseteq [0,1]^n$ (i.e. consisting of finitely many simplices) a finite, axis-parallel, cubical complex $X \subseteq [0,1]^n$ such that X is homotopy equivalent to P. In addition, if P is an n-dimensional manifold, (i.e. each point has a neighbourhood homeomorphic to R^n), then one can construct X as an n-manifold homeomorphic to P.*

Sketch of proof:
First we have to render P to an n-dimensional manifold homotopy equivalent to P. This can be achieved by triangulating the unit cube $[0,1]^n$, such that the triangulation contains P; by constructing the second normal subdivision of this triangulation and then taking only those (closed) simplices of the subdivision, which share at least one point with P.

By repeatedly cutting off small enough axis-parallel boxes from the boundary of the resulting n-dimensional polytope $P' \subseteq [0,1]^n$, it is easy to obtain a finite, axis-parallel, cubical complex $X \subseteq [0,1]^n$ homeomorphic to P' (inductively cut off first the points, then the edges, then the 2-dim. faces etc.). $\square$

In order to carry over known undecidability results from group theory and topology to fundamental groups and free spaces of kinematic scenes it suffices to prove the following main theorem:

Theorem 1 *For each finite, axis-parallel, cubical complex $X \subseteq [0,1]^n$ it is possible to construct a planar kinematic scene consisting of polygonal objects with purely translational freedom of motion, such that the free space of this scene is homotopy equivalent to X. Furthermore, if X is an n-dimensional manifold, the closure of the free space of this scene is homeomorphic to X.*

This theorem completes diagram 1: Markov's proof of the undecidability of the homeomorphism problem works with a 4-dimensional polytope in a 4-dimensioinal space and thus can be carried over to the closure of free spaces; concerning homotopy equivalence and fundamental groups we do not have to take care about the dimensions, i.e. the polytope may be embedded in a more than 4-dimensional space.

The reason why we confine our investigation to axis-parallel, cubical complexes is because they allow a recursive decomposition into lower-dimensional parts (see figure 2).

Lemma 2 *If X is an axis-parallel, cubical complex in $[0,1]^n$, $n \geq 2$, then there exists a decomposition*

$$X = X^{[1]} \times I_1 \,\dot\cup\, \ldots \dot\cup\, X^{[l]} \times I_l$$

such that $(I_1, \ldots, I_l)$ is a decomposition of $[0,1]$ into intervals and $\forall\, 1 \leq j \leq l$, $X^{[j]}$ is an axis-parallel, cubical complex in $[0,1]^{n-1}$.

By considering the intervals $I_1, \ldots, I_l$ as nodes of a tree, it is possible to describe the n-dimensional complex by a planar structure:

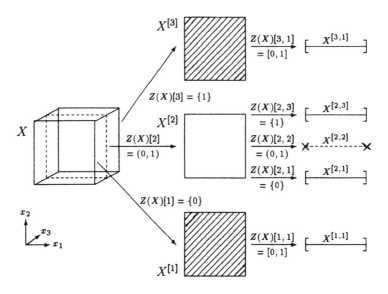

Figure 2: decomposition of $X = \partial[0,1]^3$

Definition 2 *Let $X \subseteq [0,1]^n$ be an axis-parallel, cubical complex. If $n = 1$, let $(I_1, \ldots, I_l)$ be a decomposition of $[0,1]$ with $I_j \cap X = I_j$ or $I_j \cap X = \emptyset$, $1 \leq j \leq l$; if $n > 1$, let $X = X^{[1]} \times I_1 \cup \ldots \cup X^{[l]} \times I_l$ according to lemma 2.*
A decomposition tree $Z(X)$ is a partial function from the set of all finite tupels $\mathbf{N}^*$ (paths) to the set of all subintervals of $[0,1]$, recursively defined by

$$Z(X)[j] := I_j \; ; \; 1 \leq j \leq l,$$
$$Z(X)[j, i_1, \ldots, i_r] := Z(X^{[j]})[i_1, \ldots, i_r] \, , \, 1 \leq j \leq l \, , \, r < n.$$

Any path $[i_1, \ldots, i_r]$ of length $r \leq n$ in the tree corresponds to a subbox $[0,1]^{n-r} \times Z(X)[i_1, \ldots, i_r] \times \ldots \times Z(X)[i_1, i_2] \times Z(X)[i_1] \subseteq [0,1]^n$. The intersection of this subbox with X does not "depend" on the last r variables, i.e. the projection of the intersection to the first $n - r$ variables does not change when $(x_{n-r+1}, \ldots, x_n)$ vary over $Z(X)[i_1, \ldots, i_r] \times \ldots \times Z(X)[i_1]$. This projection equals $X^{[i_1, \ldots, i_r]} := (\ldots ((X^{[i_1]})^{[i_2]}) \ldots)^{[i_r]}$, and is decomposed in the $r+1$-th recursion step according to the $n - r$-th variable.
The subboxes $Z(X)[i_1, \ldots, i_n] \times \ldots \times Z(X)[i_1]$ corresponding to all paths from the root to the leaves of the decomposition tree constitute a decomposition of the unit cube into boxes belonging either to X or to $[0,1]^n \setminus X$. Accordingly, we mark the leaves with "$\in X$" and "$\notin X$".
We will illustrate the decomposition of axis-parallel, cubical complexes by the example of the surface of the cube $X = \partial[0,1]^3$ (see figure 2). This example will be continued in the next section to illustrate the construction of scenes.
If we scan according to the last variable x_3 across the unit cube, then the intersection of the scan plane with X changes at only finitely many points, the

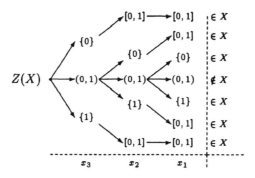

Figure 3: complete decomposition tree $Z(X)$

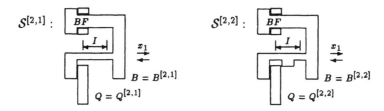

Figure 4: scenes $S^{[2,1]}$ / $S^{[2,2]}$ for the 1-dimensional free spaces $X^{[2,1]}$ / $X^{[2,2]}$

endpoints of the decomposition intervals. The intersection consists of the front side for $x_3 \in \{0\}$, of the "frame" for $x_3 \in (0,1)$ and of the back side for $x_3 \in \{1\}$. The decomposition can be continued recursively according to variable x_2. The complete decomposition tree is shown in figure 3.

4 The Construction of Scenes for Arbitrary Spaces

We will construct the scenes recursively, i.e. we will build the scene for an n-dimensional space X using the scenes for the $n-1$-dimensional spaces $X^{[1]}, \ldots, X^{[l]}$. Thus we first begin with scenes for 1-dimensional complexes, i.e. unions of closed intervals.

To construct scenes with a 1-dimensional free space like $X^{[2,1]} = [0,1]$ or $X^{[2,2]} = \{0\} \cup \{1\}$, we use a bar B, which can be moved in the horizontal direction and which is formed such that the area in which the bar may slide can be restricted to the desired intervals by the help of an obstacle Q. In figure 4, scenes for $X^{[2,1]}$ and $X^{[2,2]}$ are shown: the thickness of the obstacle Q can be chosen as less than half the length of the smallest forbidden interval. The names of the objects are indexed by the tuple describing the adequate path in the decomposition tree.

To construct a scene for a 2-dimensional free space like $X^{[2]}$, we want to use the already constructed scenes for the 1-dimensional spaces resulting from the decomposition and "activate" them according to the position of a new sliding

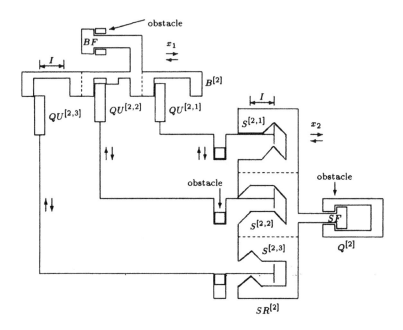

Figure 5: construction idea of a scene $\mathcal{S}^{[2]}$ for $X^{[2]}$.

object representing the value of parameter x_2. How is it possible to choose between $B^{[2,1]}$, $B^{[2,2]}$ and $B^{[2,3]}$? We first connect these bars to a new bar B, such that again one parameter x_1 suffices to describe its position (cf. figure 5). By using closed sides as bottom sides of B and top side of Q, we can make it possible that the area of allowed positions of B is enlarged to the entire interval $[0, 1]$ by moving Q downwards at an arbitrary small amount. Thus, we substitute the obstacles $Q^{[2,1]}$, $Q^{[2,2]}$ and $Q^{[2,3]}$ by vertically movable "plungers" $QU^{[2,1]}$, $QU^{[2,2]}$ and $QU^{[2,3]}$, whose position is determined by the position of a "sliding regulator" $SR^{[2]}$.

The task of $SR^{[2]}$ is to ensure that the object $QU^{[2,i]}$ is in its upmost position iff the horizontal position of the regulator corresponds to a value of parameter x_2 belonging to interval $Z(X)[2, i]$. The construction principle is shown in figure 5.

But this scene doesn't fulfill its task completely: the object $QU^{[2,2]}$ should be in its upmost position iff x_2 is in the open interval $(0, 1)$, but it really restricts the allowed positions of the big bar $B^{[2]}$ for $x_2 \in [0, 1]$.

In our construction, we cannot avoid the problem that a plunger must be in its upmost position during closed intervals of horizontal positions of its regulator. Hence at the intersection $\overline{I_1} \cap \overline{I_2}$ of two intervals $I_1 = Z(X)[i_1, \ldots, i_r]$ and $I_2 = Z(X)[i_1, \ldots, i_r + 1]$ both the plungers of the scene for $X^{[i_1, \ldots, i_r]}$ and the scene for $X^{[i_1, \ldots, i_r+1]}$ are active, thus forcing the free space to be the intersection of both spaces. Therefore, our problem only affects the border of X, i.e. some

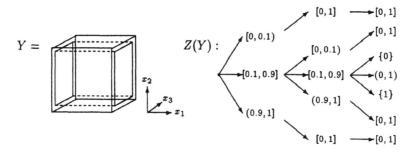

Figure 6: the blown up space Y and its decomposition tree

closed sides of X may get lost. If X is an n-dimensional manifold, then the closure of the free space naturally equals X. But if $X \subseteq [0,1]^n$ contains parts of dimension lower than n (as in our example), the topological structure may be changed.

In this case, if we are only interested in fundamental groups or homotopy type, it is possible to "cheat" by starting not with X itself but with an enlarged version of X.

We start with a decomposition tree of X that contains only open intervals or point intervals for the second up to the n-th coordinate (this is easy to achieve). The point intervals could cause problems, thus we replace them by open intervals:

$$\cdots \quad \{a_1\} \quad , \quad (a_1, a_2) \quad , \quad \{a_2\} \quad \cdots$$
$$\rightarrow \quad \cdots \quad (a_1 - \varepsilon, a_1 + \varepsilon) \quad , \quad [a_1 + \varepsilon, a_2 - \varepsilon] \quad , \quad (a_2 - \varepsilon, a_2 + \varepsilon) \quad \cdots$$

If we choose ε less than half the length of the smallest open interval in $Z(X)$, the resulting tree represents a space Y which is just a blown up version of X and especially homotopy equivalent to X:

$$X'_{(i)} := \{\, x + (0, \ldots, \overset{i}{\overset{\leftrightarrow}{0, \delta, 0}}, \ldots, 0) \; ; \; x \in X' \, , \; -\varepsilon < \delta < \varepsilon \,\} \cap [0,1]^n,$$
$$Y := (\ldots((X_{(n)})_{(n-1)})\ldots)_{(2)}.$$

With $\varepsilon = 0.1$, we get the situation shown in figure 6 for our example. If we apply our construction principle for the decomposition tree $Z(Y)$, the free space of the resulting scene is equal to Y.

Let $I_1 = Z(Y)[i_1, \ldots, i_r] = (a, b)$ and $I_2 = Z(Y)[i_1, \ldots, i_r + 1] = [b, c]$. Then $Z(X)[i_1, \ldots, i_r]$ was a point interval and hence $X^{[i_1, \ldots, i_r+1]} \subseteq X^{[i_1, \ldots, i_r]}$ (X is closed !). Hence, also $Y^{[i_1, \ldots, i_r+1]} \subseteq Y^{[i_1, \ldots, i_r]}$ and the allowed parameters $(x_1, \ldots, x_{r-1})$ at $x_r = b$ are just $Y^{[i_1, \ldots, i_r]} \cap Y^{[i_1, \ldots, i_r+1]} = Y^{[i_1, \ldots, i_r+1]}$. As b belongs to interval $I_2 = Z(Y)[i_1, \ldots, i_r + 1]$, it was correct to blow up point intervals to open intervals in order to get Y as free space.

In the last construction step of our example we will build a scene S for the entire space Y (respectively X) by using the scenes $S^{[1]}$, $S^{[2]}$ and $S^{[3]}$ for $Y^{[1]}$, $Y^{[2]}$

and $Y^{[3]}$. The first action is - analogous to step 2 - to connect the bars to one bar B. Again we will "activate" the scenes according to the parameter x_3 which is represented by the horizontal position of a new sliding regulator SR. This is achieved by moving the entire scenes (i.e. all regulators and plungers) up and down by SR in such a way that the plungers of scene $S^{[i]}$ are able to restrict the allowed positions of the bar B iff the position of SR corresponds to a value of x_3 in $\overline{Z(Y)}[i]$. For this purpose we substitute the obstacles $Q^{[i]}$ by vertically movable objects $QU^{[i]}$ whose vertical position is determined by the position of SR. We fix the vertical position of SR again by an obstacle Q - thus making further recursion possible.

By connecting the bars $B^{[1]}$, $B^{[2]}$ and $B^{[3]}$ to B we could achieve that only one single parameter x_1 determines the position of all the subbars. In this step, we also have to ensure that only one parameter x_2 determines the horizontal positions of the regulators $SR^{[1]}$, $SR^{[2]}$ and $SR^{[3]}$. As these regulators must be able to move up and down independently, a rigid connection is not possible. Thus we use telescopic connection parts, as shown in figure 7.

It is easy to see that the construction process can be continued to get scenes for spaces of arbitrary dimension, especially of dimension 4. The resulting scenes are similar to the decomposition tree, in fact instead of a recursive construction we could have built the scenes directly starting from this tree. An interval $Z(Y)[i_1, \ldots, i_r]$ corresponds to a regulator $SR^{[i_1, \ldots, i_r]}$ iff $0 \le r \le n - 2$ and to an elementary subbar $B^{[i_1, \ldots, i_r]}$ iff $r = n - 1$ (the root at depth 0 corresponds to SR). The horizontal position of all the regulators at depth r corresponds to parameter x_{n-r}, the horizontal position of the bar B corresponds to x_1. If for $r \le n - 2$ interval $Z(Y)[i_1, \ldots, i_r]$ has λ sons in the decomposition tree, then $SR^{[i_1, \ldots, i_r]}$ consists of λ parts such that the $j - th$ part forces $QU^{[i_1, \ldots, i_r, j]}$ to be in an upper position iff $x_{n-r-1} \in \overline{Z(Y)}[i_1, \ldots, i_r, j]$ (see figure 8). The proportions of the objects, especially of the telescopic connection parts, depend on their "depth" in the construction (the vertical movement of the regulators at depth r can be larger than at depth $r - 1$).

These considerations complete our construction, which leads to scenes having the properties required in the main theorem.

5   Results

In this section we will give a brief overview about undecidability results that follow from our construction and discuss them.

Of the known undecidability results from group theory, which all can be carried over to fundamental groups of kinematic scenes, we state the following (cf. [8],[9]):

It is not decidable, if

- the fundamental groups of two given kinematic scenes are isomorphic;

- the fundamental group of a given kinematic scene has a fixed given "Markov property" such as finiteness, triviality or commutativity;

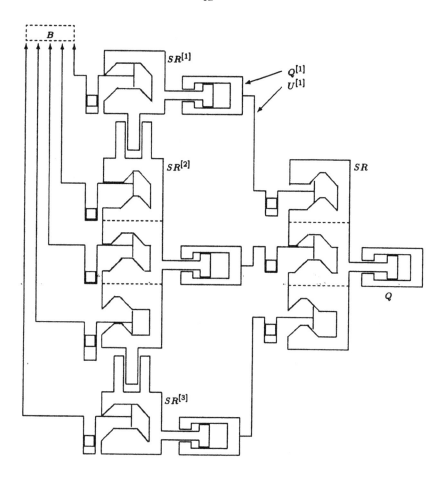

Figure 7: the scene $\mathcal{S}$ with free space Y

- the fundamental group is a non-trivial direct/free product.

Thus the fundamental group does not provide us with a useful tool to get an envision of the "structure" of an arbitrary mechanism.

The classification problem from the motivation - to compare mechanisms according to the topological structure of their free space or to construct adequate normal forms - is undecidable for the most common invariants homeomorphy and homotopy equivalence.

Concerning normal forms we have the interesting result that rotations do not increase the "topological complexity" of kinematic scenes: for each kinematic scene in which there are objects with rotational freedom it is possible to construct a "comparable" planar scene (i.e. according to homeomorphy/homotopy) in which all objects have only translational freedom of motion (to prove this, one has to use Hironaka's result ([10]) that each semi-algebraic manifold can be

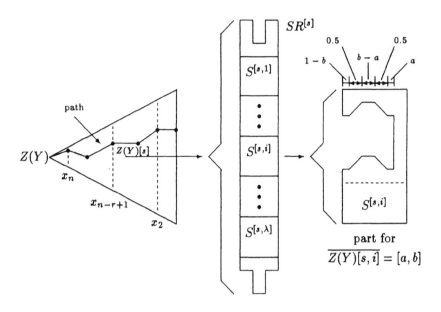

Figure 8: from the decomposition tree to the scene

triangulated and thus be transformed to a polytope).

Another interesting undecidability result concerns the word problem of funda-mental groups, which carries over to the homotopy problem of paths in the free space of a given kinematic scene: we even can state that there is a fixed scene such that it is not decidable, if a loop in its free space (i.e. a cyclic motion) is homotopic to the trivial loop staying at the base point. If we want to optimize a given path according to an optimality criterion such that the trivial loop is always more optimal than any other loop, then we cannot find the optimum in the homotopy class of this path because else we could solve the word problem for arbitrary groups. Thus deformation algorithms can never guarantee to find global optima (e.g. paths of minimal length).

Nevertheless we should not be too discouraged by all these undecidability results when we want to use "topological reasoning" to support qualitative reasoning about physical devices. For practical purposes, geometric invariants like ho-mology may be sufficient. Furthermore, the most interesting case of classifying 2-dimensional manifolds (e.g. to investigate the interactions between kinematic pairs) is possible and the development of efficient algorithms a nice topic. The investigation of special cases of kinematic scenes - e.g. the case of convex objects - seems to be interesting from both a computer scientific and a mathematical point of view, because interesting classes of groups could be discovered.

6 Acknowledgements

This paper was supported by "Graduiertenkolleg am Fachbereich Informatik der Universität des Saarlandes" (granted by Deutsche Forschungsgemeinschaft) and contains parts of the Ph.D. thesis of the author. I thank Prof. G. Hotz, who suggested the topic and who helped me in many discussions.

References

[1] *Artificial Intelligence*, Special Volume on Qualitative Reasoning about Physical Systems, Vol. 24, Nos. 1-3, North-Holland, 1984

[2] Joskowicz, Sacks : "Computational Kinematics", *Artificial Intelligence*, Vol. 51, Nos. 1-3, North-Holland, 1991

[3] Forbus, Nielsen, Faltings : "Qualitative Spatial Reasoning", *Artificial Intelligence*, Vol. 51, Nos. 1-3, North-Holland, 1991

[4] Schwartz, Sharir : " Algorithmic Motion Planning in Robotics", *Handbook of Theoretical Computer Science Volume A*, edited by Jan van Leeuwen, Elsevier 1990

[5] C.-K. Yap : "Algorithmic Motion Planning", *Algorithmic and Geometric Aspects of Robotics Volume 1*, edited by J.T. Schwartz, C.-K. Yap, Lawrence Erlbaum Associates 1987

[6] A.A. Markov : "The Problem of Homeomorphy", *Proceedings of International Congress of Mathematicians*, 1958, pp. 300-306 (in Russian)

[7] Boone, Haken, Poénaru : "On Recursively Unsolvable Problems in Topology and Their Classification", *Contributions to Mathematical Logic*, edited by H. Schmidt, K. Schutte and H.-J. Thiele, North-Holland Pub. Co., 1968, pp. 37-74

[8] M.O. Rabin : "Recursive Unsolvability of Group Theoretic Problems", *Annals of Mathematics*, Vol. 67, No. 1, January 1958, pp. 172-194

[9] Lyndon, Schupp : *Combinatorial Group Theory*, Ergebnisse Math. Grenzgebiete 89, Springer-Verlag New York Heidelberg Berlin 1980

[10] H. Hironaka : "Triangulations of algebraic sets", *Proceedings, Symposia in Pure Math*, Vol. 29, American Mathematical Society 1975, pp. 165-185

A Symbolic Approach to Interval Constraint Problems

Peter Ladkin*
Universität Bern
Länggassstrasse 51
CH-3012 Bern
ladkin@iam.unibe.ch

Alexander Reinefeld
Paderborn Center for Parallel Computing
Warburger Str. 100
D-4790 Paderborn, Germany
ar@uni-paderborn.de

Abstract

We report on a symbolic approach to solving constraint problems, which uses relation algebra. The method gives good results for problems with constraints that are relations on intervals. Problems of up to 500 variables may be solved in expected cubic time. Strong evidence is presented that significant backtracking on random problems occurs only in the range $6 \leq n.c \leq 15$, for $c \geq 0.5$, where n is the number of variables, and c is the ratio of non-trivial constraints to possible constraints in the problem. Space performance of the method is affected by the branching factor during search, and time performance by path-consistency calculations, including the calculation of compositions of relations.

1 Introduction

1.1 Synopsis

Constraint satisfaction problems, or CSPs, are well-known in AI, e.g. [Mac87]. They are a form of logic programming (see Section 2), and are usually solved by various tuned search methods through tuples of values (e.g. [Güs89]). Interval constraints form a particular class of constraint problems that *prima facie* may not be solved using these methods, since the domains of the relations are infinite. Interval constraint problems are found in planning and scheduling in AI, and solving such problems is also essential for the mixed qualitative-quantitative problems studied in [KauLad91]. Allen was the first to show that pruning techniques such as path-consistency may be used, if performed as symbolic computations with the constraint relations themselves rather than on tuples of values [All83], and later research has built on this insight (e.g. see [Lad89] for a survey, and [vBe92] for more recent work).

Ladkin and Maddux showed that the relation algebra of Tarski is an appropriate mathematical context in which to pursue a symbolic approach to constraint problems [LadMad87, LadMad88]. An *algebra of relations* is a set of binary relations closed under the Boolean operations (i.e. forming a Boolean algebra), with the additional operations of *composition* and *converse*, and containing an *identity* relation (an algebraic identity element for the composition operation). All these operations are used in constraint satisfaction techniques for binary constraints - the reader may find evidence in Sections 2 and 3. Every finite algebra

---

*Current address: Dept. of Computing Science, University of Stirling, Stirling FK9 4LA, Scotland

of relations contains a collection of *minimal non-empty* relations, called *atoms* or *atomic relations*, such that every relation in the algebra may be expressed as a sum of atoms.

In this paper, we present a design for a general algorithm architecture in which constraint problems in various different domains may be solved symbolically (Figure 12). Our design, which relies on highly pruned search, is aimed towards NP-hard problems in which some form of search is a component of a solution algorithm. Path-consistency preprocessing is carried out, followed by a search which involves pruning at various stages. Our design may be used when there is a class of constraint relations C such that (a) the composition table for members of C may be stored and accessed efficiently; (b) there is an efficient[1] method for solving problems whose constraints are all in C; (c) all constraint relations in the problem domain may be represented as unions of members of C. The last condition (c) entails that the atomic relations are members of C, and so if conditions (a) and (b) hold for the atomic relations, they form the minimal class C for which (a) - (c) hold. The algorithm takes as input a constraint problem in the domain, and returns a more restricted problem in which all constraints are relations in C that are potentially tighter than the originals. One may then apply the C-constraint-solver (condition (b)) to complete the solution. Although $C =$ *atomic relations* will work if any class C does, it is preferable to choose class C as large as possible satisfying (a) - (c), to reduce the branching factor in search and to reduce the time taken in calculations of compositions (see below).

Interval problems fit this paradigm. It is NP-complete to solve them [ViKavBe89], so search must be employed in some form in a solution algorithm (unless P=NP). There are 13 atoms (atomic relations) in the interval algebra, giving a composition table of 169 entries (144 entries are actually used since one of the atoms is an identity for composition); and problems with atomic constraints may be solved deterministically in quadratic time. There is also a larger class C with 187 members, namely the *pointisable relations*, fulfilling conditions (a) - (c). (See Section 2 for a definition of *pointisable*.) Hence we can compare performance of the algorithm design with $C =$ *atoms* with that of $C =$ *pointisables*. The pointisable relations were introduced independently in [LadMad88] and in [vBeCoh89]. A quadratic time algorithm for solving interval problems with either atomic constraints or pointisable constraints may be derived from [vBe90].

Our algorithm design assumes that the algebra of constraint relations is finite. There are important infinite classes of constraints, for example numerical constraints such as the constraints $a \leq x - y \leq b$ (for a, b real numbers) of [DeMePe91]. These classes do not satisfy condition (a) above, but nevertheless there is an efficient way to calculate the composition of two constraints, which in the case of [DeMePe91] turns out to be much faster than the methods we must use. However, in these cases there are often polynomial solution algorithms, so our algorithm design, which relies on highly pruned search, is unlikely to be as effective. Furthermore, performance of the algorithm is affected by calculation of the compositions (see below), and we have performed no experiments on problems for which this is no longer a factor.

In previous experiments on interval problems, we found that path-consistency is particularly effective as a preprocessing method, especially on large problems [ReiLad92], and that when used also for incrementally pruning the search, reduces the search almost to a linear selection in most cases [LadRei92]. Our results were reported for *completion coefficient* approximately equal to 1, and $C =$ *atoms*. The *completion coefficient* is the ratio of actual non-trivial constraints in a problem to the number of possible constraints.

We show here that the conclusions of [LadRei92] are also valid for varying values of the completion coefficient $c \geq 0.5$, namely that problems with non-trivial search lie generally

---

[1]By 'efficient' here we mean low polynomial time, e.g. quadratic time.

in the range $6 \leq n.c \leq 15$ for this range of c. We also performed experiments for $C =$ *pointisable relations*. This reduced the branching factor in search, and therefore the space requirements of the solution method, allowing us to study behaviour on consistent problems of up to 500 variables and 250,000 non-trivial constraints. (Using $C = atoms$, we had been limited by memory requirements to problems with fewer than 200 variables.) We conclude that the branching factor during search is an important space performance factor which may be influenced by judicious choice of the class C.

By comparing times taken on smaller problems with $C = atoms$ with that taken on the same problems with $C = pointisables$, we also found that a significant amount of computation time was involved in calculating the composition of constraint relations. There are 2^{13} possible interval constraint relations, so caching the entire composition table ($64m$ entries) is infeasible with current memory sizes. However, a significant time improvement was observed by caching the compositions of all pointisable relations (approximately $40k$ entries) rather than just the compositions of atomic relations (169 entries). We conclude that time performance of the algorithm design is strongly affected by the proportion of compositions which may be cached.

The computation of compositions takes place within path-consistency computations, but is not the only time bottleneck in these computations. Path-consistency is known to be cubic time [Mac77], and there is a class of examples on which iterative local algorithms must take quadratic time on parallel machines [LadMad88]. However, there is experimental evidence that iterative local algorithms generally take linear time in parallel [SusHen91]. We compute path-consistency using the relational-matrix composition algorithm of [LadMad88], explained in Section 3. The speed of path-consistency computations is essential to the performance of the algorithm design, and the raw Ladkin-Maddux algorithm may be improved by tuning. For the larger consistent problems we generated, we found that path-consistency computations dominate solution time. We conclude that improving path-consistency computations is essential for good time performance of the constraint-solution algorithm design.

Overview of the Paper. We first present some of the theoretical background for symbolic constraint problem solving. We then describe the algorithm design. Finally we summarise some previous experimental results, and present new results on interval problems.

1.2 Performance on Interval Problems

We discuss here the performance of the symbolic constraint solver on interval problems. We showed in [LadRei92] that using path-consistency as a pruning method during search results in virtually a linear selection with insignificant backtracking on significant classes of problems.

Problems with Randomly Generated Constraints. For random problems, algorithm versions return atomic reductions in less than half a second on the average, with the hardest problem taking only half a minute on a RISC workstation. The fast solution time seems to be due to (a) the pruning power of path-consistency; (b) the fact that all randomly generated interval networks of size ≥ 14 variables were found to be inconsistent, which is rapidly detected by a path-consistency computation; and (c) that small networks are consistent, and solution nodes are dense, entailing that there is little backtracking on these problems. To obtain these conclusions, we ran large-scale experiments on over 200,000 networks of sizes $3 \leq n \leq 20$. Constraint problems were generated using a distribution in which, for every individual constraint in a problem, each possible relation had equal probability of being that constraint. These results are reported in [LadRei92], and are summarised in Figure 14.

Large Consistent Problems. To solve large consistent problems, we used search on *pointisable subsets* of a constraint, rather than on atomic subrelations of the constraint. (See the definition of pointisable constraint in Section 2.) The pointisable subsets are a class of constraints that fit the general features required for application of our methods, namely (a) there are relatively few of them (less than 200), so that the composition table can be cached; (b) pointisable problems may be solved deterministically in quadratic time [vBe90]; and (c) all relations are a sum of pointisables (since atoms are pointisables). In particular, all relations are a sum of very few pointisable relations, the average being ~ 3. Consequently, choosing pointisable subrelations during search, instead of atomic subrelations, reduces the search branching factor from an average of between 6 and 7 to an average of ~ 3. Space is the more significant limitation on solution algorithm performance rather than time, and it was possible to solve problems of 500 variables using pointisable search, whereas problems of 200 variables caused significant memory problems with search using atomic subrelations. The maximal problem size is mainly restricted by the available memory space to hold the (reduced) network matrices in the depth-first search tree.

These large problems were generated using the method of [ReiLad92] (also noted in [LadRei92]), namely generating a configuration of intervals with the endpoints of each interval picked randomly from a finite collection of possible endpoints. This yields a solvable interval constraint problem with atomic constraints, and to each constraint noise was then added. At these larger sizes, using this problem generation method, it turns out that a path-consistency computation returns a solution by itself in one pass, and the solution time is expected $O(n^3)$. For networks of between 200 and 500 variables, the solution is found by the initial path-consistency computation. The ratio (average solution speed for k-variable networks): (average solution speed for m-variable networks) for $100 \leq m < k \leq 500$ is thus approximately $k^3 : m^3$.

In practice (e.g. Nökel's technical diagnosis system [Nök91]), large application problems are unlikely to exhibit the characteristics of random networks, and one may expect that domain specific knowledge can be exploited to aid in the solution process. However, domain-specific knowledge is unlikely to reduce the roughly cubic time needed to solve large problems, unless it can considerably reduce the time spent in path-consistency calculations.

Extrapolation to General Problems. The evidence from testing on interval constraint problems thus suggests that on general constraint problems fitting the conditions (a) - (c), path-consistency reduces the solution search to an almost linear selection of atomic labels on most problems; path-consistency is by itself an excellent consistency heuristic for random networks with $n.c < 6$ or $n.c > 15$ or on very large problems in general, and expected behaviour on large problems is cubic time on a serial processor.

2 Binary Constraint Networks

In this section, we introduce binary constraint problems and give some mathematical background. We use interval constraint problems as an example.

A binary constraint problem on n variables $x_1, ..., x_n$ is given by a formula

$$(P_{12}(x_1, x_2) \wedge P_{13}(x_1, x_3) \wedge \ \ldots\ldots \wedge P_{n-1,n}(x_{n-1}, x_n))$$

where the P_{ij} are predicate symbols representing relational constraints on the variables x_i taken in pairs [Mac87]. Constraints are thus binary relations. A constraint problem is *satisfied* by finding values for the variables which satisfy the logical formula. Since we shall represent binary constraint problems as graphs (e.g. Figure 3), we shall also refer to a constraint

Equals:	$Id(S) = \{\langle\langle x,y\rangle,\langle x',y'\rangle\rangle : x = x' < y = y' \in R\}$
Precedes:	$P = \{\langle\langle x,y\rangle,\langle x',y'\rangle\rangle : x < y < x' < y' \in R\}$
During:	$D = \{\langle\langle x,y\rangle,\langle x',y'\rangle\rangle : x' < x < y < y' \in R\}$
Overlaps:	$O = \{\langle\langle x,y\rangle,\langle x',y'\rangle\rangle : x < x' < y < y' \in R\}$
Meets:	$M = \{\langle\langle x,y\rangle,\langle x',y'\rangle\rangle : x < y = x' < y' \in R\}$
Starts:	$S = \{\langle\langle x,y\rangle,\langle x',y'\rangle\rangle : x = x' < y < y' \in R\}$
Finishes:	$F = \{\langle\langle x,y\rangle,\langle x',y'\rangle\rangle : x' < x < y = y' \in R\}$
Converses:	$P^{\smile}, D^{\smile}, O^{\smile}, M^{\smile}, S^{\smile}, F^{\smile}$

Figure 1: The Formal Definition of The Interval Relations

problem as a *network* whose nodes are labelled with the variables and whose edges are labelled with symbols representing the constraint relations between the nodes.

In general constraint problems, the constraints P_{ij} may be given by explicit tuples of values, or by a rule which generates explicit tuples of values (e.g. [Güs89]). Alternatively, the constraints may be given symbolically. One way is for the constraints to be specified as *unions* of distinguished relations from some small class A. (Binary relations mathematically speaking are sets of pairs, and *union* is normal set union. We shall also call it a *sum*, to conform with relation-algebraic usage.)

The properties of this distinguished subclass A that we need are that (i) relations in A are pairwise disjoint (no pair of values belongs to more than one relation); (ii) the union of all members of A is everything (every pair of values belongs to one of these relations); and (iii) all other constraint relations may be written as a union of relations in A. We call members of a class A satisfying (i) - (iii) *atomic relations*. The class of all possible unions of the atomic relations forms a Boolean algebra, of which the atomic relations are atoms [BurSan81]. Further, if this Boolean algebra is closed under composition and converse, and contains the equality relation, then this class will form a relation algebra [JònTar52].

We shall illustrate these and further concepts with the example of interval relations. An *interval* is just a pair $\langle a, b\rangle$ of real or rational numbers, such that $a < b$ [All83]. Suppose we consider just the binary relations on intervals obtained by considering the ordering (or equality) of their endpoints. There are 13 atomic relations in this class[2]. Seven of them are defined in Figure 1, and the other six are obtained by taking the *converses* (see below) of these relations[3]. The atomic relations are sets of interval pairs $\langle\langle x,y\rangle,\langle x',y'\rangle\rangle$. There are thus 2^{13} possible interval relations, obtained by taking all possible unions of these atomic relations[4]. By the equivalence of set union and logical or ($\vee$), we may write a binary constraint in the logical form

$$P_{ij}(x_i, x_j) \equiv ((x_i\, R_1\, x_j) \vee \ldots\ldots(x_i\, R_q\, x_j))$$

where the R_p are specific atomic relations from Figure 1. Thus by rewriting each P_{ij} as such a clause, we can see that an interval constraint problem is a logical formula in a restricted conjunctive normal form, in which each individual clause only has literals containing the same two free variables.

When dealing with interval relations, we shall write relations in infix, rather than prefix, form. We now consider the example constraint problem

---

[2] A general reference for interval relations is [vBen91].

[3] In most papers, including [All83], these relations are defined by pictures.

[4] Note that although the relations themselves are infinite, there are only finitely many relations obtained by this procedure.

	1'	p	p⌣	d	d⌣	o	o⌣	m	m⌣	s	s⌣	f	f⌣
1'	1'	p	p⌣	d	d⌣	o	o⌣	m	m⌣	s	s⌣	f	f⌣
p	p	p	1	u	p	p	u	p	u	p	p	u	p
p⌣	p⌣	1	p⌣	v	p⌣	v	p⌣	v⌣	p⌣	v⌣	p⌣	p⌣	p⌣
d	d	p	p⌣	d	1	u	v⌣	p	p⌣	d	v⌣	d	u
d⌣	d⌣	v	u⌣	n	d⌣	z⌣	y	z⌣	y	z⌣	d⌣	y⌣	d⌣
o	o	p	u⌣	y	v	x	n	p	y⌣	o	z⌣	y	x
o⌣	o⌣	v	p⌣	z	u⌣	n	x⌣	z⌣	p	z	x⌣	o⌣	y⌣
m	m	p	u⌣	y	p	p	y	p	a	m	m	y	p
m⌣	m⌣	v	p⌣	z	p⌣	z	p⌣	b	p⌣	z	p⌣	m⌣	m⌣
s	s	p	p⌣	d	v	x	z	p	m⌣	s	b	d	x
s⌣	s⌣	v	p⌣	z	d⌣	z⌣	o⌣	z⌣	m⌣	b	s⌣	o⌣	d⌣
f	f	p	p⌣	d	u⌣	y	x⌣	m	p⌣	d	x⌣	f	a
f⌣	f⌣	p	u⌣	y	d⌣	o	y⌣	m	y⌣	o	d⌣	a	f⌣

$x = (p+o+m)$
$y = (d+o+s)$
$z = (d+o^{\smile}+f)$
$a = (1'+f+f^{\smile})$
$b = (1'+s+s^{\smile})$
$u = (x+y) = (p+o+m+d+s)$
$v = (x+z^{\smile}) = (p+o+m+d^{\smile}+f^{\smile})$
$n = (z+z^{\smile}+b) = (y+y^{\smile}+a) = (1'+f+d+o+s+f^{\smile}+d^{\smile}+o^{\smile}+s^{\smile})$

Figure 2: The Relation Composition Table for the Interval Algebra

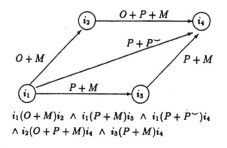

$i_1(O+M)i_2 \wedge i_1(P+M)i_3 \wedge i_1(P+P^{\smile})i_4$
$\wedge i_2(O+P+M)i_4 \wedge i_3(P+M)i_4$

Figure 3: An Interval Constraint Problem in 4 variables and its Network

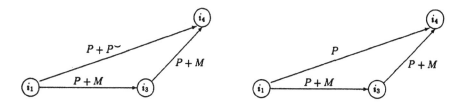

Figure 4: A Triangle Violating Path Consistency, and its Modification

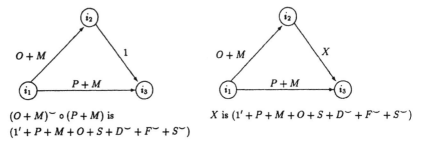

$(O+M)^\smile \circ (P+M)$ is
$(1'+P+M+O+S+D^\smile+F^\smile+S^\smile)$

X is $(1'+P+M+O+S+D^\smile+F^\smile+S^\smile)$

Figure 5: Another (Implicit) Violating Triangle and Its Modification

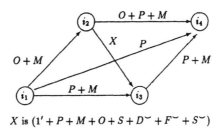

X is $(1'+P+M+O+S+D^\smile+F^\smile+S^\smile)$

Figure 6: The Path-Consistent Reduction

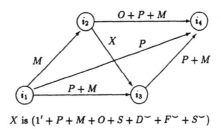

X is $(1'+P+M+O+S+D^\smile+F^\smile+S^\smile)$

Figure 7: Choice of M between i_1 and i_2

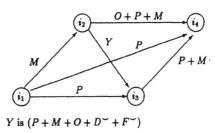

Y is $(P+M+O+D^\smile+F^\smile)$

Figure 8: Choice of P between i_1 and i_3, and reduction

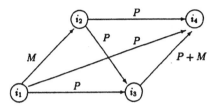

Figure 9: Choice of P between i_2 and i_3, and reduction

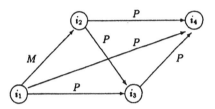

Figure 10: Final choice of P between i_3 and i_4

$$(i_1 \, O \, i_2 \; \lor \; i_1 \, M \, i_2) \; \land \; (i_1 \, P \, i_3 \; \lor \; i_1 \, M \, i_3) \; \land \; (i_1 \, P \, i_4 \; \lor \; i_1 \, P^\smile \, i_4) \; \land$$

$$(i_2 \, O \, i_4 \; \lor \; i_2 \, P \, i_4 \; \lor \; i_2 \, M \, i_4) \; \land \; (i_3 \, P \, i_4 \; \lor \; i_3 \, M \, i_4)$$

which is shown in graphical form as a constraint network in Figure 3. The relation $P_{12}(i_1, i_2)$, for example, is $(i_1 \, O \, i_2 \; \lor \; i_1 \, M \, i_2)$ in this problem.

Relation Sum. The two literals in the P_{12} clause share the same arguments (in the same order). The expression defines a binary relation between these two arguments, which is the set-theoretic union of the relations defined by each literal, since the logical operator is $\lor$. We shall denote this new relation by $(O + M)$, and the corresponding operator on relations by $+$. This definition enables us to represent the problem in a more compact form as

$$i_1 \, (O + M) \, i_2 \; \land \; i_1 \, (P + M) \, i_3 \; \land \; i_1 \, (P + P^\smile) \, i_4 \; \land \; i_2 \, (O + P + M) \, i_4 \; \land \; i_3 \, (P + M) \, i_4$$

One calculates the sum symbolically in the following manner. The sum of $(P + M + O)$ and $(O + S + F + M)$ is just the sum of atoms occurring in either, i.e. $(P + M + O + S + F)$.

Relation Intersection. Relation intersection is just ordinary set intersection. Since we work with interval constraint relations which are sums of 13 atomic relations, the symbolic calculation of this operation is simple. Because the atomic relations are pairwise disjoint (i.e. the intersection of any two of them is empty), easy set theory shows that the intersection of two relations such as $(P + M + O)$ and $(O + S + F + M)$ is just the sum of the atomic relations common to both, namely $(O + M)$.

Relation Converse. The *converse* $R^\smile$ of a relation R is defined by the equivalence

$$x R^\smile y \equiv_{def} y R x.$$

If the relation R labels the edge $[i, j]$ in a network, then the same information is conveyed by a label of $R^\smile$ on the reverse edge $[j, i]$. The converse operation is needed in path-consistency computations.

Pointisable Relations. A *pointisable relation* on intervals is a relation which may be written as a conjunction of equality and inequality relations on its endpoints. For example, let x_L, x_R denote respectively the left and right endpoints of interval x. Then $x(O + M)y$ may be written as $(x_L < y_L \wedge y_L \leq x_R)$, and thus is pointisable. However, not all formulas may be written thus, e.g. $x(P + P^\smile)y$ (a succinct formula for this relation is the disjunction $(x_R < y_L \vee y_R < x_L)$). Interval constraint problems all of whose constraints are pointisable may be written without loss of information as constraint problems whose variables are the endpoints, and whose non-trivial, non-empty constraints are one of $<, =, >, \leq, \neq, \geq$. Such problems may be solved in quadratic time [vBe90], and all path-consistent such problems are soluble [LadMad88].

There are 187 pointisable relations, defined and investigated in [LadMad88], and independently in [vBeCoh89]. Nökel's [Nök91] technical diagnosis system uses algorithms on 1-convex and 2-convex relations, which are subclasses of the pointisable relations. For further information on the relation between pointisable interval constraint problems and equivalent point networks, see [LadMad88], [KauLad91].

3   3-Consistency and Path-Consistency

In this section we describe the property of path-consistency, which turns out to be equivalent to 3-consistency. Path consistency is the highly effective pruning technique employed in our algorithm design. We give a formal representation and an algorithmic formulation of a general path-consistency algorithm which is based on relational matrices.

Composition. Given binary relations $R(x, y)$, $S(x, y)$, the *composition* $R \circ S$ of the relations is defined by

$$(R \circ S)(x, y) \Leftrightarrow (\exists z)(R(x, z) \ \& \ S(z, y))$$

The composition table of the 13 atomic relations on intervals was given in [All83], and shown in Figure 2, using auxiliary definitions of eight other (non-atomic) interval relations.

From the composition table from atoms, one calculates the composition of two arbitrary relations by using the distributive laws $R \circ (S + T) = (R \circ S) + (R \circ T)$ and $(R + S) \circ T = (R \circ T) + (S \circ T)$. Use of these distributive laws is very time-consuming, and can easily account for most of the computation time in solving constraint networks symbolically using our design. There is thus a space/time tradeoff in how much of the entire composition table to store, and how much to compute using the distributive laws.

3-Consistency. By the definition, $(a, b) \in (R \circ S)$ if and only if there is a value c such that $(a, c) \in R$ and $(c, b) \in S$. Now, for any satisfying values $a_1, ..., a_n$ for the variables $x_1, .., x_n$ we must have, by this definition, $\langle a_i, a_k \rangle \in P_{ik} \Rightarrow \langle a_i, a_k \rangle \in P_{ij} \circ P_{jk}$, for any $i, j, k \leq n$. This necessary condition may be used as a pruning technique to narrow down the potential choices of a_i and a_k: we calculate the relation $P_{ij} \circ P_{jk}$, and then intersect it with P_{ik} to form the new, potentially smaller relation (and thus tighter constraint) $P_{ik} \cap (P_{ij} \circ P_{jk})$. We then use this as the new constraint on x_i and x_k. We call this a *triangle operation*. A triangle operation on our example problem shown in Figure 3 is given in Figure 4, and another in Figure 5. We say that a triangle operation *stabilises* if the result is identical to the original label, equivalently that $(P_{ij} \circ P_{jk}) \subseteq P_{ik}$. We say that a network is *3-consistent*, or *is stable*, if every triangle operation stabilises. The stable network resulting from triangle operations on Figure 3 is shown in Figure 6. To stabilise a network, triangle operations must be performed on all possible constraints P_{ij} using all length-2 paths for every P_{ij} (which involves also using the $\smile$ operation on labels)[5]. In general, triangle operations must be iterated for each P_{ij} for the network to stabilise.

3-consistency is equivalent to path-consistency [Mon74], so we follow standard terminological abuse by conflating the two notions.

3-Consistency Is The Least Expensive Pruning Technique. The practical importance of this form of pruning for interval constraint networks is that it is the least expensive form of consistency-checking. Many authorities recommend a computationally cheaper, but less thorough, check called *arc-consistency*, for other problems. However, this cannot help for interval constraint problems, since every interval problem is guaranteed to be arc-consistent, as noted in [LadMad88].

Complexity. Path-consistency is a necessary condition for consistency but does not imply consistency [All83]. As we have noted before, path-consistency algorithms are serial cubic time [Mac77], or parallel $O(n^2 \log n)$ time, and iterative algorithms take $\theta(n^2)$ time [LadMad88], but in practice seem to run in near linear time on fast parallel machines [SusHen91, Sus91]. Path-consistency computations have been implemented in a planning system which uses interval computations [Koo88,Koo89].

A Path-Consistency Reduction Scheme. Let *intersection* of relations be denoted by '$\cdot$', and *composition* of relations by '$\circ$'. Let P_{ij} be the relational constraint between variables x_i and x_j. Then a path-consistency computation may be regarded as computing the greatest fixed point of the following sets of equations (i.e. the collection of largest possible relations r_{ij} satisfying them):

$$r_{ij} \leq P_{ij}$$

$$r_{ij} = r_{ij} \cdot \prod_k (r_{ik} \circ r_{kj})$$

where the symbol $\prod_k$ denotes the product taken over all $k \leq n$. This fixed point is the most-general path-consistent reduction of the original network. This observation leads to the representation of constraint problems by $(n \times n)$ matrices of relation symbols, where the matrix M corresponding to the constraint problem A has entries $M_{ij} = P_{ij}$. We call such a matrix a *relational matrix*. By its nature, the matrix has to represent a complete graph, so it contains an entry of **1** (the universal relation[6], i.e. the union of all 13 atoms) wherever no constraint edge appears in the original problem.

Product and *composition* for relational matrices are defined by the following schemes, which use intersection and composition defined on relations:

$$(M \cdot M')_{ij} = M_{ij} \cdot (M')_{ij}$$

$$(M \circ M')_{ij} = \prod_{k \leq n} M_{ik} \circ (M')_{kj}$$

Define also the power M^n by the usual induction, $M^1 = M$, and $M^{n+1} = (M^n) \circ M$. So, for example, $M^2 = (M \circ M)$, as we use in our path-consistency function in Figure 11.

```
function PC (var M: matrix): boolean;
    repeat
        M ← M · M²;
    until M = M²;
    return (M ≠ 0);
end;
```

Figure 11: Path-Consistency Algorithm

A Path-Consistency Algorithm. The path-consistency algorithms in the literature to date are all *iterative local* algorithms [LadMad88, revised version], differing mainly in the details of the iterations over the triangles. Our formulation in Figure 11 works on a relational matrix M, which is iteratively reduced in the **repeat** loop. If PC detects a zero edge in M it halts immediately and returns false, thereby indicating that the network is not path-consistent.

In a practical implementation, only the triangles of edges P_{ij} whose labels have changed in the previous iteration need to be recomputed. Some authors [Mac77,All83] propose a queue data structure for maintaining the triangles that must be recomputed. The computation then simply proceeds until the queue is empty. In our implementation we flip a bit in the network matrix, denoting that this element must be recomputed in the next iteration. This preserves the same edge evaluation order from one iteration to the next.

We say that a network *fails* if a path-consistency computation detects inconsistency, and it *succeeds* if it stabilises without detecting inconsistency. Succeeding networks are usually consistent, but this need not be always the case, as we show in Figure 14.

4   A Generic Solution Algorithm Architecture

Figure 12 illustrates our suggested generic algorithm design in a flow chart. Details on the path-consistency check and the backtracking search are given in the pseudo-code in Figs. 11 and 13.

At the core of any interval constraint solver is an intelligent backtracking search, which is preceeded by a preprocessing phase and followed by some postprocessing. The preprocessing usually comprises a path-consistency check (Fig. 11), but other reduction schemes are also possible. Path-consistency is a necessary pre-condition for consistency, and it often suffices to prove a network inconsistent. Using path-consistency as a preprocessing technique has the additional advantage that it returns an equivalent, syntactically *reduced network*, where many irrelevant atoms (that can not be part of a solution) are deleted. For large networks, the reduction is more pronounced than for small ones, so that only few (or none) nodes need to be expanded in the solution search.

After the preprocessing phase the network is further reduced by an intelligent backtracking search until it is known to be consistent. A network is *consistent*, if there exists a solution, or equivalently for the case of interval problems, a consistent atomic reduction [LadMad88]. Figure 13 gives pseudo code for our consistency check routine, CC. It works on a relation matrix M and is started on the first edge M_{12}, i.e. function CC is initially invoked with parameters M, $i = 1$ and $j = 2$. CC selects a sub-label $k_l \in E$ from the edge M_{ij}, where E is (in our case) either the set of pointisable labels or the set of the 13 atomic labels. If the resulting network is still path-consistent, CC recursively deepens on the next edge, and

---

[5] This is how all 3-consistency algorithms in the literature work.

[6] Called 'no info' in [All83].

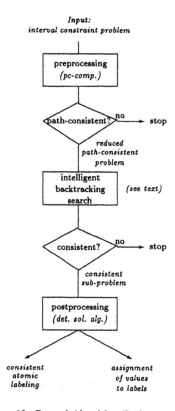

Input:
interval constraint problem

preprocessing
(pc-comp.)

path-consistent? — no → stop

reduced
path-consistent
problem

intelligent
backtracking   *(see text)*
search

consistent? — no → stop

consistent
sub-problem

postprocessing
(det. sol. alg.)

consistent   *assignment*
atomic   *of values*
labeling   *to labels*

Figure 12: General Algorithm Design

selects again a sub-label $l \in E$ on that edge. When all selected labels form a path consistent network, the whole network is consistent [LadMad88]. If the network does not succeed at some point, the algorithm backtracks and chooses another sub-label l on that label. Only when all sub-labels have been unsuccessfully tried is the network inconsistent.

Figures 7 - 10 show an actual run of the algorithm on the initial path-consistent reduction (Figure 6) of our example (Figure 3).

Each PC-call in the search reduces the network matrix M. The resulting search tree is bushy near the root and gets thinner in deeper levels. The larger the network, the more reductions are achieved. This is because in the large networks, more label compositions are intersected, which constrains the remaining label to a greater extent. In very large networks (eg. $n > 100$), the initial path-consistency check often suffices to determine a solution (or inconsistency).

Actually to obtain a solution, an assignment of values to variables, from a consistent atomic labelling, one needs additional postprocessing, a deterministic procedure of $O(n^2)$ steps, as we have mentioned.

4.1   Algorithm Improvements

Our actual implementation includes the following improvements [ReiLad92]:

```
function CC (var M: matrix; i, j: integer): boolean;
{Recursive consistency computation, starting at edge M_ij}
    M' ← M;                                          {save matrix M}
    for each sublabel l_k ∈ M_ij do begin
        M_ij ← l_k;
        if PC(M) then
            if M_ij is last edge or CC(M, next_i, next_j) then
                return (true);
        M ← M';                                      {restore M}
    end;
    return (false);                                  {no consistent atomic labeling}
end;
```

Figure 13: Consistency Computation Algorithm

	p	$p^{\smile}$	d	$d^{\smile}$	o	$o^{\smile}$	m	$m^{\smile}$	s	$s^{\smile}$	f	$f^{\smile}$	$\sum$
p	1	13	5	1	1	5	1	5	1	1	5	1	41
$p^{\smile}$	13	1	5	1	5	1	5	1	5	1	1	1	41
d	1	1	1	13	5	5	1	1	1	5	1	5	41
$d^{\smile}$	5	5	9	1	3	3	3	3	3	1	3	1	41
o	1	5	3	5	3	9	1	3	1	3	3	3	41
$o^{\smile}$	5	1	3	5	9	3	3	1	3	3	1	3	41
m	1	5	3	1	1	3	1	3	1	1	3	1	25
$m^{\smile}$	5	1	3	1	3	1	3	1	3	1	1	1	25
s	1	1	1	5	3	3	1	1	1	3	1	3	25
$s^{\smile}$	5	1	3	1	3	1	3	1	3	1	1	1	25
f	1	1	1	5	3	3	1	1	1	3	1	3	25
$f^{\smile}$	1	5	3	1	1	3	1	3	1	1	3	1	25

Table 1: Number of Atoms resulting from Label Compositions

- We changed the PC-routine so that only the relevant edges (lying on triangles that have been changed by CC) are recomputed.

- Our CC procedure skips all edges with atomic labels. Due to the previous PC computation, these edges are already known to be consistent and no further check is necessary.

- CC does not select arbitrary atoms for further exploration, but rather it selects *equals*-atoms first (if there are any). The motivation is that, when selecting an *equals* atom, the dimension of the network matrix can be reduced by joining the two nodes and intersecting the labels of all adjacent edges.

- Our algorithm selects point-meeting relations at the beginning, because these relations have only few entries in the composition table (see Table 4.1) and hence greatly reduce the size of the search space. (Point-meeting relations are the ones with intervals starting or ending at exactly the same time instance, eg. *equals, meets, starts, finishes* and their converses.)

- For the pointisable networks, we implemented a composition table that holds all possible compositions of all pointisable labels. There are 187 different pointisable labels, so the table has $187 \times 187 \approx 34k$ entries. This enhancement gave a 20-fold speedup over using

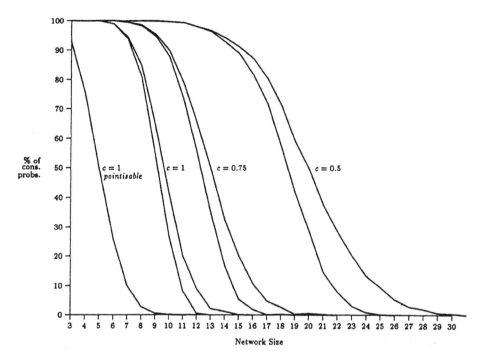

Figure 14: Percentage of Consistent Networks (1000 Probs.)

the table for atoms and using the distributive laws, on the same problems, thus showing the benefits of caching many compositions when feasible.

5 Empirical Results for Random Networks

We generated interval constraint networks where the labels are independent, identically distributed random variables drawn from a discrete probability distribution. The networks have the following properties:

- Each atom within a label occurs with the same probability.

- Each label within the network occurs with the same probability.

We excluded initial networks with zero-labels, since they are known to be inconsistent. Also, since one cannot expect constraint graphs to be complete in practice, we generated graphs with different completion coefficients ranging from $c = 1$ (complete network) to $c = 0.5$ (every second edge has a non-trivial constraint). Unconstrained edges are labeled with the universal relation 1, as explained above.

Figure 14 shows the percentage of networks of sizes $3 \leq n \leq 30$ that succeeded in the first path-consistency computation, and also those that were later found to be consistent by the solution search. Three pairs of graphs are shown (for completion coefficients of 1, 3/4 and 1/2). In each pair, the left hand graph gives the percentage of consistent networks and

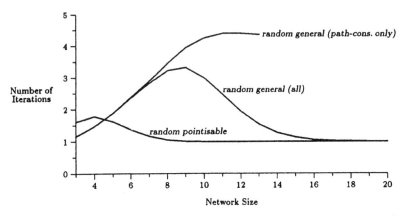

Figure 15: Iterations in Initial Path-Consistency Computation, 10,000 Probs.

the right hand graph gives the percentage of path-consistent networks. For the pointisable networks, only one graph is shown, because here path-consistency implies consistency.

All graphs show a sharp decline in the consistency rate with increasing network size. We call it the *transition range*. For complete networks ($c = 1$), the transition range lies between the size 8 networks (where roughly 20% fail) and the size 11 networks (where roughly 80% fail). Networks in the transition range took particularly long to search, in comparison with networks outside the transition range. Even so, the average solution time required within the transition range was less than half a second of CPU time on a Sun SparcStation 1. Although the problem is NP-complete, we saw no case that needed longer than half a minute to solve, amongst the 200,000 complete networks generated.

Moreover, for $c = 1$, we found that *all* complete networks that we randomly generated with more than 13 variables are inconsistent, as are all pointisable networks with more than 11 variables [LadRei92]. While at sizes 14 to 17, some small number ($\leq 1\%$) of these inconsistent networks succeed, inconsistency of large networks is usually detected right in the initial path-consistency computation.

It is interesting to note that the transition range gets wider in the networks for which $c < 1$. Also the distance between the two graphs (consistent and path-consistent) is growing, which indicates that the initial path-consistency computation is not so effective in networks with smaller values of c. (We attribute this unwanted behavior to the smaller number of intersections of relations, which is less constraining.)

Average Number of Iterations to Stabilise. Figure 15 shows the average number of iterations required to stabilise in the path-consistency computation for complete networks. The number of iterations correlates with the transition range. The peak (of 3.3 iterations) occurs in the middle of the transition range at sizes 9 and 10.

While for sizes $n \leq 7$ always the same average number of iterations are required, regardless of whether a network is path-consistent or not, in larger sizes the consistent networks need more iterations to stabilise than the inconsistent ones. Inconsistency is usually detected right in the first iteration.

Initial Label Sizes Figure 16 gives the initial distribution of the label sizes in random networks. In absolute numbers, the white bars add up to $2^{13} = 8192$, which is the total

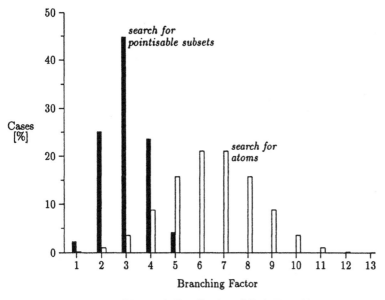

Figure 16: Distribution of Node Branchings

number of atoms over all possible labelings, ranging from $\emptyset$ to 1. Here, the numbers are given relative to the maximum, $8192 = 100\%$.

In random networks, there are an average of $13/2 = 6.5$ atoms per label before the initial path-consistency computation. This would be the average branching degree in a brute force solution search when no pc-reduction is applied. The white bars in Fig. 16 show the distribution of the label sizes over the whole range from one to thirteen atoms. The curve has the shape of a Gaussian with the peak lying at 6 and 7, as it should be.

When searching for pointisable sub-labels, the branching is done on subsets rather than single atoms. This reduces the branching degree to an average of 3.02, as can be seen in the black bars in Fig. 16. Note that this gives a worst case estimate of the search branching degree, because the frequent path-consistency checks in (and before) the search further reduce the label size.

Since the branching degree is crucial to the overall performance of any search problem, we expect the search-for-pointisable-subsets to be much more efficient than a direct search for atomic labelings. In our experiments, we noted speedups between a factor of 3 and 7 for network sizes ranging from 10 to 50, respectively. In the very large problems of size $n > 100$, path-consistency is such an effective pruning technique that further search is only seldomly required. Hence, in the very large problems, the search-for-pointisable-subsets can not speed up solution time. However, as we have noted, it is nevertheless a significant factor in space efficiency.

Average Label Reductions by Path-Consistency. We also investigated the pruning power of the path-consistency computation in the search for atomic labelings. In inconsistent networks the first iteration of the path-consistency computation almost halves the label sizes, especially when the networks are very small (≤ 7) or very large (≥ 14). The following iterations then reduce the remaining atoms further, until a zero-label is found.

For the succeeding networks, also half of the total work is done in the first iteration. In an additional experiment with large (non-random) *consistent* networks, we found that path-consistency is more effective the larger the network. In the size-80 networks, for example, path consistency reduces the average label size to merely 1.1 atoms per label. In addition, fewer iterations are required. For the details of these experiments see [ReiLad92].

Solution Density in the Search for Atomic Labelings. The effort needed to solve a given problem depends on the density of the solution nodes in the search space. As explained above, this effort is greatly reduced by the path-consistency computations, which reduces the size of the search tree. But how much does path-consistency reduce the average label size?

To answer this question, we modified our algorithm to examine the complete search space and return *all* solutions, rather than only one. Overall, we found a surprisingly high solution density. In networks of size $n < 8$ every 10th to 25th leaf node (path) is a solution. In the larger networks, more nodes must be searched to find a solution, with the peak lying at size 9-networks, where every $\approx$ 200th node contains a solution. But even in this worst case, the solution density is quite high considering the large search space.

When the path-consistency heuristic is applied in every interior node of the search tree (like in the above CC algorithm), even fewer nodes are expanded. The solution density is then at about 0.4, that is, there is a 40% probability for a node being a solution node. Unfortunately, the structure of the search tree is highly unbalanced, with the solution nodes clustering in certain subtrees. One cannot conclude that *each* interior search node has the same probability of lying on a solution path. But the given figure of 40% remains the same over the whole range of network sizes from $n = 3$ to $n = 11$.

Solution is Thus Expected Cubic Time. From the high solution density following each path-consistency computation, we can predict that the path-consistency computation dominates other aspects of the solution algorithm, and that in practice one can expect the number of path consistency computations to be almost constant. It therefore follows from [Mac77, MacFre85] that the solution procedure is *expected cubic time* for serial computations. Our statistics bear this out.

Resource Constraints. While for the larger networks a cubic time solution algorithm could be expected to be impractical in terms of *time* taken, in our experience the excessive *space* requirements of the solution search are the more restrictive constraint. Even with a simple depth-first search approach the solution process builds a search tree of depth $d = n \cdot (n - 1) \cdot 1/2$. For every interior node one (triangular) network matrix is stored on a stack, which results in a total space requirement of d^2 words. For size-100-networks with $c = 1$, for example, a maximum of 24 million memory words are needed. Note, however, that our program usually needs less storage than this, because it skips all atomic edges. In our experiments, we saw no case which needed a full search depth of $n \cdot (n - 1) \cdot 1/2$.

6 Conclusions

We have presented a general algorithm architecture for solving binary constraint problems symbolically. Any class of constraint problems is included in a relation algebra, which, when finite, contains atomic relations. Our algorithm design may be used when the composition table for the atomic relations may be stored and accessed efficiently, and when there is an efficient method for solving problems whose constraints are all atomic relations. Instead of searching on atomic relations, the method may be generalised to use any subclass C of the constraint relations for which (a) the composition table may be stored and accessed efficiently;

(b) there is an efficient method for solving problems whose constraints are all in C; (c) all constraint relations (including the atomic ones) may be represented as sums of elements of C. In this case, the algorithm searches on subrelations of a constraint that are in C, and caches the composition table for all of C. This reduces the branching factor in search, which decreases the space requirements of the algorithm, and reduces the time taken in calculating compositions of arbitrary constraints using the distributive law.

Our experiments on interval problems showed that:

- Our methods are efficient enough to solve interval constraint problems of 500 variables and more in reasonable CPU time. This includes problems in the *transition range*, which has been observed by others typically to include the hardest areas of random problems [ChKaTa91].

- Path-consistency computations *during* the search are an excellent pruning technique, considerably reducing backtracking during search even on problems in the transition range. For problems outside the transition range, search becomes almost a linear selection of atomic (or pointisable) labels for the edges.

- The computation time is dominated by path-consistency computations, which are actual cubic time ($\approx 1/6 \cdot n^3$), in particular by calculating compositions of arbitrary constraint relations from the cached composition table.

- The computation space requirements are affected by the branching factor in the search, and are a more significant resource limitation than the time requirements. Hence, more sophisticated solution search methods (like best-first searches) are not generally needed.

We conclude that the most sensitive performance factors in time and space are

- Reducing the branching factor during search (space critical);

- Caching as much of the composition table as possible (a space/time tradeoff);

- Using a fast symbolic path-consistency algorithm (time critical).

Acknowledgements

This work has benefitted from continuing discussions with Peter van Beek, who has independently verified many of the phenomena reported here. Peter also suggested experimenting with $C = pointisables$, and obtained the first results for this case, some of which appear in [vBe92]. To him, our grateful thanks for ongoing stimulation.

Much of this work was done while the second author was a member of the Fachbereich Informatik at the University of Hamburg.

References

All83: Allen, J.F., *Maintaining Knowledge about Temporal Intervals*, Comm. ACM 26 (11), November 1983, 832-843.

BurSan81: Burris, S., and Sankappanavar, H.P., *A Course in Universal Algebra*, Springer Verlag, 1981.

ChKaTa91: Cheeseman, P., Kanefsky, R., and Taylor, W.M., *Where the Really Hard Problems Are*, Proceedings of the 12th International Joint Conference on Artificial Intelligence (IJCAI-91), pp 331-337, Morgan Kaufmann 1991.

DeMePe91: Dechter, R., Meiri, I., and Pearl, J., *Temporal Constraint Networks*. Artificial Intelligence 49, 1991, 61-95.

Fre78: Freuder, E.C., *Synthesizing Constraint Expressions*, Communications of the ACM 21 (11), Nov 1978, 958-966.

Güs89 Güsgen, H.-W., *CONSAT: A System for Constraint Satisfaction*, Morgan Kaufmann/Pitman 1989.

JònTar52: Jònsson, B. and Tarski, A., *Boolean Algebras with Operators II*, American J. Mathematics 74, 1952, 127-162.

KauLad91: Kautz, H.A. and Ladkin, P.B., *Integrating Metric and Qualitative Temporal Reasoning*, Proceedings of AAAI-91, the 9th National Conference on AI, AAAI Press 1991.

Koo88: Koomen, J.A.G.M., *The TIMELOGIC Temporal Reasoning System*, Technical Report 231, University of Rochester Dept. of Computer Science, 1988.

Koo89: Koomen, J.A.G.M., *Localizing Temporal Constraint Propagation*, in Proceedings of KR89. the First International Conference on Principles of Knowledge Representation and Reasoning, pp198-202, Morgan Kaufmann 1989.

LadMad88: Ladkin, P.B., and Maddux, R.D., *On Binary Constraint Networks*, Kestrel Institute Technical Report KES.U.88.8. An extensively revised 1992 version is *On Binary Constraint Problems*, submitted for publication.

LadRei92: Ladkin, P.B., and Reinefeld, A., *Effective Solution of Qualitative Interval Constraint Problems*, Artificial Intelligence, to appear.

Mac77: Mackworth, A.K., *Consistency in Networks of Relations*, Artificial Intelligence 8, 1977, 99-118.

Mac87 : Mackworth, A.K., *Constraint Satisfaction*, in the *Encyclopedia of Artificial Intelligence*, ed. S. Shapiro, Wiley Interscience 1987.

MacFre85: Mackworth, A.K., and Freuder, E.C., *The Complexity of Some Polynomial Network Consistency Algorithms for Constraint Satisfaction Problems*, Artificial Intelligence 25, 65-74, 1985.

MohHen86: Mohr, R., and Henderson, T.C., *Arc and Path Consistency Revisited*, Artificial Intelligence 28, 1986, 225-233.

Nök91: Nökel, K., *Temporally Distributed Symptoms in Technical Diagnosis*, Lecture Notes in Artificial Intelligence 517, Springer Verlag 1991.

ReiLad92: Reinefeld, A., and Ladkin, P.B., *Fast Solution of Large Interval Constraint Networks*, in Procs. 9th Canadian Conf. on Art. Intell., AI'92, Vancouver (May 1992), pp156-162.

Sus91: Susswein, S., *Parallel Path Consistency*, MS Thesis, University of Utah, Department of Computer Science, 1991.

SusHen91: Susswein, S., Henderson, T.C., Zachary, J., Hinker, P., Hansen, C., and Marsden, G., *Parallel Path Consistency*, University of Utah, Technical Report UUCS-91-010, July 30, 1991, revised version to appear, International Journal of Parallel Programming.

vBeCoh89: van Beek, P.G., and Cohen, R., *Approximation Algorithms for Temporal Reasoning*, in Proceedings of IJCAI89, the 11th Joint Conference on Artifical Intelligence, 1291-1296, Morgan Kaufmann 1989; full version in *Computational Intelligence*, 1991.

vBe90: van Beek, P.G., *Reasoning About Qualitative Temporal Information*, in Proceedings of AAAI90, the 8th National Conference on Artificial Intelligence, pp728-734, Morgan Kaufmann 1990.

vBe92: van Beek, P.G., *Reasoning About Qualitative Temporal Information*, *Artificial Intelligence*, to appear.

vBen91: van Benthem, J.F.A.K., *The Logic of Time*, 2nd Edition, Kluwer 1991.

ViKaVB89: Vilain, M., Kautz, H., and van Beek, P.G., *Constraint Propagation Algorithms for Temporal Reasoning*, in Weld and de Kleer, eds., *Readings in Qualititative Reasoning About Physical Systems*, Morgan Kaufmann 1989.

An Algebraic Approach to
Knowledge-Based Modeling

Gerhard Schwärzler

Swiss Federal Institute of Technology Zürich
Department of Mathematics, ETH-Zentrum, CH-8092 Zürich

Abstract. In the presented approach the basic domain for modeling components of a system will be sets of first order formulas. The formation process of rules is iterated and leads to the notion of cumulative logic programs, which are identified with elements of a graph algebra. The appropriate definition of the application operation on cumulative logic programs is given. The structure of the modeled system is specified by equations, and qualitative modeling is related to the algebraic problem of solving system of equations in a graph algebra. Using the concepts of consistency and knowledge extension, an algorithm for approximating solutions to such equational systems is presented.

Keywords: qualitative modeling, model-based reasoning, expert systems, logic programming, combinatory algebra.

1 Introduction

Classical expert systems that contain high-level knowledge from experts of the field in question, suffer from this "shallow-knowledge" architecture. In a way, they are collections of formal descriptions of the experts' *experience*, and so they reason basically on the basis of this experience. Due to this architecture, these systems often fail in situations for which they have no explicitly stated knowledge, and so their success relies heavily on a good tuning for specific applications. "It is believed that many of the problems with such systems are due to their lack of underlying (deep) knowledge about their domain" [1]. There are deep knowledge approaches in the field of qualitative physics, where models are constructed from differential equations with the help of qualitative algebras [2, 3, 4]. P. Struss demonstrates in [5] that these algebras are in principle interval-arithmetic calculi. Bratko et al. [6] take an approach, which is more related to the field of expert systems, but with the application of a higher abstraction, they lose the algebraic description of the system by equations which is a key idea of modeling in a mathematical framework.

In this paper, we present an approach, where each component of the system is modeled by its own expert system, in the form of so-called Cumulative Logic Programs, which represent (may be partial) knowledge about it (chapter 3). Transformation of knowledge is defined by "higher-order" programs and the operation "apply". This operation furnishes Cumulative Logic Programs with an algebraic structure, which is related to a graph algebra (chapter 4). Chapter 5 introduces a graphical tool called "interaction graphs", specifying the overall structure of the modeled system, and which gives rise to systems of equations. Finally, in chapter 6 the concept of consistency is used to transform general systems of equations into fixpoint form, and the concept of knowledge extension will be applied for exploring non-least fixpoints, which are in may cases the main interest of the modeling effort. This outline will be given informally; formal definitions and the necessary proofs will be published in [7].

The ideas presented here basically go back to Prof. Engeler [8, 9, 10], whom I would like to thank a lot for his valuable advice and his encouragement for my research in implementing a modeling environment based on the presented approach.

2 Example

In order to explain our approach, let us consider a small, but nontrivial example[1] in the domain of car electrics (see Fig. 1). It comprises a motor which is connected via a driving-belt with an alternator. The alternating current supplied at the output of the alternator is converted into direct current by the voltage regulator which charges the battery. Connected to the battery is the ignition and injection electronics (Jetronic) and, controlled by the position of the key, the starter and a light, the latter with an intermediate switch.

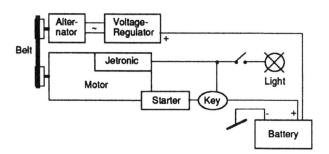

Fig. 1. Car electric system.

The model to be constructed should help us to answer questions such as:

a) What can happen, if the engine is running between 1000 and 2000 rpm?
b) The car does not start. What might be wrong?
c) Why is the light functioning, if the voltage regulator is damaged (short-circuited), but the Jetronic is not?

---

[1] A similar example is used by Price and Lee [1] to explain the difference between "shallow" and "deep-knowledge" models.

At first sight, these questions are of a very different nature. Answers to questions such as the first one are called *envisionment* in the context of qualitative modeling which means, that we are looking for all possible states of the system with certain assumptions, such as "the motor is running between 1000 and 2000 rpm". Questions of type b) ask for a diagnosis; they may be viewed as an envisionment of system on the assumptions of the observed behavior, which gives us the set of all possible faults. This set of faults may be restricted if we put some additional observations into the set of assumptions, such as "the light is not burning, although I switched it on". Finally the last question indicates that we may want an explanation of a certain behavior.

Consider the observation b) that the car does not start and the light does not burn, even if the switch is on. A "shallow-knowledge" expert system will make a diagnosis by means of a table of faults and causes. This table will contain the obvious facts that the battery may be flat or the starter and the light are damaged. If the system has been tuned for a certain period, it will consider faults which are rarely encountered, such as "the electrical contacts in the lock are damaged", but it may miss the fact that the supplying wire of the lock is disconnected.

3 Cumulative Logic Programs

As we are following a "deep knowledge" approach, we will model each of the components of the electric system by its own knowledge-base, which comprises the "first principles" of the device. Mathematically the knowledge-base is a set of formulas that denote the known properties of the component in question, and which are all simultaneously true.

Atomic knowledge is expressed by first-order formulas, and we will adopt the PROLOG syntax where the names of predicate and function symbols start with a lowercase letter, and variable names with an uppercase letter. Quantifiers are omitted and variables are implicitly all-quantified over the scope of a formula. E. g. the knowledge-base of the voltage regulator may look like

$$voltage(out, common) = high,$$
$$connected(out, wire\_red),$$
$$voltage(Contact, Contact) = 0, \dots$$

and would be read as: the voltage between the contact *out* and the common potential is high, contact *out* is connected to a red wire, and for *each contact*, the electrical potential difference to itself is zero etc.

Now, since the components of the physical system interact, the knowledge-bases modeling the components have to be related in our approach. This relation is expressed by knowledge transformations, namely: the knowledge of the influencing components is transformed into knowledge of the influenced component. We will express this dependency by a set of knowledge transformation rules, which model the interaction. E.g. in our physical example the voltage regulator and the battery interact by means of a (let's say red) wire, which forces the corresponding contacts to be on the same electrical potential. In terms of the knowledge-bases this is expressed as: if both the battery and the voltage regulator are connected to some wire W, and the electrical potential at the voltage regulator connection of W is V, then the same potential is at the battery connection of W. In other words the transformation $Wire$ takes knowledge from the battery (1) and the voltage regulator (2) and generates additional knowledge about the battery (3), which is expressed as:

$$\{connected(C_0, W)\} \qquad\qquad (1)$$
$$\rightarrow \{connected(C_1, W), voltage(C_1, common) = V\} \quad (2)$$
$$\rightarrow voltage(C_0, common) = V. \qquad\qquad (3)$$

Note that the arrows in the rule do not mean logical implication, but just separate the arguments and the result of the transformation.

A knowledge transformation F is a set of such rules, and the result of applying it to a knowledge-base X is defined by the binary operation apply ".":

$$F \cdot X = \{a\sigma : \alpha \rightarrow a \in F, \beta \subseteq X, \beta\sigma = \alpha\sigma\}$$

where σ is the most general unifier of α and β, and $\alpha \rightarrow a$ is a rule with α a finite set of formulas and a a rule or an atomic formula. If the transformation F has more than one argument (as in the wire example above), it will contain rules with multiple arrows of the form:

$$\alpha_1 \rightarrow \alpha_2 \rightarrow \ldots \rightarrow \alpha_n \rightarrow a$$

where the α_i are finite subsets of formulas, and a is a formula. The transformation F applied to the arguments $X_1, X_2, \ldots, X_n$ is then expressed by

$$F \cdot X_1 \cdot X_2 \cdot \ldots \cdot X_n.$$

To avoid clustering of parentheses, let the rule operator "$\rightarrow$" be right associative (i.e. read $\alpha \rightarrow \beta \rightarrow a$ as $\alpha \rightarrow (\beta \rightarrow a)$).

In the example above, the $Wire$ transformation contains rules of the form

$$\{a_1, \ldots, a_m\} \rightarrow \{b_1, \ldots, b_n\} \rightarrow a.$$

The knowledge transformation performed by $Wire$ is expressed by $Wire \cdot Battery \cdot Volt.Regulator$, where the subexpression $Wire \cdot Battery$ will give us a set of formulas $\{\beta_i \rightarrow a_i : i = 1 \ldots n\}$, which applied on $Volt.Regulator$ gives the intended result.

When regarding the wire as a component of the system, which indeed it is in the physical system, we have to admit transformations being dependent on other components. This observation forces us to combine the notions component and knowledge transformation into the notion $process$, and we have to admit rules about rules, rules about rules about rules, etc. This iterated rule formation process (cumulation of rules) leads us to Cumulative Logic Programs (CLP), which will serve as the modeling domain for processes, and which we will define formally below.

Let A be the set of all atomic formulas built from given sets of constants and predicate, function and variable symbols. Then $G(A)$ is defined recursively by

$$G_0(A) = A,$$
$$G_{n+1}(A) = G_n(A) \cup \{\alpha \to a : \alpha \subseteq G_n(A), \alpha \text{ finite}, a \in G(A)_n\}.$$

$G(A)$ then is the union of these sets, and the language of Cumulative Logic Programs CLP is the set of all subsets of G(A).

$$G(A) = \bigcup_n G_n(A), \qquad CLP = \mathcal{P}(G(A)).$$

Note that $G(A)$ contains rules of arbitrary nesting, e.g.

$$\{\alpha \to a, \alpha_1 \to \beta_1 \to a_1\} \to \{\alpha_2 \to a_2\} \to b_2$$

which needs the unification to be defined for (nested) rules.

4   Denotational Semantic of CLP

Before continuing with the outline of the modeling approach, we will turn our attention to CLP and give a denotational semantic. As a first step, we will expand formulas containing variables into sets of formulas, by substituting each variable with all terms built from function symbols and constants. In general, this set will be infinite. The expansion $[X]$ of a set of formulas X is then just the result of expanding each formula in X. The application operation on expanded sets $[F]$ and $[X]$ is then

$$[F] \cdot [X] = \{a : (\alpha \to a) \in [F], \alpha \subseteq [X]\},$$

which is the application operation in Engeler's graph algebra. More specific, let A be the set of all atomic formulas without variables and D_A the powerset of $G(A)$, then $\langle D_A, \cdot \rangle$ is a graph algebra, which in turn is a combinatory algebra [11].

As a second step, let us consider the logic closure of CLP (see also [13]). If we are modeling in a certain domain such as car electrics, we always have a certain theory in mind, e.g. when writing

$$low \leq voltage(in_0, common) < high$$

with the possible qualitative voltage values $0 \leq low \leq normal \leq high$, we also intend the formula

$$0 \leq voltage(in_0, common) < high,$$

which is less specific, but also true and can be inferred from the first one. The logical closure $[X]$ of the process X is the expansion $[X]$ of X plus all formulas, that can be derived from $[X]$ in the given way. With this notion of logic closure, we can precisely define the relation of exactness of knowledge. Let A, B be processes, then B contains *more knowledge* than A, if

$$[A] \subseteq [B].$$

With this relation, CLP is a complete lattice.

5 Interaction Graphs and Equations

The language CLP that has been introduced in the previous chapter serves as a language for describing the function of the processes in the modeled system. With the introduction of the application operation "·" we have defined knowledge transformation. With these two notions in mind, we will now turn our attention to the global aspects of the system and introduce a graphical language, namely *interaction graphs*. They are used for defining the structure of the system by identifying the individual processes and specifying the way in which they interact. An interaction graph consists of:

a) a labeled circle for each process
b) a network of arrows, specifying the interactions between processes,

where each arrow

- terminates at a process that is being influenced,
- has one or more shafts that originate at the influencing processes and
- goes through the process that constitutes the knowledge transformation.

E.g. the interaction graph of the voltage regulator - battery subsystem (Fig. 2) reads as: knowledge from the knowledge-bases (KB) of the battery and the voltage regulator are processes by the KB of the wire, and put into the KB of the battery.

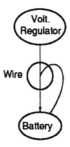

Fig. 2. Interaction graph of the voltage regulator - battery subsystem.

Figure 3 gives the complete interaction graph of the car's electric system.

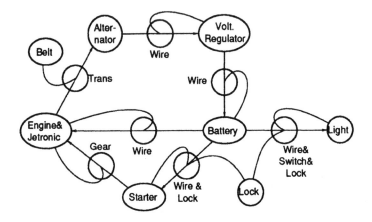

Fig. 3. Interaction graph of the car electric system

Interaction Graphs can be compared to diagrams used in other disciplines, such as data base schemes or Petri nets. Petri nets model dynamic systems where the flow of "tokens" in a network of nodes represents the flow of data, material, control etc. of a real-world system. In interaction graphs these tokens are logic programs, and are much more expressive than the binary on/off or discrete value information used in classical Petri nets. [2] In contrast to the dynamic nature of Petri nets, information once available in a node of an interaction graph, doesn't vanish, but is accumulated in it. This does not mean that our approach is unable to formulate dynamic behavior. Here, such behavior would be expressed explicitly by formulas (e.g. time dependent functions, powerseries-approximations etc.). The main difference to Petri nets and similar approaches is the reflexive nature of the underlying modeling domain, which turns mappings within the domain of knowledge-bases into objects of the domain itself.

Since we already defined the knowledge transformation mechanism with the help of the application operation, we can now translate an interaction graph into a system of equations by expressing each basic building block

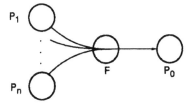

with the corresponding equation

$$P_0 = F \cdot P_1 \cdot \ldots \cdot P_n$$

---

[2] High Order Petri nets use enhanced tokens, that carry more and structured information.

which claims that the knowledge about process P_0 is completely determined by the interaction $F \cdot P_1 \cdot \ldots \cdot P_n$. Sometimes, only partial knowledge of P_0 may be determined by the interaction, i.e. $P_0 = P_0 \cup F \cdot P_1 \cdot \ldots \cdot P_n$, which can be expressed by means of a modified F that takes P_0 as an additional argument: $P_0 = F' \cdot P_0 \cdot P_1 \cdot \ldots \cdot P_n$.

Applying this transformation to the whole interaction graph gives then a system of equations, and the purpose of the modeling, e.g. answering questions such as the ones posed in chapter 2, is to solve these equations.

6 Consistency and Fixpoint Equations, Extension of Least Fixpoints

In the previous chapter, knowledge-based modeling was identified with the algebraic problem of solving equations in the domain of CLP. In the following, let us turn to the domain of the graph algebra, into which we have mapped CLP by giving the semantic $[.]$. The problem of solving systems of equations in a graph algebra is in general unsolvable (due to the fact, that combinatory algebras are universal computing devices), whereas for the important class of simultaneous fixpoint equations, an algorithm for approximating the least fixpoint is well known (note that "." is continuous).

Given $F_1, \ldots, F_n \subseteq D_A$ and variables $X_1, \ldots, X_n$ and a set of simultaneous fixpoint equations $\Sigma = \{X_i = F_i \cdot X_1 \cdot X_2 \cdot \ldots \cdot X_n : i = 1 \ldots n\}$, then the least fixpoint of Σ is:

$$X_1 = \bigcup_j X_1^j, \quad X_2 = \bigcup_j X_2^j, \quad \ldots, \quad X_n = \bigcup_j X_n^j$$

with

$$X_i^0 = \emptyset \quad \text{and}$$
$$X_i^{j+i} = F_i \cdot X_1^j \cdot X_2^j \cdot \ldots \cdot X_n^j.$$

Although the equations derived from an interaction graph have the form $P_{i_j} = \tau(P_1, \ldots, P_n)$, where $\tau(P_1, \ldots, P_n)$ is an arbitrary applicational term in $P_1, \ldots, P_n$, they need not form a system of simultaneous fixpoint equations, since a process P_{i_j} can figure twice or more on the lefthand side (Fig. 4).

This means, that we have to deal with general systems of equations. In [10] Engeler gives a semi-algorithm for solving the equation $a \cdot x = b \cdot x$ using the concepts of *consistency* and *knowledge extension*. [3] The key idea is to successively extend $x = x_0 \subseteq x_1 \subseteq \ldots$ and $y = y_0 \subseteq y_1 \subseteq \ldots$ in such a way that the union of the results from the left and the right hand side $a \cdot x \cup b \cdot y$ are consistent.

In the following we will apply the same concepts in a different framework. Let a *consistency transformation* C be defined as a CLP with:

- $[C \cdot X \cdot Y] = [X \cup Y]$ and

---

[3] Each system of equations can be reduced to the form $a \cdot x = b \cdot x$ [12].

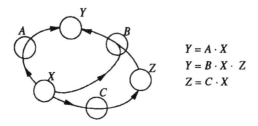

$Y = A \cdot X$

$Y = B \cdot X \cdot Z$

$Z = C \cdot X$

Fig. 4. Interaction graph with an non-fixpoint equational system

– $fail \in C \cdot X \cdot Y$ if $[\![X \cup Y]\!]$ is inconsistent,

where $fail$ is a formula specifying inconsistency (we could use $0 = 1$ as well). Consistency transformations may also perform some knowledge compression. If for example

$$low < voltage(in, common) \in X$$
$$voltage(in, common) \leq normal \in Y$$

and we again suppose the qualitative voltage values $0 \leq low \leq normal \leq high$, then $C \cdot X \cdot Y$ needs only to include $voltage(in, common) = normal$.

Suppose our model includes equations

$$Q = \tau_1(P_1, \ldots, P_n)$$
$$\ldots$$
$$Q = \tau_m(P_1, \ldots, P_n)$$

where Q is one of the P_i. Then we transform these equations into

$$Q_1 = \tau_1(P_1, \ldots, P_n)$$
$$\ldots$$
$$Q_m = \tau_m(P_1, \ldots, P_n)$$
$$Q = C \cdot Q \cdot (C \cdot Q_1 \cdot (\ldots (C \cdot Q_{m-1} \cdot Q_m) \ldots))$$

where $Q_1, \ldots, Q_m$ are new process symbols and C is the consistency transformation of an appropriate domain. Note that whereas the original set of equations claims strict equality $\tau_1(P_1, \ldots, P_n) = \ldots = \tau_m(P_1, \ldots, P_n)$, the second one just claims that the union $\tau_1(P_1, \ldots P_n) \cup \ldots \cup \tau_m(P_1, \ldots, P_n)$ is consistent.

In this fashion we transform each system of equations Σ derived from an interaction graph into the associated consistency fixpoint system Σ^*. When approximating solutions to Σ^* with the recursion given above, we need not start with $X_i^0 = \emptyset$ but can start with the knowledge already contained in the process.

In many cases, the least fixpoint of Σ^* will be reached after a few iterations, either because $fail$ has been generated at some process, or no more formulas are derivable from the consistency equations, i.e. we have reached a consistent least fixpoint, but are still faced with uncertain knowledge such as $0 \leq voltage(in, common) < high$, whereas we want to know an exact value.

To proceed further we need a second concept. Let X be a process, then $\{E_1, \ldots, E_k\}$ are *consistent extensions* of X if:

- $fail \notin C \cdot X \cdot E_i$, i.e. $X \cup E_i$ is consistent and
- for every E with $X \cup E$ consistent, there is a E_j with $E \cup E_j$ consistent, i.e. the E_i form a consistent covering of X.

E.g. a process containing $0 \leq voltage(in, common) < high$ has extensions $\{E_1, E_2\}$ with:

$$voltage(in, common) = 0 \in E_0$$
$$0 < voltage(in, common) \leq normal \in E_1.$$

We can now formulate the approximation algorithm:

(1) Compute the consistency fixpoint of the current setting.
(2) If $fail$ is in some process then halt.
(3) Extend some process P by nondeterministically choosing some extension E_i of P, and setting $P := P \cup E_i$
(4) If no such extension is possible then halt.
(5) Loop back to step (1)

The nondeterminism in step (3) reflects the fact that we are searching in the space of consistent extensions of the initial setting. It can be implemented by parallelism or by backtracking (the latter is applied in the current implementation).

7 Conclusion

The presented approach is a contribution to both artificial intelligence *and* computer algebra. On the one hand, by combining concepts from logic programming and qualitative modeling, we believe that it is a promising tool for building "deep-knowledge" expert systems. The combination of these concepts is performed in a clear and well-understood mathematical framework. By making knowledge-bases to objects that are manipulated by computer, we introduce computer algebra concepts to artificial intelligence.

On the other hand, computer algebra may benefit from our approach. If we think of the components of the system as standing for real values, functions, or transformations on functions, mathematical equations can be formulated by interaction graphs. Mathematical objects then are not represented by symbols, closed formulas, algorithms, or limits of approximations etc., but by collections of knowledge about them, which may in fact include these established representations. E.g. we could think of representing the sine function by a knowledge-base containing

- an algorithm for computing the floating point approximation of the sine function at some point,
- knowledge about its derivative,
- knowledge that it is continuous, periodic etc.

The approximation algorithm of chapter 6 has been implemented in a prototypical modeling environment [7]. An extended example application for diagnosis of a technical system is currently being built within this environment by the diploma student E. Burkhard.

References

1. Price, C., Lee M.: "Applications of deep knowledge". *Artificial Intelligence in engineering*, 1988, Vol. 3, No. 1, pp.1-7.
2. de Kleer, J., and Brown, J.S.: "A Qualitative Physics Based on Confluences" *Artificial Intelligence* **24**, (1984), p. 7–83.
3. Forbus, K.D.: "Qualitative Process Theory". *Artificial Intelligence* **24**, (1984), p. 85–168.
4. Kuipers, B.: "Qualitative Simulation" *Artificial Intelligence* **29**, (1986), p. 289–338.
5. Struss, P.: "Mathematical aspects of qualitative modelling" *Artificial Intelligence in Engineering* (1988), pp. 156–169.
6. Bratko, I., Mosetic, I., Lavrac, N.: "Automatic Synthesis and Compression of Cardiological Knowledge" *Machine Intelligence* **11**, (1988), pp. 435–454.
7. Schwärzler G.: "Knowledge-Based Modeling of Cooperative Processes" PhD Thesis, ETH Zürich, to appear 1992.
8. Engeler, E.: "Modelling of Cooperative Processes" Report No. 86-06, Math. Dept. ETH Zürich, 1986.
9. Engeler, E.: "Cumulative Logic Programs and Modelling" *Logic Colloquium '86*, Drake, F., Truss, J., (Editors), North Holland, 1988.
10. Engeler, E.: "Sketch of a New Discipline of Modelling" Report Math. Dept. ETH Zürich, 1988.
11. Engeler, E.: "Algebras and Combinators" *Algebra Universalis*, (1981), p 389-392.
12. Engeler, E.: "Equations in Combinatory Algebras". *Proceedings of "Logic of Programs '83", SLNCS 164*, (1984).
13. Aberer K.: "Combinatory Differential Fields and Constructive Analysis" PhD Thesis No. 9357, ETH Zürich, (1991).

Intelligent User Interfaces for Symbolic and Numeric Computation

Michael E. Clarkson,
Centre for Earth and Space Science,
York University,
North York, Ontario, Canada, M3J 1P3

and,
Centre de Morphologie Mathématique,
35 Rue Saint–Honoré,
77305 Fontainebelau, France

Abstract. We have implemented an intelligent user interface for the symbolic algebra system Macsyma, using a rule–based expert system based on Emycin. The system acts as an intelligent front–end, communicating with a back–end Macsyma sub–process. The rule base of the system has been extended to include knowledge about numerical solutions to problems that Macsyma is unable to solve algebraically, or on request by the user. The system incorporates an on–line documentation brower, and facilities to review and explain the results of a consultation. The system has been written using portable object–oriented programming techniques, based on CLOS and CLX.

Keywords: Intelligent User Interface, Expert Systems, Symbolic Computation

General purpose computer algebra (CA) systems such as Reduce, Macsyma, SMP, Maple and Mathematica provide a very wide range of techniques to help scientists solve problems in mathematics, and a great body of code has been contributed for a wide range of advanced applications [1–3]. However, our observation of the users of all of these systems is that they have proven to be very difficult to use. These systems are very large, and simply learning the proper command syntax can be a major hurdle, especially when there are usually several different ways to solve any given problem. In addition, for inexperienced users there is often the temptation to try and solve a problem with brute–force, without applying any of the finesse that mathematics requires.

There is also a degree of difficulty for the beginner that arises from the lack of semantic uniqueness that is inherent in mathematics. For example, factoring an expression usually makes it smaller, or reduces it in size:

$$\text{FACTOR}(x^6 + 3 x^4 + 3 x^2 + 1) \;\to\; (x^2 + 1)^3$$

However, in some cases factoring an expression makes it bigger:

$$\text{FACTOR}(x^6 - 1) \;\to\; (x - 1)(x + 1)(x^2 - x + 1)(x^2 + x + 1)$$

Thus the intuitive concept of "reducing" or "making smaller" does not always correspond to a unique mathematical counterpart. Furthermore, the most straightforward approach from a mathematical point of view may be less than optimal from a CA point of view. Often the former might be characterised as a "brute–force" approach, lacking in any kind of subtlety or optimization. Although provably correct, the mathematical approach may not be successful on a CA system because of limitations of time or memory resources. This is especially critical in computer algebra, because many of the algebraic algorithms grow exponentially or even factorially with increasing size. This internal "expression swell" can be fatal to large brute–force computations.

Macsyma is a particularly difficult system to use from a user's point of view. It is a large CA system, with over 1000 global variables and functions defined and documented. It was developed by a large number of people over a ten year period, and grew without the benefit of a coherent naming scheme for the functions and variables. At the same time, **Macsyma** is probably the most powerful of the generally available CA systems, and with the requisite skill, it can be an excellent system to solve problems in the physical sciences.

In fact, the skills needed to operate an CA system like **Macsyma** are in themselves a considerable body of knowledge. One has to have good understanding of mathematics to know how to approach any non–trivial problem, as well as extensive training on how to implement the chosen approach. Further consideration must also be given to selecting implementation strategies based on efficiency. Given the structured nature of this knowledge and expertise, it immediately suggests the idea of combining CA systems with rule–based expert systems in order to provide an intelligent guide to aid in mathematical problem solving.

Rule–Based Expert Systems

In a certain sense, any computer language could have been used to write the system, just as any computer language can be used to write a symbolic algebra system (including Cobol! [4]). There is nothing magical about using a rule–based expert system to write the user interface; it is simply a very high level computer language with a predefined flow of control. Although there has been a great deal of hyperbole in recent years describing the capacities of rule–based expert systems, it is clear that they offer some very useful properties for organizing any body of knowledge that can be structured into rules. If combined with a CA system, rule–based expert systems offer a number of features described here.

- Expert systems can provide a consistent user interface, that can greatly ease the current problems of human–computer interaction. In many rule–based systems, the user is presented with a menu of choices in order to define the goal. At any particular decision point, all of the relevant possibilities to choose from are presented, so it is easy to define the goal precisely. These menus can be state dependent, so that the menu items presented for selection depend on the answer to the questions that have been asked before. This means that only a minimum amount of information needs to be solicited from the user.

- A structured knowledge base can take advantage of exponential growth to traverse a large body of knowledge by asking only a very few questions. In theory, if each menu presented to the user contains 5 choices, one can distinguish between 125 different possibilities with only 3 questions, 625 possibilities with 4 questions, and so on. This means that layering an expert system on top of a CA system can organize a large body of knowledge without unduly slowing down the user.

- Once the knowledge base structure has been written, additional rules can be added easily, always increasing the usefulness and completeness of the system as time goes on. Knowledge coded into rule–based expert systems is usually easily modified and changed. Macsyma is a Lisp based program, and both Macsyma and Lisp tend to be idiomatic in how they are programmed: it is possible to capture many the common mathematical or programming idioms into rules, and extend the system to incorporate idioms for speed, efficiency or completeness.

- Expert systems provide a mechanism to review a consultation. This provides a way to explore alternatives to solving a problem, without having to have the CA system carry out the operation. One can backup and rerun a consultation with different choices, and the user can call up an explanation of why a particular rule was invoked. In this sense, the knowledge coded into a rule–based system is much more accessible than knowledge hard–coded into the CA system.

- An expert system for computer algebra could correlate a particular function with the associated flags or variables that may effect the particular function's action. The choice of which flags or global variables are relevant usually depends on the context and form of the expression. With a small amount of parsing of the expression's structure and contents, we are able to eliminate a large number of flags that may effect a given command, and can highlight variables that are most likely to have an important effect.

- An expert system for computer algebra can examine the data which is to be passed to the CA system, to check for consistency and correctness, such as basic datatype and range checks.

- Expert systems can provide an excellent teaching tool, especially when they are integrated with good documentation. The knowledge base can be include active pointers to the appropriate sections from the User's or Reference manual, to be available at any point in the consultation. In addition an explanation of the consultation's reasoning can be saved or printed for later study and analysis.

Some parts of Macsyma already have decision structures that are similar to rule–based knowledge. A large base of mathematical knowledge that is "hard-coded" into the program, sometimes using heuristics and other times using deterministic algorithmic methods, but in almost all cases the knowledge is inaccessible to the user. Usually these structures are found at the top level of a Macsyma function, and dispatch the expression to sub–functions based on the expression type.

If we take for example **Macsyma**'s **LAPLACE** command, it has a top–level **cond** statement that dispatches auxiliary Lisp functions **laptimes**, **lapext** etc. in the appropriate cases. This structure can be considered to be equivalent to a set of **if-then-else** "rules", that might be embodied in the consultation rules described above. However, there would be no point in removing these "hard–coded" rules and replace them with rule–based rules, for the following reasons. Firstly, there would be a severe performance penalty for doing so. The simple Lisp conditional statement is a very efficient way of selecting between mutually exclusive alternatives. Although we wish to make parts of the mathematical knowledge base accessible to the user, it is wise to be selective as to just how much knowledge is made available. Secondly, we wish to make as few changes to the source code of **Macsyma** as possible. This avoids the problem of maintaining the code to the CA system itself. As we have referred to previously, **Macsyma** is huge, and we have no wish to maintain the CA system as well as the rule–based expert system. Thirdly, most of this "hard–coded" knowledge is in the form of mutually exclusive cases as in the example above, and hence there is little benefit to be derived from reordering the rules. It would be an inefficient use of an expert system to apply it to these kinds of cases. And finally, we wish to use the expert system to deal with cases where there is interaction between the rules, such as the correlation of the form of an expression, with the goal of the consultation. This kind of correlation using "meta–knowledge" can only be carried out using a high level knowledge structure, and would prohibitively expensive to implement directly into a CA system, because of the multitude of cases that have to be allowed for.

We feel that a "shallow" coupling of the two expert systems is preferable to designing a monolithic hybrid system [5]. We will use the expert system for the parts that it can do best: to provide a consistent user–interface that allows the user to succinctly specify his goal; to correlate high–level knowledge about the CA system syntax is needed to solve the problem, and to perform rule–based error checking and optimization of the **Macsyma** input. The rule–based system is essentially separate from the CA system, and the two sub–systems will run on different CPU's, or as different processes on the same CPU. By using separate packages for the rule–based system and for the CA system, we can use existing rule–based and CA systems, so that they do not have to be written from scratch. The two sub–systems can be maintained separately, avoiding the potential nightmare of having to maintain a specially modified CA system.

Current Implementation

The **Praxis** knowledge base is made up of a set of frames and sub–frames grouped into a decision tree. These frames are lexical environments where the environment of a sub–frame is contained in the lexical scope of its parent. Not only do frames make it easier to organize the information of a large knowledge base in a modular way, but they minimize the amount of information that the system needs to keep in memory during a consultation. The rules and parameters of a frame are loaded only when the frame is entered for the first time (instantiated), so large knowledge bases can be constructed without severely degrading the system response times. Currently, Praxis consists of 35 frames with about 500 rules.

Emycin

The system is based on an implementation of the Emycin expert system [6]. Pmycin is a simple yet powerful rule–based expert system, that has backward chaining or forward chaining rules. It handles conditional reasoning with uncertain data using the Dempster–Shafer theory of evidence. The optimization of the flow of a consultation is aided through rule utilities, parameter utilities, precedent rules, and frame initial data. Parameter can acquire values either through assignment by the rules, or by active values, default values and parameter methods. Because of the access to underlying Lisp or foreign C functions, user–defined functions can be added to make the system extensible. Also, the traditional Emycin analysis and English translation of the elements of the knowledge base, and the reasons for a specific conclusion are available.

In order to take advantage of the abstraction and simplification that comes from using object–oriented programming techniques, Pmycin is written in the Common Lisp Object System (CLOS) [7]. Frames, rules, parameters, operators, and variables are all CLOS classes; tracing, translation, and help documentation are implemented as generic functions.

The graphical user interface to Pmycin is written using Picasso [8,9], which is an object–oriented graphical user interface development system. The application framework uses high–level graphical objects, which use inheritance and generic functions to simplify the creation of instances of a class. Picasso is written in Common Lisp, using CLOS and CLX [10].

We have found that the programming ease and maintenance of the system has been greatly enhanced by using object–oriented programming techniques, and estimate that the overall expert system and user interface code is about one–third of the number of lines of what it would otherwise have been. The entire expert system, including the windowing user–interface code is less than 5000 lines of code. Also, as CLOS and CLX available for most computers that run Lisp, the system is very portable. The main problem is of course the one to which we referred to earlier, namely the size and speed of the underlying system. The size of Picasso is comparable to that of Macsyma itself, and the Common Lisp core image before loading the expert system or knowledge base, but including Picasso, CLOS and CLX, is 16 Mbytes.

Consulting the Expert System

The consultation begins by prompting the user for a definition of what he or she wants to do, which becomes the goal of the consultation. This is done by presenting a series of multiple choice menus that will try to progressively define the type of mathematical operation that he or she wishes to perform. The rules then determine the type of mathematical operation that is desired by the user, and the best way of carrying out that operation using the symbolic algebra system. Finally, the system generates the specific syntax of the Macsyma command, and may perform source–level optimizations to improve the efficiency of the result. More than one possible solution to the problem at hand may be given, and the user is presented with a choice amongst the alternatives, and an indication of the certainty with which the expert system concluded each answer.

Having completed a consultation with the knowledge base, the system should have arrived at a command to carry out what the user wants to accomplish. This command is then dispatched to a back–end Macsyma process for execution. A remote GNU Emacs server is established separately from the expert system Lisp process. This Emacs process asynchronously listens to a TCP/IP port for incoming Emacs Lisp commands, and executes them. In this case, the expert system opens a network stream connection to the Emacs server process, and dispatches the Emacs code necessary to submit the generated syntax to a subprocess buffer running Macsyma. This buffer allows the user to dispatch to Macsyma the command generated by the expert system, or carry out a normal interactive computer algebra session with Macsyma. As the expert system and Macsyma process operate asynchronously, the user can continue to query the expert system while the Macsyma process is busy with a calculation. By making a minor modification to the top–level Macsyma read–eval–print loop, and using the process sentinel functions of GNU Emacs, the Lisp list form of the result of the evaluation is passed back to the expert system process, so that it can determine if the generated command was successful. This also allows the results of a limited form of structure analysis of the expression to be used during the expert system inferencing.

Numerical Linkage

In a sense, Praxis can be viewed as a system of automatic programming [11], where the expert system acts as a Programmer's Apprentice that uses high–level commands (in this case menu–driven) to generate valid syntax for a target programming language (in this case, Macsyma). Similar approaches have been adopted for the semi–automatic generation of code for scientific–problem solving using numerical subroutine libraries [12,13]. These latter systems generate FORTRAN code that links to large numerical libraries such as NAG, IMSL or NBS CMLIB, and use as a basis for the organization of the mathematical knowledge, the GAMS framework for the management of scientific software [14].

At a recent conference on programming environments for high–level scientific problem solving [15], it became apparent that the range of applied mathematics represented by the GAMS taxonomy, and by the structure of knowledge in Praxis(Figure 2) is very similar, and that there is a very high degree of overlap. The idea presented itself of combining the two structures so that the domain of the Praxis consultation process would be extended to include numerical as well

```
A -- Arithmetic, error analysis
B -- Number theory
C -- Elementary and special functions (search also class L5)
D -- Linear Algebra
E -- Interpolation
F -- Solution of nonlinear equations
G -- Optimization (search also classes K, L8)
H -- Differentiation, integration
I -- Differential and integral equations
J -- Integral transforms
K -- Approximation (search also class L8)
L -- Statistics, probability
M -- Simulation, stochastic modeling (search also classes L6 and L10)
N -- Data handling (search also class L2)
O -- Symbolic computation
P -- Computational geometry (search also classes G and Q)
Q -- Graphics (search also class L3)
R -- Service routines
S -- Software development tools
Z -- Other
```

Fig. 1. GAMS Primary Categories

as symbolic computing.

The organization of the GAMS hierachy is somewhat arbitrary, and perhaps less structured than that of the knowledge base frames. This might be expected as the latter was written at the same time as a restructuring of the Macsyma reference manual was carried out. The symbiotic nature of the two processes of structuring the knowledge base and organizing the manual was very fruitful in enforcing some kind of order on the structure.

In the GAMS taxonomy, the entire spectrum of numerical subroutines from a subroutine library such as IMSL is represented, including areas such as graphics, service routines and software development tools. At this stage in its development, these areas are not contained in the Praxis knowledge base tree. So for the sake of this discussion, let us narrow the focus a little to the area of overlap between the two structures, and concentrate on the facilities for carrying out applied mathematics. We will therefore eliminate from current consideration the follow-ing GAMS categories: G Optimization, L Statistics, probability, M Simulation, stochastic modeling, N Data handling, O Symbolic computation, Q Graphics, R Service routines, S Software development tools, and Z Other.

In the GAMS taxonomy, the category for Symbolic Computation is structure-less, and in effect all computer algebra is lumped into this category. At the risk of perpetration a similar error, we shall place GAMS as a category in our frame structure, but as we shall see below, we will create the necessary links between the symbolic and numerical structures. In comparing the resulting structure with Figure 2, we see, as could be expected, a very large degree of overlap between the two structures. Perhaps in the future it may be possible to improve on the organization of the GAMS taxonomy, but for the moment, it has the status of something akin to a standard. Therefore, rather than departing from the GAMS structure, or sacrificing the extra organization of the knowledge base frames, we will adopt a spirit of what the French would call "cohabitation" and support both structures in one rules base.

```
GOAL                        - the "parent" frame
    LANGUAGE
        MATH-FUNCTIONS          - Mathematical Functions
    MANIPULATE
        EVALUATE                - Evaluate Expressions
        PART                    - Select a Part of an Expression
        SUBSTITUTE              - Substitute a Part of an Expression
        LISTS                   - List Manipulation
        MAP                     - Map functions on expressions
    SIMPLIFY
        EXPAND                  - Expand Expressions
        FACTOR                  - Factor Expressions
        COMBINE                 - Combine Expressions
    MATRIX
        MAT-BASIC               - Basic Matrix Operations
        MAT-INFO                - Matrix Information
        MAT-DEFINE              - Define Matrices
        MAT-OPERATE             - Operate on Matrices
        MAT-MANIPULATE          - Manipulate Matrices
        MAT-FLAGS               - Variables that effect Matrices
    SERIES
        SUM                     - Summation
        TAYLOR                  - Taylor Series
    CALCULUS
        DIFF                    - Differentiate Expressions
        INTEGRATE               - Integrate Expressions
        LAPLACE                 - Take Laplace Transforms
    SOLVE
        SOLVE-EQN               - Solve Equations
        SOLVE-POLY              - Solve Polynomials and their roots
        SOLVE-DE                - Solve Differential Equations
    SYSTEM
        HELP                    - Getting Help on a Macsyma Command
        SAVE                    - Saving Macsyma Expressions to a file
        TRACE                   - Tracing the Execution of a Function
    GAMS
```

Fig. 2. Praxis Knowledge Base Frames

Towards this end, we proceed in the following manner: we create a top-level sub-frame called **GAMS**, which has the entire GAMS taxonomy in its original form. As we shall detail below (§), the sub-frame contains the rules and parameters necessary to traverse the tree, and select any subroutine classified by the taxonomy that is available from the IMSL library. On the other hand, the "symbolic" Praxis is expanded to include pointers to the **GAMS** frame, which are when a symbolic solution to a problem cannot be found, or when a numerical option is selected from one of the menus presented to the user.

The GAMS Frame

Using automatic programming techniques, a frame for the GAMS hierachy was automatically generated from a list representation of the GAMS structure. Parameters were generated to provide the user with menu selections corresponding to each level of the structure, and rules were generated so that the user can traverse the tree to select a particular category of interest.

Once the expert system rules have determined which GAMS category the problem lies in, a subprocess of the GNU Emacs server consults the IMSL help program `imsl.idf`, which deposits in a temporary Emacs buffer a listing of all IMSL routines (if any) that or of the relevant GAMS classification.

1. a purely "symbolic" answer, which is a fragment of **Macsyma** code, that is then dispatched to the back–end **Macsyma** process. This result will be obtained when the rule–base contains a purely symbolic solution, and the user opts for symbolic choices in the menus.
2. a purely "numeric" answer, which is a fragment of FORTRAN code, that is subsequently compiled in an Emacs buffer and linked to the numerical library. This result will be obtained when the rule–base fails to obtain purely symbolic solution, when a symbolic solution is tried, but **Macsyma** fails to arrive at a solution, or when the user opts for numeric choices in the menus.
3. a hybrid "symbolic–numeric" answer, which is a fragment of FORTRAN code that is compiled and linked to the numerical library, and then dynamically linked into the running **Macsyma** process using the foreign–function interface available in most Common Lisps, to make the numerical function or subroutine available to the current **Macsyma** session. This result will be obtained under considerations similar to 2) above, expect that the user is presented with the choice of creating a self–contained FORTRAN program, or loading the numerical code into the **Macsyma** process.

We have implemented a second sub-process buffer of the Emacs process that will retrieve the documentation of the relevant FORTRAN function from the IMSL library, using the IMSL interactive help program `imslidf`. The results from this search are reformatted by Emacs into a prototype FORTRAN program in the case of 2) above, and are working on the case 3) above in a manner very similar to that which was done in **Naglink** [16].

Conclusion

We are hopeful that a port of the rule base to Reduce, Maple, Mathematica or Axiom could be carried out without major changes to the system. It should be even more straight–forward to port the numerical part of the system to other numerical libraries such as NAG, as only the external GAMS category search program needs changing, and the small portion of GNU Emacs code that formats the result into an executable program. This would raise the possibility of a very desirable prospect, which is a system independent interface for both symbolic and numeric computation.

References

1. V. Ellen Golden, editor. *Proceedings of the 1984 MACSYMA Users' Conference*, General Electric, Schenectady N.Y, 1984.
2. J. Calmet and J. A. van Hultzen. Computer algebra applications. In B. Buchberger G. E. Collins and R. Loos, editors, *Computer Algebra, Symbolic and Algebraic Computation*, page 245. Springer Verlag, 1983.
3. M. E. Clarkson. A bibliography of computer algebra references. Artificial Intelligence and Image Analysis Laboratory Technical Report AI/IA-89/2, Institute for Space and Terrestrial Science, 108A Farquharson Bldg., 4700 Keele Street, North York, Ontario, M3J 1P3, 1989.
4. John Fitch, P. Herbert, and A. C. Norman. Design features of COBALG. In *SYMSAC '76: Proceedings of the 1976 ACM Symposium on Symbolic and Algebraic Computation*, page 185, New York, 1976. Association for Computing Machinery.
5. J. S. Kowalik, editor. *Coupling Symbolic and Numerical Computing in Expert Systems*, New York, 1986. Elsevier Science.
6. Bruce G. Buchanan and Edward H. Shortliffe, editors. *Rule-Based Expert Systems: the MYCIN experiments of the Stanford Heuristic Programming Project*. Addison-Welsley, Reading, Masschusetts, 1984.
7. S. Keene. *Object-Oriented Programming in COMMON LISP*. Adisson-Wesley, 1988.
8. P. Schank, J. Constan, C. Liu, L. Rowe, S. Seitz, and B. Smith. PICASSO Reference Manual. Technical report, Computer Science Division, University of California Berkeley, 1990.
9. L. Rowe, J. Constan, B. Smith, S. Seitz, and C. Liu. The PICASSO application framework. Technical report, Computer Science Division, University of California Berkeley, 1990.
10. R. W. Scheifler and O. LaMott. CLX programmer's reference. Technical report, Texas Instuments, 1989.
11. R. C. Waters. The programmer's apprentice: A session with KBEmacs. *IEEE Transactions on Software Engineering*, 11(11):1296, 1985.
12. R. F. Boisvert. Toward an intelligent system for mathematical software selection. In Einarsson [18], page 51.
13. P. Buis, W. Dyksen, and J. Korb. Fortran interface blocks as an interface description language for remote procedure call. In Einarsson [18], page 116.
14. R. F. Boisvert, S. E. Howe, and D. K. Kahaner. Gams: A framework for the management of scientific software. *ACM Transactions on Mathematical Software*, 11(4):313, 1985.

15. M. E. Clarkson. Expert systems as an intelligent user interface for symbolic algebra. In Einarsson [18], page 137.

16. K.A. Broughan, G. Keady, T. D. Robb, M. G. Richardson, and M. C. Dewar. Some symbolic computing links to the NAG numeric library. *SIGSAM Bulletin*, page 28, June 1991.

17. M. E. Clarkson. Praxis: An expert system for macsyma. In *International Symposium on Design and Implementation of Symbolic Computation Systems*, volume 429 of *Lecture Notes in Computer Science*, page 264, Berlin-Heidelberg-New York, 1990. Springer-Verlag.

18. B. Einarsson, editor. *Working Conference on Programming Environments for High–Level Scientific Problem Solving (Karlsruhe, Germany)*, Bergen Scientific Centre, 23-27 September 1991. International Federation for Information Processing, IBM.

The Progress Towards an Intelligent Assistant — A Discussion Paper

Greg Butler

Centre Interuniversitaire en Calcul Mathématique Algébrique
Department of Computer Science, Concordia University
Montreal PQ H3G 1M8 Canada
gregb@cs.concordia.ca

Abstract. Powerful computer workstations, communication networks, algebraic and numeric software systems are changing the way mathematicians work. Expert systems already provide user-friendly interfaces to several software tools, but will AI play a still larger role in the future? We open a discussion on the interaction of AI and Symbolic Mathematical Computation, and pose many questions which should be addressed.

1 Introduction

We open a discussion on the interaction of Artificial Intelligence and Symbolic Mathematical Computation, which is the theme of this conference and the subject of the following panel discussion. I propose to provide a focus for the discussion by posing the fundamental question:

Will developments in Artificial Intelligence, computer workstations, and software systems provide mathematicians with an automated intelligent research assistant?

This question really asks whether today's proven techniques of AI — and the promising experimental techniques now being investigated — are sufficient to intelligently support the work of a research mathematician. Here "intelligently" means at the level of a human research assistant. If not, then how much support could these techniques provide to a working mathematician?

As mathematicians are already heavy users of numeric and symbolic computation, especially in their exploratory work, this question also presumes that such intelligent assistance would interact with tools for Symbolic Mathematical Computation.

While a definitive answer to the above question is not yet known, there has been progress in a number of areas that offers promise. Some positive signs are

- PRAXIS [15, 16], an expert system interface for Macsyma;

- IRENA [17, 18], an expert interface, which uses Reduce, to the NAG library of numerical software;

- advanced simulation systems [1], and qualitative physical modelling [28];

- Scratchpad/Axiom [23], a rigorous computer algebra system;

- the Mantra system [12] for hybrid knowledge representation in mathematics and symbolic computation; and

- multi-paradigm programming languages, such as LIFE [2].

In particular, the initial impact of Artificial Intelligence has come from intelligent user interfaces rather than from theorem provers capable of reasoning as well as human mathematicians. Exploratory mathematics has benefitted from the availability of computer algebra systems [9, 14, 31]; from numerical software libraries, such as the NAG library; and, to a lesser extent, from theorem provers [4, 32]. The study of dynamic systems and chaos relies heavily on computer experiments [30]. Inexperienced and casual users of software are guided by expert system interfaces that hide the idiosyncracies and the myriad of parameter settings of numerical libraries [17, 18], of computer algebra systems [15], and of simulation packages [1]. Some interfaces [1] even plan computer experiments and interpret the results.

Advances in programming languages, be they general- or special-purpose, offer perhaps the greatest promise of complete and flexible integration of the various techniques from Artificial Intelligence with those of Symbolic Mathematical Computation (and other areas). For example, the Scratchpad language [23] has a type system offering the rigour to precisely express the mathematical requirements of a computation. Furthermore, the knowledge representation system Mantra is being applied [12] to state and check the semantic requirements of Scratchpad algorithms. Another example, the language LIFE [2], successfully provides a semantics which combines the programming paradigms of logic programming, functional programming, equations, and inheritance. Hence, it is conceivable to program systems combining reasoning, database queries, simplification, and function evaluation in a framework with the convenience, modularity, and re-usability of an object-oriented type system.

First we will present some background material, then pose several issues and questions raised by the fundamental question above, and then we conclude.

2   Some Background

Computer algebra [5], includes both special-purpose and general-purpose computer algebra systems. Significant systems in terms of future trends are Maple, Mathematica, Cayley, and Scratchpad. Maple adopts a 'kernel plus libraries' software architecture and uses hashing extensively. It is compact and fast, and its "algebra engine" is being used as a component of problem solving environments. Mathematica has sophisticated 3D graphics and active hypermedia as part of its user interface. Cayley is attempting to integrate database and deductive facilities with the algorithmic aspects of a computer algebra system, and its user language also aims for easy extensibility through user definition of new algebraic domains. Scratchpad offers strong typing, and extensibility through user definition of categories, domains, and packages. It also has an active hypermedia user interface with 3D graphics.

The traditional view of "scientific computation" is that of numerical computation and simulation. Several libraries, such as NAG, EISPACK, and LINPACK, have been developed as tools. The vast amount of data produced during numerical modelling has stressed the importance of scientific visualisation [21], which has been ably supported by the Supercomputer Centres in the USA. The user interfaces of Mathematica and Scratchpad are also evidence of the role of visualisation in experimental mathematics. A number of simulation packages are mentioned in [1, 30], and [30] also points out the role of non-numeric simulation. Statistical packages are often used to analyse the significance of the results from modelling and simulation.

Collections of raw data and processed data are a vital part of scientific computation. The task of classifying all examples with certain properties is one important task of mathematicians, and the resulting catalogues must be disseminated and made accessible. The issues involved in such databases are different from those of traditional commercial tasks — where the issues have been largely resolved by relational databases — and are still unresolved. Scientific databases are discussed in [20], mathematical databases are discussed in [10, 11], while [29] is a good introduction to deductive and object-oriented databases and knowledge based systems.

Some notable advances in automated theorem proving include the ability to generate theorem provers from descriptions of the logic, inference rules, and strategies [22], the use of meta-level reasoning or plans [8, 6, 27, 7], and the use of algebraic computation in geometric theorem proving [24]. Related issues such as the representation of mathematical knowledge and interactive proof verifiers were pioneered by Automath [19].

Languages are particularly important. The description of new domains of study and methods for reasoning and computation requires a language. Languages also control the input to meta-level tools such as compiler generators and theorem prover generators. So languages are required to express the problem, the model, axioms, theorems, search queries, plans, strategies, and meta-level reasoning. A number of "universal" languages have been developed for these roles, including logic, set theory, graph theory, universal algebra, and category theory. While the languages need a solid semantic foundation, they should accommodate the usual informality of everyday mathematics (and infer the omitted details, as is done in type inference systems, where possible).

Sophisticated software such as numerical libraries and computer algebra systems require a very experienced user to get the best out of them. There are many ways to solve the same problem and the tradeoffs in performance between them can cover several magnitudes of time and space usage. It has been demonstrated that this expertise can be captured in a rule based expert system. Praxis [15] is an expert system interface to the integration facilities of Macsyma. IRENA [17, 18] uses a rule based expert system and the computer algebra system Reduce to set the appropriate parameter settings, condition the equations, and select the relevant Fortran subroutines of the NAG library in order to solve a problem.

Geometric theorem provers can automatically generate an algebraic description of a problem from a pictorial representation that is input using a WIMP interface. The *Bifurcation Interpreter*, described in [1], can generate a numerical procedure describing the behaviour of an electrical circuit from a Lisp description of the circuit. The results of numerical experiments of a dynamic system may be automatically analysed by *PLR*, which combines geometric reasoning, computer algebra, and inequality reasoning to determine whether trajectories cross regions in a phase space, and by *KAM*, which uses computer vision techniques to interpret the phase space and generate numerical experiments which explore the phase space. *KAM* has a grammar describing permissible phase spaces.

As well as the above examples of the automatic control and interpretation of simulation experiments, there has been other significant work on meta-level reasoning and planning. PRESS [8, 6, 27] solves algebraic problems at the level of final year high school students, by selecting and appropriate manipulation technique from a number of possibilities. The OYSTER-CLAM system [7] provides a language for describing plans for solving problems by induction, and reasons with these plans as it solves the problem.

3 Questions

Rather than tackle the fundamental question above directly, let us look at the issues or questions it raises. The questions fall into three categories:

- Requirements of mathematicians,

- Techniques of AI, and

- Pragmatics.

These categories address the basic concerns of what do we want, do we have the tools to build what we want, and do we have he resources to build what we want.

3.1 Requirements of Mathematicians

The basic questions are:

> *What are the requirements of an intelligent assistant for mathematicians?*
> *Are they the same, or different, for those for general engineers and scientists?*

Are the requirements of the mathematical community homogeneous?
e.g. for applications of mathematics, for researchers in applied math-
ematics, for researchers in pure mathematics, for fluid dynamicists,
for logicians, for group theorists?

The nature of research mathematics is to push situations to the limit in order to ascertain the precise circumstances under which something is valid. This often involves exploring circumstances which have no physical analogue and therefore must be described and discussed in an abstract symbolic manner. The novel circumstances also require the development of a language — new concepts, symbols, formalisms, and pictorial representations — for communication and reasoning about the abstract domain. The properties and limitations of both the domain and the language are the subject of mathematical research.

Can these requirements be satisfied by proven techniques from Arti-
ficial Intelligence?
Can we at least identify the current use of "intelligent" tools in the
different mathematical disciplines?

3.2 Techniques from Artificial Intelligence

There is one basic question that must be addressed in any discussion on the role of Artificial Intelligence:

What do we mean by "intelligent"?

or we might side-step that philosophical issue by forming a consensus of techniques which we agree lie in AI.

List those techniques that lie in Artificial Intelligence, and are (or
may be) relevant to Symbolic Mathematical Computation?

What seems or does not seem intelligent often depends on our understanding of the task. For example, to a lay person, the following tasks from computer algebra may seem "intelligent" but are only algorithmic (though the algorithms rely on very deep mathematics).

- factoring a 100-digit integer,

- integration of a function,

- geometric theorem proving, and

- determining the composition factors of a permutation group.

On the other hand, there have been notable successes within the AI community to mathematical problem solving. Most of these demonstrated new techniques and their application, but most of the time the applications have not been "industrial strength" nor delivered new mathematical results:

- Gelernter's geometry theorem prover,

- PRESS applying meta-level reasoning, and

- Lenat's AM.

The question is whether these techniques can be scaled up. (Slagle's SIN and Moses' SAINT systems for integration, using pattern matching techniques, have largely been superseded by algorithmic algebraic approaches to integration.)

> What are the "industrial strength" techniques in Artificial Intelligence?

The best-known examples come from expert systems, some automated reasoning systems (for verification of software and hardware), and some "intelligent" simulation systems.

> Do they include techniques from the fields of expert databases, knowledge-bases, intelligent tutoring systems, inductive learning and data mining?

> Would we be content with an "intelligent" user-friendly interface, or do we demand more "intelligence" in an intelligent assistant?

3.3 Pragmatics

Unfortunately, someone has to do the work and pay the bills, so the pragmatic considerations of creating such intelligent assistants must be considered.

> What demand would there be for intelligent assistants?
> Will commercial software companies create/market such software?
> Are we talking about one system, or many systems with well-defined interfaces?
> Is there a role for scientific organisations, individual research groups, or individual researchers?
> Can the research community (of mathematicians? or computer scientists? or both?) do it on its own?

4 Conclusion

The panel discussion concluded that there were "industrial-strength" techniques from the areas of expert database systems (see [25]); machine learning (in particular, the ID3 algorithm of Quinlan [26], which induces a decision tree from classified data samples); and user modelling (see [3]).

The prospects for an automated intelligent research assistant for mathematicians are good. User interfaces already use expert system technology. There are good prospects for AI to deliver with advances in planning, strategies, automated theorem proving and machine learning. However, it is unclear whether we will ever achieve deep and creative automated reasoning on a par with human mathematicians.

But to be truly flexible and extensible requires the ability to create new domains, languages, formal systems, inference and computational methods within the new domains. This requires a meta-level approach, such as theorem prover generators, which draw on our deep understanding and modelling of established domains, languages, reasoning techniques and computational paradigms.

References

[1] H. Abelson, M. Eisenberg, M. Halfant, J. Katzenelson, E. Sacks, G.J. Sussman, J. Wisdom, and K. Yip, *Intelligence in scientific computing*, CACM **32**, 5 (1989) 546–562.

[2] H. Aït-Kaci and A. Podelski, *An overview of LIFE*, Lecture Notes in Computer Science **504**, Springer-Verlag, 1991, pp.42–58.

[3] G. A. Boy, **Intelligent Assistant Systems**, Knowledge-Based Systems, volume 6, J. Boose and B. Gaines (eds), Academic Press, London, 1991.

[4] R.S. Boyer and J.S. Moore, **A Computational Logic**, Academic Press, New York, 1979.

[5] B. Buchberger, G.E. Collins, and R. Loos, **Computer Algebra : Symbolic and Algebraic Computation**, Springer-Verlag, Wien, 1982.

[6] A. Bundy, **The Computer Modelling of Mathematical Reasoning**, Academic Press, London, 1983.

[7] A. Bundy, *The use of explicit plans to guide inductive proofs*, **9th International Conference on Automated Deduction**, E. Lusk and R. Overbeek (eds), Springer LNCS **310**, 1988, pp. 111–120.

[8] A. Bundy and B. Welham, *Using meta-level inference for selective application of multiple rewrite rule sets in algebraic manipulation*, Artificial Intelligence **16** (1981) 189–211.

[9] G. Butler and J.J. Cannon, *Cayley, version 4: the user language*, **Symbolic and Algebraic Computation**, P. Gianni (ed.), Springer LNCS **358**, 1989, pp. 456–466.

[10] G. Butler and S.S. Iyer, *Deductive mathematical databases — a case study*, **Statistical and Scientific Database Management**, Z. Michalewicz (ed.), Springer LNCS **420**, 1990, pp. 50–64.

[11] G. Butler, S.S. Iyer, and S.H. Ley, *A deductive database of the groups of order dividing 128*, **ISSAC 91**, ACM, New York, 1991, pp.210–218.

[12] J. Calmet, G. Bittencourt, and I.A. Tjandra, *Mantra: A shell for hybrid knowledge representation*, Proceedings of the Third International Conference on Tools for Artificial Intelligence, San Jose, USA, November 5–8, 1991, IEEE Computer Society Press.

[13] J. Calmet, K. Homann, and I.A. Tjandra, *Unified domains and abstract computational structures*, these proceedings.

[14] B.W. Char, K.O. Geddes, W.M. Gentleman, and G.H. Gonnet, *The design of Maple: a compact, portable and powerful computer algebra system*, **Computer Algebra**, J.A. van Hulzen (ed.), Springer LNCS **162**, 1983, pp.101–115.

[15] M. Clarkson, *Praxis: a rule-based expert system for Macsyma*, **Design and Implementation of Symbolic Computation Systems**, A. Miola (ed.), Springer LNCS **429**, 1990, pp. 264–265.

[16] M. Clarkson, *Intelligent user interface for symbolic and numeric computation*, these proceedings.

[17] M.C. Dewar, *IRENA — An integrated symbolic and numerical computation environment*, **SYMSAC 89**, ACM, New York, 1989, pp.171–179.

[18] M.C. Dewar and M.G. Richardson, *Reconciling symbolic and numeric computation in a practical setting*, **Design and Implementation of Symbolic Computation Systems**, A. Miola (ed.), Springer LNCS **429**, 1990, pp.195–204.

[19] N.G. de Bruijn, *A survey of the project Automath*, **Essays in Combinatory Logic, Lambda Calculus, and Formalism**, J.P. Seldin and J.R. Hindley (eds), Academic Press, 1980, pp. 589–606.

[20] J.C. French, A.K. Jones, and J.L. Pfaltz, *Summary of the final report of the NSF workshop on scientific database management*, SIGMOD Record **19**, 4 (1990) 32–40.

[21] K.A. Frenkel, *The art and science of visualising data*, CACM **31**, 2 (1988) 110–121.

[22] M. Grundy, **Theorem Prover Generation using Refutation Procedures**, Ph. D. Thesis, University of Sydney, 1990, 129 pages.

[23] R.D. Jenks, R.S. Sutor, and S.M. Watt, *Scratchpad II: an abstract datatype system for mathematical computation*, **Trends in Computer Algebra**, R. Janssen (ed.), Springer LNCS **296**, 1988, pp.12–37.

[24] D. Kapur and J.L. Mundy (eds), **Special Volume on Geometric Reasoning**, Artificial Intelligence **37** 1–3 (1988).

[25] L. Kerschberg, *A multiple paradigm approach to query optimization: Integrating historical, structural and behavioral information sources*, these proceedings.

[26] J.R. Quinlan, *Simplifying decision trees*, Proceedings of the First Knowledge Acquisition for Knowledge-Bases Systems Workshop, Banff, Canada, November 1986.

[27] L. Sterling, A. Bundy, L. Byrd, R. O'Keefe and B. Silver, *Solving symbolic equations with PRESS*, Journal of Symbolic Computation **7** (1989) 71–84.

[28] P. Struss, *Qualitative modelling*, these proceedings.

[29] J.D. Ullman, **Database and Knowledge-Based Systems, Volumes I and II**, Computer Science Press, Rockdale, MD, 1989.

[30] S. Wolfram, *Computer software in science and mathematics*, Scientific American **251** (September 1984) 188–203.

[31] S. Wolfram, **Mathematica: A System for Doing Mathematics**, Addison-Wesley, 1988.

[32] L. Wos, R. Overbeek, E. Lusk and R.J. Boyle, **Automated Reasoning: Introduction and Applications**, Prentice Hall, Englewood Cliffs, 1984.

On Mathematical Modeling in Robotics

J. Pfalzgraf *

RISC-Linz

Johannes Kepler University

A-4040 Linz, Austria

E-mail: jpfalzgr@risc.uni-linz.ac.at

Abstract

Some selected applications of mathematical modeling methods in the AI field robotics are presented in survey style by examples of geometric reasoning, topological reasoning and so-called fibered logical spaces for logical reasoning in robotics. The main perspective is on interaction and combination of different fields and methods from symbolic mathematical computation and AI and the mutual stimulation given by the various disciplines.

1 Introductory Remarks

The field of robotics is one of the traditional areas of artificial intelligence. Robotics is highly interdisciplinary and the problems are ranging from difficult theoretical questions to hard engineering problems like in sensor fusion (symbolic and subsymbolic processing), for example.

The tendency to build autonomous robots immediately leads to difficult problems in cognitive sciences.

To make it short, robotics is an area where many disciplines come together, in particular, it is an ideal field where AI and symbolic mathematical computation and mathematics in general can interact and cooperate in a variety of ways.

With that in mind, we want to discuss in this contribution applications of mathematical and symbolic computational modeling in robotics from different perspectives like:

geometry ('geometric reasoning'), topology ('topological reasoning'), and logical aspects, more precisely, a geometric-logical view namely the use of "logical fiberings" in robot multi-tasking. We illustrate our idea in this direction by a simple robot scenario example.

---

*sponsored by the Austrian Ministry of Science and Research (BMWF), ESPRIT BRA 3125 "MEDLAR"

Part of these considerations are topics of work in the ESPRIT project MEDLAR [1] (Mechanizing Deduction in the Logics of Practical Reasonig), where the work of the RISC-Linz group is on mathematical foundations, (constructive algebraic methods in) geometric reasoning, automated theorem proving, and robotics scenario modelling (under the aspect of applications of various reasoning methods).

In a case study ([34]) a list was compiled with characteristic problems arising in robotics.

Some of these problem fields are typical application areas of symbolic computation methods, as a sample we are discussing below among others the *inverse kinematics problem*.

We would like to point out, that the topics presented in this article are a personal selection, we cannot give a complete overview here of existing methods and results in this permanently increasing field. Thus, the title of this work should be read as "some selected topics in mathematical modeling in robotics", or so.

This article is written in a survey style and symbolic mathematical details are mostly omitted – we want to discuss this in more detail in another place.
But, nevertheless, we hope to be able to provide here a certain spectrum of informations – corresponding to the intentions of the conference – and some suggestions and stimulations for further reading.

Concerning the *literature references* included in this text we point out, too, that they are by no means exhaustive; in some respect we quote some representative titles and refer to the extensive literature meanwhile available in the large area of robotics.

2   A Mathematical Model of the Robot

We now briefly discuss a possiblity to establish a mathematical model of a robot, the *robot map*.

2.1   A Remark on Kinematics

Roughly spoken, a robot as we are considering it is a system of rigid parts (links, sub-arms) linked by joints and this as a whole forms a robot arm which in many cases is an open kinematic chain (tree like structure).
The desription of the motion of such a system is studied in the classical field of kinematics (in classical mechanics as a branch of physics and engineering and also a part of classical geometry).
In general, kinematics is the study of the motion of rigid bodies without considering forces and momentums - (the latter are considered in robot dynamics).

---

[1] the MEDLAR group of RISC-Linz is sponsored by the Austrian Ministry of Science and Research (BMWF), ESPRIT BRA 3125. This Basic Research Action is continuing as BRA 6471, MEDLAR II

Technically the most popular types of joints are *rotational (revolute)* and *translational (prismatic)* joints, respectively.

In a rotational joint the local motion is decribed by the rotation around an axis, thus an angle of rotation is the corresponding parameter for mathematical description of the motion.

In a translational joint the local motion is described by sliding along a certain distance in the direction of a real straight line.

Thus, mathematically, the corresponding local parameters, namely rotation angles and translational parameters (distance) , respectively, can be modelled as elements of a full unit circle (denoted by S^1) in the corresponding rotation plane and as elements of a suitable straight line (denoted by $\mathbb{R}$), respectively. This will be formally described below.

Fore more details about kinematics and theoretical results cf. e.g. [1], [28], [23] and classical texts on kinematics.

2.2 The Robot Map

We assume that the robot has t translational and r rotational joints.

The aim is to model the robot as a mathematical map from joint space (configuration space) $\mathcal{C}$ to work space $\mathcal{W}$ (cf. e.g. [24], [8] and also [12]) : $R : \mathcal{C} \to \mathcal{W}$

Counting the joints, we have $n = r + t$ *degrees of freedom* (DOF $= n = r + t$). Thinking about the largest possible *joint space* (where all smaller joint spaces can be embedded) we come to the following considerations:

the parameter value of a translational joint runs through a real interval, we model this more generally by the whole real line $\mathbb{R}$ (generalized case), rotational joint values can analogously be interpreted as points of the unit circle S^1 in the real plane; recall that $S^1 = \{z = e^{i\phi} | 0 < \phi \leq 2\pi\}$ and a rotation through angle ϕ corresponds to a multiplication of a 2-dimensional vector by $e^{i\phi}$ in the corresponding "rotation plane" (technically, we use the complex numbers represention here).

Summarizing, the manifold $\mathcal{C}$ corresponding to r rotational and t translational joints is in the most general sense the manifold: $(\mathbb{R} \times \ldots \times \mathbb{R}) \times (S^1 \times \ldots \times S^1)$, where the first cartesian product has t factors, and the second r factors ($\mathbb{R}^t \times (S^1)^r$), for short.

Mathematically, this is a nice (differentiable) manifold which even has the structure of an abelian group.

Notational convention: the cartesian product $(S^1)^r$ is also called *r-dimensional torus* $\mathbb{T}^r$ with $\mathbb{T} = S^1$.

Turning to the *work space*: what is the complete information necessary to describe the location of the end effector (EE) of a robot?

In order to provide the complete information about the EE movement, we need a *position vector* $\vec{a} = (a_1, a_2, a_3) \in \mathbb{R}^3$ and the *orientation* of the EE. Subsequently, we abbreviate $\vec{a}$ by a, for short.

The orientation in Euclidean 3-space can be uniquely determined by an orthonormal frame; the variety of all frames can be described by the group of all proper rotations, the *special orthogonal group*:

$SO_3(\mathbb{R}) = \{A | A$ is an orthogonal 3×3 matrix with det $A = +1\}$.

Hence the whole work space manifold, describing all possible EE locations in the most general case, is $\mathbb{R}^3 \times SO_3(\mathbb{R})$, the Euclidean motion group. Again, this is a nice differentiable manifold. It is of dimension 6, this is why 6-DOF robots are so widespread in practical applications.

Thus, in the most general case, a robot can mathematically be described by a (differentiable) map: $R : \mathbb{R}^t \times (S^1)^r \to \mathbb{R}^3 \times SO_3(\mathbb{R})$, the *robot map*.

With this translation into mathematical terms it is possible to use mathematical methods in a natural way to study robot problems. In particular, it is straight forward now with this notation to define the kinematic problem, the inverse kinematic problem, the notion of a singularity, etc..

The mechanical structure of the robot determines its workspace structure (geometry and topology). Technically relevant robot structures frequently have cube, cylinder, ball, or torus-like workspace structures.

With the above notation, let $R : C \to W$, in general, denote the robot map describing a selected robot model (forward kinematic) with joint space C and work space W, $q \in C$ and $x \in W$, denote the corresponding coordinate vectors (n-dimensional and m-dimensional, respectively).

The dimension (of the manifold) C, i.e. the number of free parameters of C, is defined to be the degree of freedom (DOF) of the robot manipulator. Thus, DOF $= \dim(C) = n$.

A *redundant manipulator* is defined as a robot having more DOF than is necessary in comparison with the dimension of the workspace, i.e.: $\dim(C) = n > m = \dim(W)$.

Redundant manipulators are of increasing interest for complex industrial applications where difficult EE movements have to be performed.

A *singular (joint) configuration* (singular point of joint space) is a point $q \in C$ in which the derivative $d(R)(q)$ — the Jacobian matrix in that point — does not have maximal rank.

This makes it mathematically plausible that in such configurations there are non reachable points. Or the velocity can have very high values near a singular point (cf. e.g. [11]).

Technically such configurations cause problems. The manipulator can move in a locked position then singularity means the EE *cannot* be moved along a small path segment with only small changes of the joint parameters (this would be possible in non singular points).

2.3 Inverse Kinematics

The *inverse kinematics problem* can be formulated as follows:

Given a point $x = (a, A) \in \mathcal{W}$ in work space specifying an EE location. What is a suitable joint configuration $q = (q_1, ..., q_n) \in \mathcal{C}$ which enables the robot to point to that EE location ?

Mathematically this amounts to finding a joint space element $q = (q_1, ..., q_n)$ such that q is mapped to x, i.e. $R(q) = x$.

Forward (direct) kinematics deals with:

Given a joint configuration q determine the corresponding location of the EE in work space, i.e. calculate $x = R(q)$.

A classical approach widely used for the formal expression of the EE coordinates in terms of the individual joint coordinates is the method by Denavit and Hardenberg, cf. e.g. [30], [11].

It uses four geometric invariants derived from the geometry of the mechanics of a robot arm.

We do not go into more details here and refer to the literature (cf. e.g., [11, 30]). This approach is often used to find solutions to the inverse kinematics problem.

REMARK

In the SMART and SAVE robot and NC simulation systems of RISC-Linz algorithms for collision checking and collision detection are implemented to manage the generation of collision free trajectories. There are also possibilities for simulation of robot multi-tasking problems available. Cf. [27], [18] (this literature is also available in the RISC-Linz Report series).

3 General Remarks on Geometric and Topological Reasoning

In the very interesting article Hopcroft [21] mentioned the importance of mathematical modeling of continuous objects and motions for "electronic prototyping" in advanced engineering.

He points out that there is a future need of such continuous mathematical and symbolical methods in addition to the discrete methods widely used in computer science.

Geometric reasoning (GR) has become a growing field where various disciplines are involved. It is of increasing importance for industrial application areas like CAD, CAM, CIM.

There are different interpretations of GR in the literature like (cf. e.g. [7]) :
GR is automated geometry theorem proving or GR is intelligent reasoning about geometric objects or GR is algorithmic problem solving for geometric objects.

From a mathematical, symbolic computation point of view the constructive algebraic methods applied in GR are of particular interest (cf. [7]). Especially those constructive methods capable of solving problems which can be formulated in algebraic geometry involving systems of polynomial equations.

Or, more general, polynomial inequalities which are suitable to model varieties of real physical environments including solid bodies with algebraic surfaces and algebraic curves as boundaries. This is of relevance, e.g., in the geometric modeling of robot cells including obstacles.

The mathematical discipline dealing with polynomial inequalities for modelling geometric objects is semi-algebraic geometry (cf. e.g. [3]).

In the frame of the ESPRIT project MEDLAR the RISC-Linz group makes, among others, major contributions in the area of algebraic methods for GR with applications to robotics and automated theorem proving. The main methods from constructive algebraic and symbolic computation are

Gröbner Bases ([5] Cylindrical Algebraic Decomposition ([10], cf. also [20]). Characteristic Sets ([40], cf. also [39]).

Cf. [39] for more details.

REMARK on Literature

In connection with the previous remarks we want to mention here, *among others*, the book on Geometric Reasoning [22]. Interesting articles are contained in [4] on "Geometry and Robotics". Of further interest for reading is "Some Topological Problems in Robotics" in [2] and articles quoted in the references in [2] on redundant manipulators (especially by C.W.Wampler, Baker and Wampler).

Again we point out that the references we included is only a personal selection and should provide a certain background information with suggestions for further reading.

4   Aspects of Geometric Reasoning

The field of geometric reasoning which covers a variety of different branches contains many problems well suited for applications of symbolic computation approaches. Especially, computer algebra methods and systems come to application.

We present here sample solutions to selected problems in robotics which were subject of work in our local research group in the MEDLAR project.

4.1   Symbolic Computation and Inverse Kinematics

A major motivation to use methods from symbolic mathematical computation an computer algebra systems to treat inverse kinematics problems in robotics is the aim to obtain general closed form solutions with algebraic expressions.

That means: typical parameters of a manipulator like lenghts of links and cosines/sines of joint angles are symbolically treated as variables (not as fixed numerical constants).

This general representation then allows to treat many specific cases (a whole robot class is represented) by merely specifying explicitly the numerical values of the corresponding parameters, respectively.

This is to be seen in completion to the pure numeric approach and in some respect can be an improvement (if the complete symbolic representation is not too complex to handle).

More general, in that way it is possible, in principle, to build up a library for symbolically represented robot types on the basis of computer algebra systems. This has been done at RISC-Linz in form of an implementation, cf. [19].

In the frame of MEDLAR the symbolic computational approach to robot kinematics was studied cf. [39] and the performance of these symbolic methods was discussed - especially the application of Gröbner bases and characteristic sets.

There are examples where this approach works well, cf. e.g. the 2-rotational joint robot treated in great generality using Gröbner bases in [6].

Also more complex robot types like a 6−DOF (rotational) robot of PUMA type which are in industrial use can be treated.

This was done by W.T. Wu - cf. [39] - applying modified characteristic sets methods.

More symbolic computational details on this treatment of inverse kinematics will be discussed in [33].

4.2 Singularity Detection

As mentioned, there is a need to know the singular configurations of a robot system in order to by-pass these singular points.

Working with numerical methods, in the neighbourhood of a singular point the problem of path tracking can be numerically intractable ("explode").

Hence it is useful to have an explicit symbolic representation of the complete set of singularities of a given robot motion along a path.

This is a typical situation where symbolic computation methods can be applied.

For a planar algebraic curve $f(x, y) = 0$, the method of Gröbner bases can be applied directly to find the singular points, namely a singular point is a common solution of the polynomials
$f(x, y) = 0$, $f_x(x, y) = 0$, $f_y(x, y) = 0$, where f_x , f_y are the partial derivatives, respectively.

The singular points can thus be expressed explicitly using Gröbner bases implemented in a computer algebra system − cf. [6], [7] for more details and an example (cf. also [39] and [33]).

REMARK

We would like to mention briefly here a very interesting application of classical (projective) geometry to problems arising in the classification of singular configurations of so-called *parallel manipulators* − cf. the article by J.P.Merlet ([4]) − a more detailed summary of that will be discussed in [33].

It shows how classical results from geometry can help to clarify difficult problems where other approaches did not succeed to solve them completely.

5 Aspects of Topological Reasoning

5.1 Remark on the General Inverse Kinematics Solution

A general formulation of the *path tracking task* can be formulated in terms of topology as follows:

Let $\mathbf{x} : [0,1] \to \mathcal{W}$, $\mathbf{x}(t) = (a(t), A(t))$ be a smooth path in the workspace $\mathcal{W} \subseteq \mathbb{R}^3 \times SO_3(\mathbb{R})$ along which the EE has to move.

Such a path reasonably should be a continuously differentiable curve - this is technically desirable such that the robot motion is without jerks.

Problem: find a suitable joint control function $\mathbf{q} : [0,1] \to \mathcal{C}$ such that for every $t \in [0,1]$ the equation $\mathsf{R}(\mathbf{q}(t)) = \mathbf{x}(t)$ holds (cf. [24]).

Another *more general question* is the following:
is it always possibel to find a global solution to an inverse kinematics problem ?

We will discuss this in more details in [33] to illustrate how classical concepts from algebraic topology can help us to find a qualitative answer to such a problem rather quickly (cf. [24].

5.2 Singularities of the Robot Map

In this section we would like to report about an example of topological reasoning which shows how rather theoretical results with initially no expected applications can under certain circumstances immediately provide substantial aid in applied fields after suitable reformulation of the corresponding problem (here: an application of a deep result from the theory of fiber bundles to a problem in robotics).

We briefly describe an interesting project where a topologist could contribute to a problem in robotics (cf. [16, 17]).
The problem can be shortly sketched as follows.
We are considering a 6 DOF robot mathematically presented by the robot map
$\mathsf{R} : \mathcal{C} \to \mathbb{R} \times SO_3(\mathbb{R})$ with $\mathsf{R}(q) = (a, A)$ (we use the full 6-dimensional workspace here).
Let the first three joint parameters of $q = (q_1, ..., q_6)$ be used for positioning $a \in \mathbb{R}^3$ of the EE, then there are the last three joint coordinates left for the orientation A of the EE.

Focusing on the orientation task now and considering only rotational joints we have the robot map
$\mathsf{R} : \mathbb{T}^3 \to SO_3(\mathbb{R})$, recall : $\mathbb{T}^3 = S^1 \times S^1 \times S^1$ (the 3-dimensional torus)

Now, dealing with the problem of singularities, the idea was (cf. D.H.Gottlieb, loc. cit.) to add another link (a fourth link for the orientation problem), i.e. to increase the DOF, expecting that there are now enough degrees of freedom to avoid the existence of singular configurations.

The corresponding robot map for the orientation problem is then $R : \mathbb{T}^4 \rightarrow SO_3(\mathbb{R})$

Of course, if still singularities would arise one could add another link, and so on.

But does this really help to avoid singular configurations in general ?
Which analytical calculation gives a definitive answer how much DOFs we need for being sure that no singularities are existing any longer ?

Actually, no mattter how many additional joints are added there can always appear singularities, in general, that means for a robot map, with $k \in \mathbb{N}$ $R : \mathbb{T}^k \rightarrow SO_3(\mathbb{R})$.

The answer can be given with the help of a theorem from fiber bundle theory, cf. [16, 17].

Thus, a topological reasoning argument can help to clarify this robotics problem and stop trying to find a solution with adding new links. Instead, one is now led to find particular solutions for the singularity problem (by-passing them in a neighborhood, or others).

We discuss the above example in more detail in [33].

6   Logical Reasoning With Fibered Systems

Subsequently, we introduce a novel approach for constructing a large class of many-valued logical systems applying the concept of "fibered structures".

Our introduction is informal not going into details here. We only want to give in a descriptive way the rough ideas of the concept, provide some background information and indicate in a small application involving a simple robotics scenario in which direction we are looking concerning possible applications of such "logical fiberings" in the field of robotics.

The notion of a *logical fibering* was introduced in [31] as a concept for mathematical modeling of a system of (possibly different) logical spaces (the fibers) residing over a base manifold (index system) forming as a whole the logical fibering. Such structures are fibered spaces where the typical fibers are themselves given logical spaces.

The idea behind this concept is to bring different structures together in one notion, the fibers carry a certain structure (in our case a logical structure) they are put together over a base manifold (index system). With the notion of a logical fibering we express the coexistence of various logical loci residing over an indexed system (base space).

Logical fiberings are linked to the labelled deductive systems (LDS), introduced by D.Gabbay ([15]), from a semantical point of view (forming certain semantical models for LDS).

Due to lack of space we cannot present more technical details about the notion of logical fiberings here. For more comments and background information about such fiberings we refer to [31]. A compressed exposition will be included in [33].

In the realm of MEDLAR we are studying applications of logical fiberings for modeling "logical controllers" in robot multi-tasking. A first approach to this will be presented subsequently in form of a particular example, as mentioned before.

In that example we apply a non-classical bivariate operation which arises naturally in a logical fibering, called *transjunction*. Some informations about this are given in the following considerations.

6.1 Remarks about Transjunctions

Let $\mathcal{L}^I$ denote a "parallel system" (cf. [31]) consisting of "local" 2−valued subsystems (the fibers) $L_i, i \in I$, where I is the index set. In such a logical fibering the following situation arises naturally for bivariate operations: a "local pair" of logical values (x_i, y_i) in $L_i \times L_i$, $i \in I$ can be mapped into different subsystems $L_j, L_k, \dots$.

The four possible local truth value input pairs from $\Omega_i \times \Omega_i$ (Ω_i consists of the two local truth values T_i, F_i) can be mapped to maximally four different subsystems.

That means that such bivariate operations distribute images over different subsystems — *a new feature*.

Such bivariate operations are called *transjunctions*.

More details are discussed in [31], section 8, where we give a classification of transjunctions.

We give a concrete example of a (conjunctional) transjunction by merely showing its truth value matrix.

$\wedge_t$	T_i	F_i
T_i	T_α	F_β
F_i	F_γ	F_δ

Generally spoken, a transjunction can be described by such a $2 \times 2 - T, F -$ *pattern* followed by a *distribution* of the $T, F - values$ over (maximally four different) value sets $\Omega_\alpha, \Omega_\beta, \Omega_\gamma, \Omega_\delta$ corresponding to subsystems $L_\alpha, L_\beta, L_\gamma, L_\delta$, where $\Omega_\alpha = \{T_\alpha, F_\alpha\}$, etc..

6.2 Example: A Logical Controller

The aim of the following considerations is to make an experiment with a transjunction applying it to a simple assembly task where three robots are interacting (cooperating).

It is a simple scenario which we chose and it should be a basis for extensions to more complex cases; for example we can extend it by further cell components of various types involving different logical control formulas.

We want to illustrate how the whole scenario can be modeled logically by a suitable simple logical fibering. In particular, the logical operation of a transjunction can possibly be applied in the sense of a "logical controller" of the corresponding assembly process.

We start with a very simple experiment and construct an easy sample assembly task involving 3 logical subsystems, each associated with one of the 3 robots involved, respectively.

Non-formal Description

Let $\mathcal{R}_0, \mathcal{R}_1, \mathcal{R}_2$ denote the 3 acting robots performing the following task.

$\mathcal{R}_0$ receives a *work piece* (e.g. a block with a hole) A, it has *to be checked* (for example by an optical sensor) whether A *is already on the work table of* $\mathcal{R}_0$ (we express this logically by the predicate "*A is well-positioned*").

We do not discuss the positioning problem from a technical point of view here. Then $\mathcal{R}_0$ also waits for a *bolt B* to be inserted into the work piece A (or a screw to be screwed into it, for example — these technical details should not be relevant here).

If these conditions are fulfilled - i.e. "*the pair (A, B) is well-positioned*" on the table of the robot system $\mathcal{R}_0$, then $\mathcal{R}_0$ *grips the bolt and inserts it in A*: performing the assembly task.

We assume that there can occur *faults in this process*, e.g. work pieces A, B with $\mathcal{R}_0$ can be missing - or being not well-positioned, respectively.

The surveillance of the process will be *logically modeled* by (local) "logical controllers" L_0, L_1, L_2 corresponding to the robots $\mathcal{R}_0, \mathcal{R}_1, \mathcal{R}_2$, respectively.

These L_0, L_1, L_2 are the *local logical subsystems* of the whole logical fibering $\mathcal{L}$ forming the *global logical model* of the scenario ("*fibered logical controller*").

We describe the logical control and communication between the robots (logical subsystems) as follows — corresponding to the four possibilities (in robot system $\mathcal{R}_0$) of A, B being "well-positioned" or not.

Let x_0, y_0 be variables in the logical system L_0 assigned to the predicates
x_0 : "A is well-positioned", y_0 : "B is well-positioned".
Let $\Omega_i = \{T_i, F_i\}$, for $i = 0, 1, 2$, denote the local truth values (logical fibering notation).

Evaluating the sensor application, we assume that in system $\mathcal{R}_0$ the position of A, B is checked simultaneously, i.e. we evaluate eval(x_0, y_0), hence there are 4 possibilities of pairs of truth values expressing the positioning of (A, B): (T_0, T_0), (T_0, F_0), (F_0, T_0), (F_0, F_0).

The whole assembly process is now described and logically controlled according to these 4 possible cases.

To this end we introduce the *conjunctive transjunction* $\Theta = x_0 \wedge_t y_0$ defined by the following truth table. (It operates on the four value pairs).

$\wedge_t$	T_0	F_0
T_0	T_0	F_1
F_0	F_2	F_1

The indices correspond to the numbering of the subsystems (robots), respectively.

A short descrition how this should work according to the four pairs:

(T_0, T_0) : the logical controller (we call it Θ) invokes a T_0 in subsystem L_0; this induces the "local" action: robot $\mathcal{R}_0$ performs the assembly.

(T_0, F_0) : Θ invokes a F_1 in subsystem L_1 (a "fault message" from $\mathcal{R}_0$) ; this induces the "local" action: robot $\mathcal{R}_1$ provides $\mathcal{R}_0$ with work piece B (well-positioned).

(F_0, T_0) : Θ invokes a F_2 in subsystem L_2; this induces the "local" action: robot $\mathcal{R}_2$ provides $\mathcal{R}_0$ with work piece A (well-positioned).

(F_0, F_0) : Θ invokes a F_1 in subsystem L_1, then the situation of (T_0, F_0) is reached and the system reacts as described for that case.

Summarizing (Principle of Process Control), the cooperating robot system is modeled following the principle that each specific robot system $\mathcal{R}_i$ performs the corresponding action in accordance with the evaluated situation (in $\mathcal{R}_0$). That means, depending on the semantical meaning of the sensor evaluation procedure $eval(x_0, y_0)$ (in L_0) the suitable action is initialized in the corresponding subsystem $\mathcal{R}_i$, respectively.

In a certain sense we can interpret that kind of modeling (demonstrated by this simple scenario) as a *conversion principle* - converting numerical data (produced by the sensor) into logical values (from the *subsymbolic to the symbolic processing level*).

The chosen simple example wanted to show that the "logical control problem" can be modelled with the help of a particular bivariate logical operation which typically arises in logical fiberings, named *transjunction*. More details will be subject of discussion in [33].

We would like to point out here, that there are many possibilities to vary and extend this scenario and this is where the "toolkit" can help much (cf. [12]). One can make modifications according to which reasoning method one wants to test and apply.

7 Remarks on Path Planning

In recent years extensive work has been done in the field of robotics path planning.
This is an area where many complex problems arise around the central question to find algorithms and complexity bounds for the construction of a collision free path joining a starting point A and a destination point B in an environment containing obstacles.
These "landscapes" are sometimes modelled as polyhedral lanscapes or as (differentiable manifolds or as algebraic curves and surfaces or semi-algebraic sets). Obstacles and objects in a 2D environment are sometimes modelled using conic sections (mainly line and circle segments) to describe their shape.
In [36], for example, the find path problem is treated for a 3D polyhedral landscape.

A general idea for treating this hard problem is to reduce the dimension of the space in which the path has to be considered. This is also desirable for reducing the complexity of the hard search problems which arise in path planning.

Reduction of dimension can for example mean to try to move along the 1-dimensional skeleton (the edges) of a polyhedron (simplex), i.e. work on the 1-dimensional graph represented by the edges (1-dimensional cells) and the corners 0-dimensional cells) of the whole complex.

Another approach (compliant motion planning) using manifolds searches for paths lying in submanifolds (e.g. intersections of manifolds) and borders of manifolds.

In [9] the notion of a retraction from topology is applied, among others.

Various algorithmic methods from symbolic mathematical computation are applied.

The large field of computational geometry (cf. [29], [35], [14]) is widely used.

We mention, for example, the method of Voronoi diagrams for constructing collision free paths, cf. e.g. [41]; cf. [38] for Voronoi diagrams in 3D.

There are approaches where obstacles are modelled as repelling potentials to achieve obstacle avoidance for moving particles, cf. e.g. [25, 26],

In Schwartz and Sharir: "On the Piano Mover's Problem II" (in [37]) the theory of real closed fields is applied to give a general theoretical treatment of the decision problem whether a continuous path exists between a start and destination point. They work with refinements of methods introduced by Tarski (1951) and Collins (1975), (cf. [37], loc.cit.).

The cylindrical algebraic decomposition method and quantifier elimination [10] can be applied in collision detection, cf. also [20].

Many more interesting contributions to that field of robotics (it is a rich source of interesting difficult problems) which are in the spirit of AI and symbolic mathematical computation can be found in the literature.

We cannot go into more details here and refer to the numerous publications. As sample references, and by no means exhausting, we include the titles mentioned in this section.

8   Remarks on Connectionist Approaches

A highly interdisciplinary and very active field in AI where mathematical and symbolic computation methods are of increasing importance and relevance is undoubtedly the large field of artificial neural networks (ANN)/ connectionist systems.

We make only some remarks here about ANN in robotics. For extensive discussions of this topic we refer to the literature and the corresponding contributions in the many conferences held on ANN and related topics.

One possible aspect of the role of ANN in robotics can be roughly described as follows.

Modelling robots with many degrees of freedom (e.g. so-called redundant manipulators) the problems in inverse kinematics and in dynamics can be analytically very hard and complex or even intractable with the current mathematical tools. This is a point where ANN approaches offer the chance to overcome the difficulty, i.e. to get a step further towards a practical solution having the hard problems in mind.

We understand this in the following sense:
use a suitably devised ANN to implement the inverse kinematics model of the robot under consideration by training the ANN with this model of the manipulator using significant training data generated as samples by the real robot.
In the ideal case the trained ANN - representing now the robot symbolically - is capable to generalize i.e. can correctly react to situations which have not been trained before.
Some promising results exist in the literature. And much research is being done in this direction. Of course, the massive parallelism which is inherently contained in ANN approaches can be exploited and this is of great relevance for realtime requirements in robotics applications - deployment of dedicated ANN hardware is of special interest here (work on that is in progress in various places).

The connectionism group at RISC-Linz is working on ANN applications in robotics. We aim at a systematic study of such methods using an integrated simulation environment incorporating powerful robotics/NC and ANN simulation tools (cf. [32] for a short description).

Concerning theoretical models, we are interested in modularized ANN approaches this is very promising and flexible. It is intended to apply mathematical methods which are in the interaction of geometry and net theory ("geometric net theory") - we refer here to the interesting "call for mathematical contributions in ANN theory" by [13] where the application of various mathematical disciplines like geometry, topology, graph theory, knot theory ... is suggested.

There is a variety of further possible deployments of ANN in the area of robotics - this can be an interesting subject of considerations elsewhere.

9 Concluding Remarks

A main background motivation of this contribution was the interaction and cooperation of AI and symbolic mathematical computation (bringing different fields and methods together).
In a survey style, we presented a personal selection of some examples which we find quite suitable to illustrate these aspects.

From the viewpoint of today mathematics we would describe the potential coming from that classical area as follows:

there is a rich source of results and methods already available which can either be applied directly or needs to be suitably reformulated and adapted in order to use it for the treatment of an AI problem.

On the other hand, there is a need for the development of new mathematical methods which are capable to support the work on the many hard problems in AI.

And, of course, there is always the promising cooperation of methods from different disciplines like AI and (symbolic) mathematical computation which provide considerable problem solving capabilities.

The selected information material we presented here should illustrate the above mentioned aspects and, hopefully, contains some suggestions valuable for the reader.

References

[1] J. Angeles. *Rational Kinematics*. Springer, New York-Berlin, 1988.

[2] D.R. Baker. Some topological problems in robotics. *Mathematical Intelligencer*, 12:66–76, 1990.

[3] J. Bochnak, M. Coste, and M-F. Roy. *Géometrie algébrique réelle*. Springer Verlag, 1987.

[4] J.-D. Boissonat and J.-P. Laumond (eds.). *Geometry and Robotics*. Springer Verlag, 1989. Lecture Notes in Computer Science 391.

[5] B. Buchberger. Gröbner bases: An algorithmic method in polynomial ideal theory. In *Multidimensional Systems Theory (N.K.Bose, ed.)*, pages 184–232. D.Reidel Publ. Comp., Dordrecht-Boston-Lancaster, 1985.

[6] B. Buchberger. Applications of Gröbner bases in non-linear computational geometry. *Mathematical Aspects of Scientific Software (J.R.Rice, ed.)*, 14:59–87, 1987.

[7] B. Buchberger, G. Collins, and B. Kutzler. Algebraic methods for geometric reasoning. *Ann. Rev. Comput. Sci.*, 3:85–119, 1988.

[8] J.W. Burdick. On the inverse kinematics of redundant manipulators: Characterization of the self-motion manifolds. *IEEE*, pages 264–270, 1989.

[9] J.F. Canny. *The Complexity of Robot Motion Planning*. MIT Press, Cambridge Massachusetts and London, 1988.

[10] G.E. Collins. Quantifier elimination for the elementary theory of real closed fields by cylindrical algebraic decomposition. *Lecture Notes In Computer Science*, 33:134–183, 1975.

[11] J.J. Craig. *Introduction to Robotics*. Addison-Wesley Publ. Co., 1986.

[12] F. Dargam, J. Pfalzgraf, V. Stahl, and K. Stokkermans. Towards a toolkit for benchmark scenarios in robot multi-tasking. Technical Report 91-45.0, RISC-Linz, J. Kepler University, Linz, Austria, 1991.

[13] R. Eckmiller. Concerning the emerging role of geometry in neuroinformatics. In *Parallel Processing in Neural Systems and Computers, Eckmiller, Hartmann, Hauske (eds.). Proceedings, Elsevier Sc. Publ. (North-Holland)*, 1990.

[14] H. Edelsbrunner. Algorithms in Combinatorial Geometry. Springer Verlag, 1987.

[15] D. Gabbay. *Labelled Deductive Systems, Part I*. CIS, University of Munich, 1990. CIS-Bericht-90-22.

[16] D.H. Gottlieb. Robots and fibre bundles. *Bulletin de la Société Mathématique de Belgique*, 38:219–223, 1987.

[17] D.H. Gottlieb. Topology and the robot arm. *Acta Applicandae Mathematicae*, 10:1–5, 1987.

[18] J. Heinzelreiter and H. Mayr. Machining simulation and verification by efficient dynamic modeling. In *23rd ISATA, Int. Symp. on Automotive Technology and Automation, Dec. 3-7, Vienna, Austria*, 1990.

[19] P. Hintenaus. The inverse kinematics system – installation guide, user's manual, program documentation. RISC-Linz Report no 87-18.0, RISC-Linz, J. Kepler University, Linz, Austria, 1987.

[20] H. Hong. Improvements in CAD–based quantifier elimination. PhD Thesis, Ohio State University, 1990.

[21] J.E. Hopcroft. The impact of robotics on computer science. *Communications of the ACM*, 29:486–498, 1986.

[22] D. Kapur and J.L. Mundy. *Geometric Reasoning*. MIT Press, Cambridge Massachusetts and London, 1989. Special Issue of AI.

[23] A. Karger and J. Novák. *Space Kinematics and Lie Groups*. Gordon and Breach Science Publishers, New York-London-Paris-Montreux, 1985.

[24] U. Karras. On mathematics in robotics. Lecture Notes (in German), DMV Seminar on Mathematics in Robotics; held in Blaubeuren, November, 1988.

[25] O. Khatib. Real-time obstacle avoidance for manipulators and mobile robots. *Internat. J. of Robotics Research*, 5:90–98, 1986.

[26] D.E. Koditschek. Exact robot navigation by means of potential functions: Some topological considerations. In *Proceed. IEEE Int. Conf. on Robotics and Automation*, 1987.

[27] H. Mayr, M. Held, and H. Öllinger. SMART- A Universal System for the Simulation of Robot and Machining Tasks. In *CAPE'89, Tokyo, Japan,* 1989.

[28] J.M. McCarthy. *An Introduction to Theoretical Kinematics.* MIT Press, Cambridge Massachusetts and London, 1990.

[29] K. Mehlhorn. *Data Structures and Algorithms, Vol. 1-3.* Springer Verlag, 1984.

[30] R.P. Paul. *Robot Manipulators.* MIT Press, Cambridge Massachusetts and London, 1982.

[31] J. Pfalzgraf. Logical fiberings and polycontextural systems. In *Fundamentals of Artificial Intelligence Research, Ph.Jorrand, J.Kelemen (eds.). Lecture Notes in Computer Science 535, Subseries in AI, Springer Verlag,* 1991.

[32] J. Pfalzgraf. Neural Networks in Robotics Simulation. In: Symbolic Computation Tools for Technological Applications, J. Heinzelreiter et.al. In *Proceedings IFAC Symposium on Robotics Control (SYROCO'91), Vienna, Austria,* 1991.

[33] J. Pfalzgraf. On geometric and topological reasoning in robotics. to be submitted to Annals of Math. and AI, 1992.

[34] J. Pfalzgraf, K. Stokkermans, and D. Wang. The robotics benchmark. Proc. 12-month MEDLAR Workshop (Weinberg Castle, Austria, November 4-7), 1990.

[35] F.P. Preparata and M.I. Shamos. *Computational Geometry.* Springer Verlag, 1985.

[36] J.H. Reif and J.A. Storer. 3-dimensional shortest paths in the presence of polyhedral obstacles. In *Mathematical Foundations of Computer Science 1988. Proceedings. Lecture Notes in Computer Science 324,* 1988.

[37] J.T. Schwartz, M. Sharir, and J. Hopcroft. *Planning, Geometry, and Complexity of Robot Motion.* Ablex Publishing, Norwood New Jersey, 1987.

[38] S. Stifter. An axiomatic approach to voronoi-diagrams in 3D. *J. of Computer and System Sciences,* 43:361–379, 1991.

[39] D.M. Wang. Reasoning about geometric problems using algebraic methods. proc. medlar 24-month review workshop, grenoble, december 1991. Technical Report 91-51.0, RISC-Linz, J. Kepler University, Linz, Austria, 1991.

[40] W.T. Wu. Basic principles of mechanical theorem proving in elementary geometries. *J. Automated Reasoning,* 2:221–252, 1986.

[41] C-K. Yap. An O(n log n) algorithm for the voronoi diagram of a set of simple curve segments. *Discrete Comput. Geom.,* 2:365–393, 1987.

Gröbner Bases: Strategies and Applications

Eric Monfroy *

ECRC
European Computer-industry Research Centre
Arabellastr. 17, D-8000 Munich 81, Germany

Abstract. Many problems of computational geometry involve non linear polynomials. Many of these problems can be solved with easy algorithms after transforming the polynomial set involved in their specification into Gröbner bases form. Gröbner bases of a system of polynomials are canonical finite sets of multivariate polynomials which define the same algebraic structure as the initial polynomial system. But the computation of Gröbner bases requires a large amount of time and space. However some strategies can be introduced to improve the computation. In this paper, we shortly introduce basic notions of Gröbner bases. Then we propose an algorithm for their computation which is based on a completion procedure for rewrite systems. This algorithm is extended with an orthogonal set of selection strategies which improve each sub-algorithms (reduction, inter reduction, critical pair choice). At last we discuss the use of the strategies applied to some examples (as a restriction of the well known piano mover's problem).

1 Introduction

Different fields of applications as computational geometry [1], robot motion planning ([2], [3], [4]), free space computation, Voronoï diagram ([5]), involve non linear geometrical objects. These kinds of problems which require polynomials to be formulated, can be solved with simple algorithms using Gröbner bases properties [6]. Gröbner bases can also be used to tackle many problems in polynomial ideal theory and algebraic geometry.

This paper starts with a brief summary (section 2) of the notion of Gröbner bases. A set of polynomials is an arbitrary specification of an algebraic structure which can be seen as an equational theory. The Gröbner base of this set of polynomials is a finite set of polynomials which defines the same equational theory. When polynomials are viewed as rewrite rules, the Gröbner base is like a canonical rewriting system for the equational theory. In section 3 we propose an algorithm for Gröbner bases computation which is based on a standard completion procedure [7]. This algorithm provides a framework for introducing and testing selection and choice strategies to improve the computation. We have elaborated an orthogonal set of strategies (section 4) which aim at avoiding worthless computation (as critical pair reduced to 0) and at determining the operation leading the most quickly to the Gröbner base. We then present a short outline (section 5) of the implementation we used to treat the applications (requiring non linear polynomials to be stated) described in section 7.

---

* Partially supported by the CHIC ESPRIT Project 5291

The two first applications aim at showing the need of strategies to speed up Gröbner bases computation and the need of criteria (see section 6) for comparing different sets of strategies. In the last application (section 7.3) we demonstrate the use of Gröbner bases for solving AI problems. It is dealing with robot motion planning, and the well known problem of the "piano mover": we want to determine whether an object can move through a right angled corridor. This application not only demonstrates that Gröbner bases are a powerful and valuable tool for problems involving non linear polynomial systems, but also underscores the gain of efficiency and the speed up provided by the selection strategies.

2 Gröbner Bases and solving algebraic equations: overview

2.1 Gröbner Bases

The Gröbner Base computation [8] of a set of polynomials doesn't lead to an explicit form of the solutions, but it gives a representation of the system from which roots may often be derived in an easy way or gives some information about the solutions and their existence. This is due to the following properties of Gröbner bases:

Property 1 Let F be a set of polynomials, and G=Gröbner_Base(F).
Then Ideal(G)=Ideal(F), which implies G and F have the same set of solutions.

The basic notion of Gröbner bases theory is *polynomial reduction* which depends on a partial ordering on polynomials. The most often used admissible orderings are *lexicographic ordering* and *total degree ordering* (these orderings are totally specified by an ordering on the indeterminates). With one of these orderings, Buchberger's algorithm for the computation of Gröbner base is described as a bases completion procedure. This procedure is based on two principal operations:

Definition 1 Reduction. g reduces to g_1 with respect to f if:
$f = c_1.m_1 + f_1$ with $c_1.m_1$ head term of f under the fixed ordering and c_1 a coefficient
$g = \ldots + c.m.m_1 + \ldots$ with m a monomial and c a coefficient
Then $g1 = \ldots - \frac{c}{c_1}.m.f_1 + \ldots$.

Definition 2 S_polynomial. Let $p_1 = c_1.m_1 + f_1$ and $p_2 = c_2.m_2 + f_2$ two non constant polynomials with $c_1.m_1$ head term of p_1 and $c_2.m_2$ head term of p_2 under the fixed ordering.
Let $m = LCM(m_1, m_2) = s_1.m_1 = s_2.m_2$, where s_1 and s_2 are monomials.
Then the polynomial $p = s_1.f_1 - \frac{c_1}{c_2}.s_2.f_2$ is called S_polynomial of p_1 and p_2.

2.2 Buchberger's Algorithm

Computation of the Gröbner base of S_0 (set of polynomials) ([9])
    -1- constructs the S_polynomials of all pairs of polynomials of S_0
    -2- adds the reduced (with respect to polynomials in S_0) non zero S_polynomials to S_0 to form S_1

-3- iterates the steps 1 and 2 on S_1 until a fixed point is reached
S_k is a Gröbner base of S_0.

The complexity of this algorithm is exponential in the worst case. Fortunately, this method can be modified and some criteria introduced to improve the performance of the algorithm ([10], [8]). Furthermore the use of strategies ([10], [11]) can also speed up the computation and decrease the number of performed operations (reductions, critical pair computation, etc).

2.3 Solving Algebraic equations

This section utilises the fact that the ideal generated by a set of polynomials F is equivalent to the one generated by its Gröbner base $GB(F)$, i.e. the solutions of F are the solutions of $GB(F)$. The first theorem defines the consistency of a system F of polynomials over $K[x_1, \ldots, x_n]$, i.e. the existence of solutions of F over an algebraic extension of K. The second theorem is a criterion which gives some properties of $GB(F)$ when F has finitely many solutions.

Theorem 3 Solvability of F. *Let $GB(F)$ be the reduced Gröbner base of F. F is solvable (has some solutions) iff $1 \notin GB(F)$:*

Theorem 4 Finitely many solutions [8]. *Let $GB(F)$ be the reduced Gröbner base of F. F has finitely many solutions iff $\forall i \in [1, n]$, $\exists k$ such that $(x_i)^k$ is the head term of a polynomial of $GB(F)$.*

When the system has finitely many solutions, two different algorithms can be used in order to compute the solutions.

The first algorithm is based on the construction of univariate polynomials p_i for each variables x_i of the problem. If F has finitely many solutions (see theorem 4), the set of roots of p_i includes the possible values of x_i in the solutions of F.

The second algorithm is based on theorem 4 and the use of a lexicographic ordering. We can deduce the form of $GB(F)$ when the Gröbner base has been computed using lexicographic ordering.

Property 2 Let $GB(F)$ be the reduced Gröbner base of F with respect to the lexicographic ordering and the precedence: $x_n > x_{n-1} > \ldots > x_1$. If F has finitely many solutions, then $GB(F) = \{p_1, \ldots, p_l, \ldots, p_k, \ldots, p_m, \ldots, p_p\}$ with $p_1 \in K[x_1]$, $p_2, \ldots, p_l \in K[x_1, x_2], p_{l+1}, \ldots, p_k \in K[x_1, x_2, x_3], \ldots, p_m, \ldots, p_p \in K[x_1, x_2, \ldots, x_n]$

The second algorithm is based on the form of the lexicographic Gröbner base (see property 2.3): the set of polynomials contains an univariate polynomial p_1 in x_1 from which it's possible to derive the roots of x_1. This gives an inclusion of the solutions of x_1. The values are propagated to p_2, and permit the computation of the roots of x_2, or a restriction (with respect to the solutions of the complete set of polynomials) of the solutions of x_1, and so on till the last polynomial.

3 Gröbner Bases computation as a completion procedure

As said in the introduction (section 1), the Knuth-Bendix procedure for the completion of rewrite systems and the Buchberger algorithm for Gröbner bases computation are very similar. In this section, we present an algorithm for computing Gröbner bases of sets of polynomials which is based on a standard completion procedure for rewrite systems. This algorithm aims at providing a framework in which orthogonal sets of strategies can be introduced in order to improve the computation.

We only give the intuitive idea of the equivalence between Buchberger's algorithm and Knuth-Bendix procedure. For more formal and theoretical proofs of the equivalence of Gröbner bases computation and completion procedure, the reader may refer to [12] and [13].

The intuitive idea is based on the equivalence of the basic operations of both algorithms. But before, we have to show the equivalence between the use of rules and the use of polynomials. We present these similarities as a set of remarks. $HT(p)$ represents the head term of the polynomial p with respect to the chosen ordering.

Remark Polynomial and Rule. Looking for the roots of a polynomial p is equivalent to looking for the solutions of the equation $p = 0$. This allows polynomials to be treated as equations. So a polynomial $p = HT(p) + p_1$ can be represented as the equation $HT(p) = -p_1$ which can be viewed as the rewrite rule $HT(p) \rightarrow -p_1$.

Remark Reduction. For Gröbner base, the reduction of a polynomial p is the replacement of a monomial m (or part of a monomial) of p which is equal to the head term of a polynomial $q = HT(q) + q_1$ by q_1. So if we consider q as a rewrite rule (remark 3), then reducing p with q is equivalent to rewriting p with q: the monomial m of p is rewritten in q_1 by using the rule q.

Remark Critical Pair. The computation of the S_polynomial of two polynomials $p = HT(p) + p_1$ and $q = HT(q) + q_1$, is equivalent to the computation of the critical pair obtained from p and q viewed as rewrite rules. The overlapping term t is $lcm(HT(p), HT(q))$. The critical pair is the equation which right hand side is t rewritten using p as a rule and the left hand side t rewritten using q. Then the right hand side minus the left hand side is equal to 0, and this difference is the S_polynomial of p and q (using remark 3).

Using the previous remarks, the Buchberger's algorithm for Gröbner base computation (section 2.2) can be viewed as a completion procedure [7].

To improve Gröbner bases computation with some strategies, the algorithm must be as general and "modular" as possible. In this way strategies may be introduced at every step of the computation, selecting which inference or which choice will lead the most quickly to the solution. A completion procedure possesses these properties and therefore we choose a completion procedure to compute Gröbner base of a set of polynomials.

3.1 Completion for Gröbner Bases

This procedure is a standard completion procedure which has been modified to be more efficient for polynomials. When Gröbner Bases are computed using rewrite system techniques, the representation of the current system during the completion may be modified. Instead of inferring from a pair (E_i, R_i) to a pair (E_{i+1}, R_{i+1}) (E_j set of equalities, R_j set of rewrite rules), it's possible to consider only one set P_i of polynomials. To do so, polynomials are always monic (head coefficient equal to 1) and "sorted" (w.r.t. the chosen ordering) so they can be viewed as rules.

In the standard completion cases of failure are due to the impossibility of orienting an equation or to the creation of an infinite set of rules. These cases have been removed from this procedure since they can't appear with polynomials.

The following Gröbner Bases computation algorithm provides the framework for the succeeding sections. It applies the reduction (section 3.3) and inter reduction (section 3.2) as soon as possible (every time a new non zero equation is introduced, or an equation is modified). Although it complicates the algorithm, this choice has been made to obtain a "minimal" set of polynomials at each step of the computation. By keeping the polynomials as reduced as possible, it is hoped that the complexity of the computation may be reduced (no parameterization for this choice has been realized).

```
Groebner_Bases(Set_Poly)
    Set_Poly := Inter_Reduce(Set_Poly)
    Set_CP := Init_CP(Set_Poly)
    While ∃ Critical_Pair ∈ Set_CP
        CP := Choose_CP(Set_CP)
        Set_CP := Set_CP \ CP
        S_Poly := Compute_CP(CP)
        New_Eq := Reduce(S_Poly, Set_Poly)
        If Valid(New_Eq, Set_Poly)
            Then
                Set_CP := Add_New_CP(New_Eq, Set_Poly)
                Set_Poly := Set_Poly ∪ {New_Eq}
                Set_Poly := Inter_Reduce(Set_Poly)
            Else
                True
        End If
    End While
    Return(Set_Poly)
End Groebner_Bases
```

Init_CP(Set_Poly) : computes possible CP between all pairs of polynomials. This function is only used once during the initialization phase.

Add_New_CP(Eq, Set_Poly) is a function which computes the possible critical pairs between Eq and each equation of Set_Poly. Each time a new equation is created, all new possible CP are added to the current set of CP that can be computed.

Valid(Poly, Set_Poly) is a boolean function which is true if Poly is not equal to 0, and Poly doesn't already exist in Set_Poly. So, the polynomial issued from a critical pair will be added to the system only under these conditions.

Choose_CP(Set_CP). This function chooses a CP in the set of possible CP using some strategies (see section 4.3).

Compute_CP(CP) Computation of the chosen CP (or S_Polynomial) as described in definition 2.

3.2 Inter-reduction

The inter-reduction ensures that all dependencies between equations are removed. That means that after inter-reduction, no equation can reduce an other equation. This is done in the following way.

There are two different sets of polynomials, one already inter reduced (IR), the other not. At each step of the computation, a non IR equation is chosen, reduced modulo the set of IR polynomials. If this equation can reduce IR polynomials then they are removed from the IR set, and added to the not IR set. Then the equation is introduced in the IR set.

The IR set of polynomials is initialized to $\emptyset$, but it can also be initialized with the last computed and reduced CP (see section 4.1).

```
Inter_Reduction(Set_Poly)
    Reduced_Poly := ∅
    While Set_Poly ≠ ∅
        Poly := Choose_Poly_Inter_Reduce(Set_Poly)
        Set_Poly := Set_Poly \ {Poly}
        New_Poly := Reduce(Poly, Reduced_Poly)
        If New_Poly ≠ 0
            Then
                Divisible := Reducible(New_Poly,Reduced_Poly)
                Reduced_Poly := Reduced_Poly \ Divisible
                Reduced_Poly := Reduced_Poly ∪ { New_Poly }
                Set_Poly := Set_Poly ∪ Divisible
            Else
                True
        End If
    End While
    Return(Reduced_Poly)
End Inter_Reduction
```

Reduced_Poly is the current set of polynomials already inter reduced.

Set_Poly is the current set of polynomials not already inter reduced.

Reducible(Poly,Set_Poly) is a function which finds every polynomials of Set_Poly which can be reduced by the polynomial Poly.

Choose_Poly_Inter_Reduce(Set_Poly) chooses a polynomial (in the set of non already inter reduced polynomials) to be reduced by each polynomial of the set Reduced_Poly (exposed in section 4.1).

3.3 Reduce

This function reduces a polynomial Poly modulo a set of polynomials Set_Poly. At the end of the computation, Poly is in normal form with respect to Set_Poly (no equation of Set_Poly can be applied to Poly).

```
Reduce(Poly,Set_Poly)
    While ∃ P ∈ Set_Poly and P reduces Poly
        Poly_Red := Choose_Poly_Reduce (Poly,Set_Poly)
        Poly := Reduction(Poly,Poly_Red)
```

End While
Return(Poly)
End Reduce

Choose_Poly_Reduce(Poly,Set_Poly) chooses a polynomial in the set Set_Poly to reduce Poly (exposed in section 4.2).

3.4 Conclusion

Choose_CP, Choose_Poly_Reduce and Choose_Poly_Inter_Reduce, are selection strategies that are described in the next section. This representation allows strategies to be introduced for all sub algorithms to improve the efficiency of the computation.

4 Selection Strategies

For a given ordering on terms, the efficiency of Gröbner Bases computation depends on the efficiency of all sub algorithms, especially of "reduce", "inter reduce" and the critical pair selection. This is the reason why strategies have been developed to improve these sub algorithms. They aim at reducing worthless computation (as critical pairs reduced to 0) and at selecting the operation which lead the quickest to Gröbner base.

The Gröbner Bases computation is also strongly dependent on variable ordering and term ordering. For term ordering, the most often used orderings are total degree , lexicographical or inverse lexicographical ordering. But for variable ordering, every permutation of the indeterminates is a possible ordering. Some strategies have been studied to find the best variable and term ordering for a given set of polynomials.

An orthogonal set of strategies has been isolated, each with a number of alternative choices. For each sub-algorithm of the completion procedure a set of strategies is offered. Each strategy of one sub-algorithm is independent from those of the other sub-algorithms. This means that the strategies can be freely combined.

4.1 Strategies for "inter-reduction"

These strategies define the ordering on which polynomials must be chosen to be reduced by the already IR set. Several techniques may be applied:

Ordering on time: the first equation to be introduced in the IR set is the last modified or computed one, and so on until they have all been inter reduced.

Smallest head term: the polynomials are ordered (with the chosen ordering on term) ascendantly with respect to their head term. Then they are introduced in the IR set in this way.

Greatest head term: same strategy than the previous one, but descending ordering of the polynomials.

Mixed Ordering: the first equation chosen is the last one computed (last non 0 reduced CP), then smallest or greatest head term ordering is applied.

4.2 Strategies for "reduce"

When a polynomial has to be reduced, the way reductions are applied influences the complexity of the computation. A strategy defining the reduction application ordering should lead to the most reduced polynomial (with respect to the term ordering). The following techniques aim to define two points:

1 the *ordering on reductions* used to reduce a polynomial p. The general idea is how to choose in a set of polynomials S (each p_i of S is able to reduce p) the polynomial p_i which will permit p to be reduced the most. When the choice is done and p is reduced with p_i, then S is recomputed, and the procedure is iterated. This process terminates when S is empty.

2 the *ordering on monomials to apply the reduction* inside the polynomial. This defines which monomial of p will be reduced first.

Ordering on Reductions

Last modified polynomial: the chosen p_i of S is the last modified or computed polynomial.

Minimal head term: the chosen p_i is the polynomial of S with the smallest head term with respect to the ordering.

Maximal head term: idem previous point but with maximal head term.

Maximal reduction: the chosen p_i is the polynomial with the maximal difference between the head term and the second term. This leads to the maximal reduction in one step.

Ordering on Monomials to be Reduced

Head term of the polynomial: the head term of the polynomial is first chosen to be reduced modulo S. Then one of the previous strategies is used to determine which polynomial will first reduce this monomial. The algorithm is a variant from the general algorithm given in Section 3.3.

```
Reduce(p)
    Result := 0
    While p ≠ 0        /* p = HT(p) + Rest(p) */
        S := Set_Reduction(HT(p))
        If S ≠ ∅
            Then
                Poly_Red := Choose_Poly(Set_Poly)
                ht := Reduction (HT(p),Poly_Red)
                p := ht + Rest(P)
            Else
                Result := Result + HT(p)
                p := Rest(p)
        End If
    End While
    Return(Result)
End Reduce
```

Order of reduction equation first: only the strategies of point 1 are used. The chosen reduction polynomial can be applied on any of the monomials of p. So the set S is larger, but not specialized on a particular monomial. The corresponding algorithm is the general algorithm "reduce".

4.3 Strategies for critical pair choice

After "inter-reduce", a critical pair is computed. But generally, there is a collection of possible computable CP. Some of them lead to a polynomial which reduce to 0, or some of them to a redundant polynomial (polynomial already in the current system). To remove these worthless cases, criteria may be introduced (see below). But even with these techniques, some strategies must be added to choose the "best" CP to compute. The aim of these strategies is to find which CP gives the smallest non zero polynomial and leads most quickly to Gröbner Bases.

Criteria: [14]

1. Let p_1 and p_2 be two polynomials.
   If $\mathrm{LCM}(\mathrm{HT}(p_1), \mathrm{HT}(p_2)) = \mathrm{HT}(p_1)\,\mathrm{HT}(p_2)$ then the corresponding CP will be reduced to 0.
2. Let p_1, p_2 and p_3 be three polynomials.
   If $\mathrm{HT}(p_3)$ divide $\mathrm{LCM}(\mathrm{HT}(p_1), \mathrm{HT}(p_2))$ then the CP between p_1 and p_3, p_2 and p_3 will be reduced to 0.

Strategies:

Last found possible CP: the CP chosen is the last detected one (after a CP computation or a polynomial update).

CP with smallest LCM: this strategy chooses the CP with respect to the smallest LCM between the to concerned polynomial head terms.

CP with smallest S_Polynomial: in this case, all the possible CP are computed. Then the CP which leads to the smallest ordering is chosen to be reduced and then introduced in the system. The idea of this strategy is to introduce in the system a polynomial as small as possible. It's the same idea than the previous technique, but the CP with the smallest LCM won't always give the smallest S_polynomial. This strategy is more powerful, but also more costly.

To reduce the extra cost, when a choice has been made, all the information collected must be memorized for the next choice. But every CP attached to a modified polynomial (during inter-reduction) must be computed again.

4.4 Strategies for variable ordering

This is one of the most influential points for Gröbner Bases complexity. In fact, the computation is strongly dependent on the chosen permutation of variables. For example, if there exists an ordering on variables which make the set of polynomials (in normal form with respect to the lexicographic ordering) nearly triangular, then the computation time may be very short.

Boege and all variable ordering: (see [15]) this strategy is used to find the best variable ordering. However, with some examples a better variable ordering may be found. This strategy is based on "reduced univariate polynomials". A "reduced univariate polynomials" for a polynomial $f(x_1, \ldots, x_r) = \sum a_i.x_1^{e(i,1)} \ldots x_r^{e(i,r)}$ (where $a_{(i)} \neq 0$) $\in K[x_1, \ldots, x_r]$

is $p_i(x_i) = g(1, \ldots, 1, x_i, 1, \ldots, 1) \in N[x_i]$ where $g = \sum x_1^{e(i,1)} \ldots x_r^{e(i,r)}$

Then, the reduced polynomial for a set of polynomials is the sum over all reduced polynomials corresponding to the elements of the set.

An ordering for univariate polynomials is defined by:

$h(x) > 0$ iff the leading coefficient of h is > 0

$h(x) > k(x)$ iff $h(x) - k(x) > 0$

With this ordering on univariate polynomials, the variable ordering is the "best" one if:

$p_1(x) \geq \ldots \geq p_r(x)$.

When the "best" ordering has been found (w.r.t. the input system), it can happen that after some reductions, the set of polynomials has considerably changed: the properties which influenced the choice have disappeared. Then, applying again these strategies may reveal that another variable ordering is better.

4.5 Strategies for term ordering

Different term orderings may be used to compute Gröbner Bases of a set of polynomials as total degree ordering and lexicographic ordering. The complexity of the algorithm depends on the chosen ordering, and is generally greater when using lexicographic ordering. But with that latter ordering, the algorithm tends to have different behaviours (due to the form of the input systems) of which it's possible to take advantage to develop strategies based on the properties of the input set of polynomials. Furthermore the computation of the finitely many solutions of the problem is easier to perform with the lexicographic ordering [10], [15] (the Gröbner bases is triangular with respect to the variables).

No work was done on this point during this study on Gröbner Bases. Every strategy and algorithm was based on the lexicographic ordering as this ordering is more suited for the computation of the finitely many solutions of polynomial systems.

5 Implementation

The implementation has been made in Prolog with SEPIA ([16]), using a link to Mathematica ([17]) to accomplish various operations as the rational calculus (for the coefficients of the polynomials) and root computation.

This choice was made because the non linear solver is embedded in a new constraint logic programming system under development at ECRC [18]. This system allows problems to be stated in a natural way, with non linear constraints arising during the evaluation of the logic program. Examples of the language in use are presented in [19]. The Gröbner base computation is based on an algorithm of standard completion [20]. However our implementation allows different strategies at each level (reduction, inter reduction, critical pair choice, variable ordering) to be dynamically combined. This implementation and the different used algorithms are described in [11].

Two different algorithms for computing the finitely many solutions of the input system were implemented using the method of univariate polynomials and an elimination method. More details about the implementation and the algorithms can be found in [21].

In order to extend the range of possible applications, polynomial inequalities have been introduced. They are only used to validate the solutions of the input system after their computation (in the future, we want to use them more actively). They also provide the user with the possibility of introducing square roots of polynomials, by adding a new variable and an inequality.

In the next sections, we discuss the results obtained with the strategies.

6 Criteria for Comparison

These strategies do not yield any theoretically guaranteed improvement on any broad classes of polynomials. We therefore tested them to obtain empirical evidence of their real value [11].

As our purpose was to compare some different strategies and algorithms, we did not implement usual methods that speed up the computation (factorization, improved rational calculus). Thus it is not easy to compare the performances of our implementation with existing algebra systems which are dedicated to this kind of computation.

Thus we propose a set of criteria which permitted us to compare the strategies. These criteria are independent from the system, the implementation language, the machine and the memory. They are based on basic operations of Gröbner bases computation.

The complexity of the computation can be evaluated with the following criteria:

number of critical pairs computed: this number represents the polynomials added to the initial system.

number of critical pairs reduced to zero: this number must be as small as possible, because every CP reduced to 0 was not worth computing. If a CP is reduced to zero, this means that this polynomial is redundant and that the information it contains was already in the current set of polynomials. One of the aims of the selection of CP is to reduce the number of such useless computation.

number of critical pairs leading to a polynomial of the current system: this must be as small as possible for the same reasons that the previous point.

number of reductions: this is to count how many times the reduction sub algorithm has been called. If reductions are applied in the right order, this number can be reduced, and with it the number of steps in the algorithm.

number of rational operations

The previous criteria helped us comparing the different strategies applied to several input sets of polynomials (see section 7). Thus we propose the following strategies combination which generally gives better results than the other ones:

Boege and all technique to determine the variable ordering
CP with the smallest LCM for the critical pair choice
Reduction of the head term of polynomials first, with application of the polynomial with **maximal head term** to reduce equations
Mixed ordering (last computed CP combined with smallest head term ordering) for inter reduction sub algorithm

7 Examples and Results

Our strategies have been tested successfully on many of the examples tackled in the thesis of Czapor [10].
For different problems different combinations of strategies yield the best performance. Typically, for large problems, a combination of strategies can be found that yields a better performance than the version 1.2 of Mathematica, our commercial system available at ECRC. These results have been obtained despite the use of a Prolog library for infinite precision rational calculations (from O'Keefe). A precise comparison with current commercial systems based on more complete information would be of considerable interest.

This section shows that applying different strategies may considerably modify the execution of the computation in term of efficiency, and performed operations. The two first examples demonstrate the need of strategies, and the last one demonstrate the use of Gröbner bases (and the introduction of inequalities) for solving AI problems.

7.1 Example 1

The input polynomial system is (see [10]):
$$zx^2 - (1/2)x - y^2$$
$$zy^2 + 2x + (1/2)$$
$$z - 16x^2 - 4xy^2 - 1$$

This example is now treated with three different sets of strategies:

```
Set 1 Strategy for Inter Reduction:
          simple strategy
          last modified equation
      Strategy for Reduction:
          ordering on reduction equations first
          ascending equation (last modification)
      Strategy for CP Choice:
          last computed cp
      Strategy for variable ordering:
          random ordering:
          variable ordering:  z > x > y
Set 2 Strategy for Inter Reduction:
          mixed choice
```

```
        smallest head term (ordering)
    Strategy for reduction:
        ordering on reducible equation first
        maximum head term (ordering)
    Strategy for CP Choice:
        cp with smallest LCM (ordering)
    Strategy for variable ordering:
        ordering chosen by the user:
        variable ordering:  z > y > x

Set 2 Strategy for Inter Reduction:
        mixed choice
        smallest head term (ordering)
    Strategy for reduction:
        ordering on reducible equation first
        maximum head term (ordering)
    Strategy for CP Choice:
        cp with smallest LCM (ordering)
    Strategy for variable ordering:
        random ordering:
        variable ordering:  z > x > y
```

The first set is "without" any strategy, this means, always performing the most obvious operation. This is now the obtained results:

	Set 1	Set 2	Set 3
User CPU Time (in s.)	40.117	5.27	35.61
Operations:			
Reductions	65	24	54
0_reductions	1	1	1
CP computed	1	1	1
CP reduced to 0	0	0	0
CP already existing	0	0	0
Arithmetic:			
Number of additions	339	102	278
Number of subtractions	71	28	60
Number of multiplications	438	151	360
Number of divisions	70	41	70
Total number of operations	918	322	768

If we use some strategies (i.e. not computing the last detected operation to perform), the computation time can be reduced by 80%, due to less performed reductions, and less rational operations (difference between set 1 and 2). Set 2 and set 3 only differ for the variable ordering. We can see that this influences the performance of the computation.

7.2 Example 2

This more complex example was suggested by Alex Sa'ndor. It is a non linear geometrical problem involving slopping ladders (for more details about this application see [22]).

The input set is the following:

$$a^2 + y^2 - p^2$$
$$b^2 + z^2 - l^2$$
$$a^2 - 800 - l^2 + 60l$$
$$b^2 - 300 - p^2 + 40p$$
$$y^2 - 300 + 20y + x^2$$
$$z^2 - 800 + 20z + x^2$$
$$x - a - b$$
$$30b - lb - al$$
$$20a - ap - bp$$
$$10p - 20y + yp$$
$$10l - 30z + lz$$

we give the results for two sets of strategies:

```
Set 1 Strategy for Inter Reduction:
          mixed choice
          smallest head term (ordering)
      Strategy for reduction:
          ordering on reducible equation first
          maximum head term (ordering)
      Strategy for CP Choice:
          cp with smallest LCM (ordering)
      Strategy for variable ordering:
          ordering chosen by the user:
          variable ordering:  x > a > b > y > z > l > p
Set 2 Strategy for Inter Reduction:
          mixed choice
          smallest head term (ordering)
      Strategy for reduction:
          ordering on reducible equation first
          maximum head term (ordering)
      Strategy for CP Choice:
          cp with smallest S_Poly (ordering)
      Strategy for variable ordering:
          ordering chosen by the user:
          variable ordering:  x > a > b > y > z > l > p
```

The result is the following:

	Set 1	Set 2
User CPU Time (in s.)	35.7	4149.93
Operations:		
Reductions	209	156
0_reductions	13	0
CP computed	14	1
CP reduced to 0	5	0
CP already existing	2	0
Arithmetic:		
Number of additions	889	2121
Number of subtractions	289	254
Number of multiplications	1300	2653
Number of divisions	132	520
Total number of operations	2610	5548

The two different runs of this example show that changing only one strategy may modify the efficiency. The second run was stopped during the computation without reaching the Grobner base of the system. The first computed critical pair (which leaded to a lot of reduction and arithmetic operations) gave back a more complex system where polynomials contained lots of monomials with non trivial rational coefficients. This example was also computed with Mathematica, without having the result after 12 days of computation.

The next section exposes some applications treated using our implementation with a lexicographic ordering on terms.

7.3 Rectangle in a right angled corridor

This problem is a restriction of the well known *"piano mover's problem"* [4]. The problem [23] is to move a rectangle R of length l and width L through a right angled corridor (b is the width of the horizontal corridor, a is the width of the vertical corridor). R has to go from an initial position I in the vertical corridor to a position F in the vertical corridor. The difficulty lies in moving through the angle (see Figure 1).

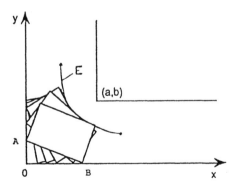

Fig. 1: Rectangle in a right angled corridor

Method We have chosen the following movement: R slices in the angle with one point of contact with the x-axis (B) and a second point of contact with the y-axis (A) (If there is a path, then this path is valid). We can determine the upper envelop E of the moving rectangle as a parametric equation:

$$x(t) = lt^3 + L\sqrt{1 - t^2} \qquad with \qquad t_1 = (1/2 - (9l^2 - 4L^2)^{1/2}/(6l))^{1/2}$$
$$y(t) = Lt + l(1 - t^2)^{(3/2)} \qquad\qquad t_2 = (1/2 + (9l^2 - 4L^2)^{1/2}/(6l))^{1/2}$$
$$t_1 \le t \le t_2$$

The Problem We want to determine the minimal width b of the horizontal corridor when knowing the dimensions l and L of the rectangle R and the width a of the vertical corridor. Thus we fix $l = 2$, $L = 1$ and $a = 3/2$.

Now, determining the minimal width of b consists in computing the collision between two moving points (see Figure 2): the first one following the envelop E, the second one having a vertical trajectory of parametric equations $x = 3/2$ and $y = rt$.

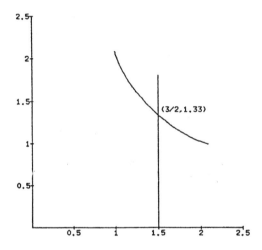

Fig. 2: Collision

In order to remove square roots, we add a new variable w, a new equation and a new inequation stating that w is the square root of $1 - t^2$. The polynomials $x - a$, $y - b$ define the collision points. Thus we obtain the following system of polynomials and inequations:

$$x - lt^3 - Lw \quad t \ge (1/2 - 2^{1/2}/3)^{1/2}$$
$$w^2 - 1 + t^2 \quad t \le (1/2 + 2^{1/2}/3)^{1/2}$$
$$y - Lt - lw^3 \quad w > 0$$
$$b - rt$$
$$2a - 3$$
$$L - 1$$
$$l - 2$$
$$x - a$$
$$y - b$$

We can now compute the Gröbner base of this polynomial system using different sets of strategies.

The most efficient computation is obtained with the set of strategies described in section 6, with an ordering chosen by the user $(a > b > l > L > x > y > w > r > t)$ instead of an ordering computed with Boege's heuristic (section 4.4). With this set of strategies, only one critical pair is computed and 27 reductions are performed. Together this leads to 169 arithmetic operations on rational numbers. The Gröbner base we obtain is the following set:

$$a - (3/2)$$
$$b - 4t^5 + 4t^3 + 3t^2 - t - 3$$
$$l - 2$$
$$L - 1$$
$$x - (3/2)$$
$$y - 4t^5 + 4t^3 + 3t^2 - t - 3$$
$$w + 2t^3 - (3/2)$$
$$t^6 - (3/2)t^3 + (1/4)t^2 + (5/16)$$
$$r + (48/5)t^5 - 4t^4 - (52/5)t^2 + (27/5)t - 1$$

We also tried some other sets of strategies with less efficient results. For instance, a modification of the ordering $(w > t > r > y > b > x > a > L > p)$ leads to compute 8 critical pairs among which 4 are reduced to 0 (worthless critical pairs) and 247 reductions. That means together 3892 arithmetic operations. Furthermore the Gröbner base is less readable: polynomials of higher degrees, larger rational numbers

In order to estimate the speed up due to a strategy, we fixed the ordering and changed one strategy at a time. Choosing a "bad" strategy can increase the number of reductions by a third.

By changing only the strategy on inter-reduction (mixed choice with largest head term ordering) it is no more possible to compute the Gröbner base in an acceptable time. This is due to the increasing number of critical pairs, and the growth of rational numbers.

Computation of the finitely many solutions There is one real solution to this Gröbner base which is computed using an elimination method or univariate polynomials: $w = 0.6640$, $x = a = 1.5$, $y = b = 1.3334$, $r = 1.7833$, $t = 0.7477$

Computing the Gröbner base of the problem stated with polynomials, and determining the solutions, permitted us to find the minimal width $b = 1.33$ of the horizontal corridor which allows the rectangle going through.

Comments The flexibility of our system is obvious. We can make different treatments of this problem (as determining if the object can go through when knowing all the parameters (decision procedure) or finding some relation between the parameters) without modifying its specification, but only the input of Gröbner bases computation.

The use of hybrid constraints for square roots was necessary to exclude imaginary solutions appearing when squaring the complete polynomials.

8 Conclusion

We have briefly exposed a summary about Gröbner bases, their properties and improvements based on strategies. We used the algorithms and strategies of [11] and [21] in order to implement a system computing Gröbner bases and the finitely many solutions of systems of polynomial equations. This implementation permitted us to treat some applications described in [22]. As shown in section 7, the strategies drastically improve the Gröbner bases computation and allows a larger range of problems to be solved.

The introduction of hybrid constraints (inequations) not only facilitates the specification of problems (see section 7.3 as example), but also enlarges the field of applications that can be treated.

We have tested several examples in order to evaluate the system and its capacities, but only three of them are presented in this paper. The presented applications (especially the piano mover's problem 7.3) show how simply a problem can be specified with polynomial equations and solved using Gröbner bases method. Our goal was to demonstrate that Gröbner bases are a powerful and valuable tool for the treatment of problems involving non linear polynomial systems.

However, this application and the other ones in [22] are still quite academic. That is the reason why next steps of this work will be to experiment with real applications in business and science.

References

1. F. P. Preparata and M. I. Shamos. *Computational Geometry: An Introduction.* Springer-Verlag, 1985.
2. H. Amet and R. Schott. Déplacement Exact d'un Polygone au Sein d'un Ensemble de Polygones. In *7ème Congrés AFCET Reconnaissance des Formes et Intelligence Artificielle, Paris*, pages 749–763, 1989.
3. J.-C. Latombe. *Robot Motion Planning.* Kluwer Academic Publishers, Boston/ Dordrecht/ London, 1991.
4. S. R. Maddila and C. K. Yap. Moving a Polygon arround the Corner in a Corridor. In *Proceedings of the ACM Symposium on Computational Geometry*, pages 187–182, Yorktown Heights, NY, 1986.
5. H. Amet, Boissonnat J.-D., and R. Schott. Calcul Dynamique du Diagramme de Voronoï d'un Ensemble de Segments. In *Journées de Géométrie Algorithmique, INRIA Sophia-Antipolis (France)*, pages 143–153, June 1990.
6. B. Buchberger. Applications of Gröbner Bases in Non-Linear Computational Geometry. In Deepak Kapur and Joseph L. Mundy, editors, *Geometric Reasoning*, pages 413–446. MIT Press, 1989.
7. D. E. Knuth and P. B. Bendix. Simple word problems in universal algebras. In J. Leech, editor, *Computational Problems in Abstract Algebra*, pages 263–297. Oxford, Pergamon Press, 1970.
8. B. Buchberger. Gröbner Bases: an Algorithmic Method in Polynomial Ideal Theory. In N. K. Bose Ed., editor, *Multidimensional Systems theory*, pages 184–232. D. Reidel Publishing Company, Dordrecht - Boston - Lancaster, 1985.
9. B. Buchberger. Some Properties of Gröbner Bases for Polynomial Ideals. *J.A.C.M. Sigsam Bull*, 10(4):19–24, 1976.

10. S. R. Czapor. *Gröbner Basis Methods for Solving Algebraic Equations*. PhD thesis, University of Waterloo, Canada, November 1989.

11. E. Monfroy. Strategies for Computing Gröbner Bases of a set of polynomials. Internal Report 92-2i, ECRC (European Computer-industry Research Centre), Munich (Germany), 1992.

12. B. Buchberger. History and Basic Features of the Critical-Pair/Completion Procedure. *Journal of Symbolic Computation*, 3:3-38, 1987.

13. F. Winkler. Knuth-Bendix Procedure and Buchberger Algorithm - A Synthesis. In *Proceedings of International Symposium on Symbolic and Algebraic Computation (ISSAC'89)*, pages 55-67, 1989.

14. B. Buchberger and F. Winkler. Miscellaneous Results on the Construction of Gröbner Bases for Polynomial Ideals. Technical Report 137, Univ. of Linz, Math. Inst., 1979.

15. W. Boege, R. Gebauer, and H. Kredel. Some Examples for Solving Systems of Algebraic Equations by Calculating Gröbner Bases. *Journal of Symbolic Computation*, 1:83-98, 1986.

16. M. Meier, A. Aggoun, D. Chan, P. Dufresne, R. Enders, D. De Villeneuve, A. Herold, P. Kay, B. Perez, E. Van Rossum, and J. Schimpf. SEPIA - an extendible Prolog system. In *Proceedings of the 11th World Computer Congress IFIP'89*, San Francisco, August 1989.

17. S. Wolfram. *Mathematica: A System for Doing Mathematics by Computer*, 1988.

18. E. Monfroy. Non Linear Constraints: a Language and a Solver. Technical report ECRC-92, ECRC (European Computer-industry Research Centre), Munich (Germany), 1992. To appear.

19. E. Monfroy. Specification of Geometrical Constraints. Technical report ECRC-92-31, ECRC (European Computer-industry Research Centre), Munich (Germany), 1992.

20. C. Kirchner and H. Kirchner. Rewriting: Theory and Applications. Working Paper for a D.E.A. lecture at the University of Nancy I, France, 1989.

21. E. Monfroy. Solutions of Algebraic Equations using Gröbner Bases. Internal Report 92-3i, ECRC (European Computer-industry Research Centre), Munich (Germany), 1992.

22. E. Monfroy. Applications of Gröbner Bases. Internal Report 92-4i, ECRC (European Computer-industry Research Centre), Munich (Germany), 1992.

23. E. Monfroy. Spécification Opérationnelle des Contraintes Géométriques: ELIOS. Rapport de DEA CRIN 90-R-130, CRIN Nancy, 1990.

Heuristic Search Strategies
for
Cylindrical Algebraic Decomposition

Hoon Hong

Research Institute for Symbolic Computation
Johannes Kepler University
A-4040 Linz, Austria
e-mail: hhong@risc.uni-linz.ac.at

Abstract. In this paper, we report how certain AI techniques can be used to speed up an algebraic algorithm for deciding the satisfiability of a system of polynomial equations, dis-equations, and inequalities. The algebraic algorithm (a restructured version of Cylindrical Algebraic Decomposition) is non-deterministic, in the sense that it can often achieve the same goal, but following different paths requiring different amounts of computing times. Obviously one wishes to follow the least time-consuming path. However, in practice it is not possible to determine such an optimal path. Thus it naturally renders itself to the heuristic search techniques of AI. In particular we experimented with Best-First strategy. The experimental results indicate that such AI techniques can often help in speeding up the algebraic method, sometimes dramatically.

1 Introduction

In this paper, we report our experience that certain AI techniques can be used to speed up an algebraic algorithm for deciding the satisfiability of polynomial constraints. Precisely put, the satisfiability problem is as follows:

Given: a system of polynomial equations, inequalities, and dis-equations with integer coefficients.

Task: decide whether there exists a common real solution of the system.

As trivial examples, the first system below has a common real solution, but the second system does not.

$$\{ \ x^3 + 2y = z^2, \ x > 0, \ y \geq 0, \ z \neq 0 \ \}$$

$$\{ \ x^3 + 2y = z^2, \ x < 0, \ y \leq 0, \ z \neq 0 \ \}$$

The satisfiability problem is a classic, but non-trivial problem of great importance in mathematics, with applications in various areas such as robot motion planning, geometric modeling, geometric theorem proving/disproving, termination proof of term rewrite systems, constraint logic programming, etc.

A special case with only polynomial equalities has been extensively studied during last two centuries under the name *elimination theory*. A breakthrough in the general case has been made by Tarski [34] who gave the first algorithm for solving this problem.[1] Following Tarski's work, various improvements and new algorithms have been devised and analyzed [33, 9, 17, 6, 10, 1, 28, 3, 5, 13, 15, 14, 8, 16, 30, 31, 32, 18, 19, 26, 20, 11, 25, 7, 24, 21, 22, 23]. (See [2] for an extensive bibliography on this subject.)

Among these, the following three algorithms are most well-known: Collins [10], Grigor'ev [15, 14], and Renegar [30, 31, 32]. Let n be the number of variables in a given system. The worst case computing times of the three algorithms are respectively: doubly exponential (2^{2^n}), quadratically exponential (2^{n^2}), and linearly exponential (2^n). Thus for sufficiently large input, Renegar's algorithms is best among them. But as analyzed in [21], for small inputs which can be decided in a tractable amount of computing time (say, less than a million years on currently available machines), Collins' algorithm is best among them.

However Collins' algorithm itself is still too time-consuming to be used for practical purposes. The main purpose of this paper is to show that Collins' algorithm can be very often speeded up, sometimes dramatically, by using certain techniques developed in AI . In the remaining of this introduction, we will sketch the main idea, of course without any claim to complete precision.

Let n be again the number of the variables occurring in a given system. Collins' method, called "Cylindrical Algebraic Decomposition", consists of two main stages:

(1) Construct a finite set of points in the n-dimensional real space, such that if the system is satisfiable, then at least one point in the set satisfies it. The points are called "sample points".
(2) Check whether at least one sample point satisfies the system. If so, the system is satisfiable, otherwise, it is un-satisfiable.

The stage (2) is relatively simple; one only needs to evaluate the signs of the polynomials on the sample points. However the stage (1) is non-trivial and also can be extremely time-consuming, thus we need to cut down the computation time of this stage.

We begin by viewing the set of sample points as the *search space* in which a "treasure" lies. A treasure we want to find is obviously a sample point which satisfies the system. If we find a treasure, then we know that the system was worthwhile to search (satisfiable), otherwise worthless (un-satisfiable).

Once we view the problem in this way, we are naturally led to the standard search technique developed in AI: *generate and test*. In other words, we generate one node at a time and test if it is a treasure. For our problem, we generate one

---

[1] Actually, Tarski's and the subsequent works deal with even more general problem, namely quantifier elimination for the first order theory of real closed fields. But in this paper we will be only concerned with the satisfiability problem.

sample point at a time, and check immediately if it satisfies the system. If so, we can conclude that the system is satisfiable, without having to generate the remaining sample points. In a sense, we are pruning the remaining portion of the search space. So here is the revised algorithm based on this idea:

(1) Generate a new sample point. Check whether it satisfies the system. If so, the system is satisfiable and we are done. Otherwise, drop to (2).

(2) If no more sample point can be generated, the system is un-satisfiable and we are done. Otherwise, loop to (1).

But then, as usual in most space search problems, a question naturally arises: *In what order should the sample points be generated?* Obviously, we always get the same result no matter what orderings are used. But the computation times can be different, thus we wish to find the ordering with the least computation time. However, in practice it is not possible to determine such an optimal ordering. Thus we are naturally led to various *heuristics* developed in AI. We experimented with the *Best-First* search strategy. The experimental results show that such heuristic search technique can often speed up the whole computation time, sometime dramatically.

We believe that various other algebraic algorithms developed in computer algebra are also amenable to such AI techniques, and we plan to investigate such possibilities.

The structure of this paper is as follows: In Section 2 we give a more precise description of Collins' algorithm. In Section 3 we elaborate on the idea sketched above. We first show how to restructure Collins' algorithm so that the technique of generate and test can be applied. Then we describe a heuristic search strategy. In Section 4 we report various experimental results.

Though the familiarity with Collins' algorithm would be helpful, it is not required in understanding the remaining sections. The same remark goes with the familiarity with heuristic search techniques. For the details of Collins' method, see [10, 4, 20]. For an overview of heuristic search strategies used in AI, see the excellent monograph [29].

2 Algebraic Algorithm

In this section, we give a brief overview of Collins' method. We describe only the super-structure of the algorithm, abstracting away all the irrelevant details. We will not provide any correctness proof, but merely present the algorithm in the quickest way possible. (For the details, see [10, 4, 20].) First we introduce several conventions and definitions in order to facilitate the subsequent discussion.

In the following, let Φ always stand for the formula $\phi_1 \wedge \ldots \wedge \phi_m$ where each ϕ_i is one of the following forms $f_i = 0$, $f_i > 0$, $f_i < 0$, $f_i \neq 0$, $f_i \geq 0$, and $f_i \leq 0$ where again $f_i \in \mathbf{Z}[x_1, \ldots, x_n]$. Also let F always stand for the set $\{f_1, \ldots, f_m\}$. (Without losing generality, we assume that each f_i is distinct.)

Definition 1 Decomposition. An F–*decomposition* of $\mathbf{R}^n$ is a finite partition D of $\mathbf{R}^n$ into connected subsets called "cells" such that each polynomial of F has a constant sign throughout each cell of D. $\square$

Definition 2 Sample Point Set. An F–*sample point set* of $\mathbf{R}^n$ is a finite set $S = \{s_1, \ldots, s_l\}$ of points of $\mathbf{R}^n$ such that there exists an F-decomposition $D = \{d_1, \ldots, d_l\}$ of $\mathbf{R}^n$ where $s_i \in d_i$ for every i. $\square$

Theorem 3. Let S be an F-sample point set of $\mathbf{R}^n$ Then we have the following

$$(\exists \bar{x} \in \mathbf{R}^n)\Phi(\bar{x}) \Longleftrightarrow (\exists s \in S)\Phi(s). \ \square$$

This theorem tells us that once we have computed an F-sample point set, the "infinite" search space is reduced to a "finite" one. Here is the main algorithm **Decide** based on this theorem:

Algorithm Decide.
(1) [Samples.] Construct an F-sample point set S of $\mathbf{R}^n$
(2) [Check.] Check whether at least one sample point of S satisfies the formula Φ. If so, Φ is satisfiable, otherwise it is un-satisfiable. $\square$

The second step is relatively easy; one only needs to evaluate the truth of Φ on each sample point of S. Thus now we focus on how the first step is done. In order to hide irrelevant details, we assume that the following three algorithms are available as black boxes:

- **Sample1** Given a finite set G of polynomials in $\mathbf{Z}[x_1]$, this algorithm constructs a G-sample point set of $\mathbf{R}$.

- **Project** Given a finite set F of polynomials in $\mathbf{Z}[x_1, \ldots, x_n]$, this algorithm constructs a "certain" finite set F' of polynomials in $\mathbf{Z}[x_1, \ldots, x_{n-1}]$.

- **Lift** Given a point $s' \in \mathbf{R}^{n-1}$ and a set F, this algorithm constructs a "certain" finite set of points in $\mathbf{R}^n$ such that they all lie on the vertical line passing through s' (that is, on $\{s'\} \times \mathbf{R}$).

We make precise of the word "certain" in the above descriptions. Let F be a finite set of polynomials in $\mathbf{Z}[x_1, \ldots, x_n]$, let F' be **Project**(F), and let S' be a F'-sample point set of $\mathbf{R}^{n-1}$. Then the set $\bigcup_{s' \in S'} \mathbf{Lift}(s', F)$ is an F-sample point set of $\mathbf{R}^n$. Now we give a recursive algorithm **Sample** which constructs an F-sample point set of $\mathbf{R}^n$.

Algorithm Sample.
(1) [Base.] If $n = 1$, then return the result of **Sample1**(F).
(2) [Project.] Set F' to be **Project**(F).
(3) [Recurse.] Recursively call **Sample** with F', obtaining an F'-sample point set S' of $\mathbf{R}^{n-1}$.
(4) [Lift.] Return the set $\bigcup_{s' \in S'} \mathbf{Lift}(s', F)$. $\square$

Below we trace this algorithm on a simple example formula. The purpose of such an exercise is to illustrate how the algorithm work, and also to prepare for the next section where we will restructure the algorithm.

Example 1: We would like to decide whether the following formula is satisfiable:

$$\Phi \ \equiv \ x_1 x_2 - 1 < 0 \ \wedge \ x_1 + x_2 - 1 = 0.$$

Let $F = \{ x_1 x_2 - 1, x_1 + x_2 - 1 \}$. First we need to construct an F-sample point set of $\mathbf{R}^2$. Here is the trace of the algorithm **Sample**.

- **Project** on F gives $F' = \{x_1, x_1^2 - x_1 + 1\}$.
- **Sample1** on F' gives $S' = \{(-1), (0), (1)\}$.
- **Lift** on (-1) gives $S_1 = \{(-1, -2), (-1, -1), (-1, 0), (-1, 2), (-1, 3)\}$.
- **Lift** on (0) gives $S_2 = \{(0, 0), (0, 1), (0, 2)\}$.
- **Lift** on (1) gives $S_3 = \{(1, -1), (1, 0), (1, 1/2), (1, 1), (1, 2)\}$.
- Combining them, we get $S = S_1 \cup S_2 \cup S_3$.

Next we check each point of S whether it satisfies the formula Φ. We find that $(-1, 2)$ satisfies Φ. Thus we conclude that Φ is satisfiable. □

This concludes the description of Collins' algorithm. One should not be misled by the simplicity of the above description, thinking that the whole algorithm is also simple. It is not. The complicated parts have been encapsulated into the above three assumed algorithms: **Sample1**, **Project**, **Lift**. These algorithms require various algebraic computations such as factorization, greatest common divisors, (sub) resultants, (sub) discriminant, square-free bases, real root isolation for multivariate polynomials with either rational number coefficients or real algebraic number coefficients.

3 Heuristic Search

In this section, we restructure the algorithm presented in the last section so that the scheme of generate and test can be applied. After having done so, we will discuss a heuristic search strategy.

In order to obtain initial insight, let us go back to the example in the last section. Note that Φ was found to be satisfiable on $(-1, 2)$, which is a sample point obtained by lifting the sample point (-1). Thus we did not have to lift the other two sample points of $\mathbf{R}$, namely (0) and (1). But the algorithm, anyways, lifted all of the three sample points of $\mathbf{R}$. Using the terminology of AI, it generated the whole search space before carrying out any testing of nodes.

Clearly we can do better by applying the AI scheme of *generate and test* where one node is generated at a time and tested immediately for being a desired goal. In order to allow such a scheme, we need to restructure the original algorithm. For that purpose, let us first pin down several basic notions:

Definition 4 Search Space. Let Φ be a formula. Then its *search space* Ω is a set of all the sample points generated by the original algorithm. Precisely, we define $\Omega = \bigcup_{i=1}^{n} S_i$ where each S_i is the set of sample points of $\mathbf{R}^i$ generated by the original algorithm. □

Definition 5 Goal. Let Φ be a formula, and let Ω be its search space. We say $s \in \Omega$ is a *goal* iff it satisfies the formula Φ. □

Definition 6 Child. Let Ω be a search space, and let s and s' be its two elements. We say s is a *child* of s' iff s is obtained by lifting s'. □

Proposition 7 Search Tree. Let Ω be a search space, and R be the child relation on Ω^2. Then the structure $< \Omega, R >$ is a *tree*. □

Now what the new algorithm should do is clear. It should construct the search space incrementally, while looking for a goal. A search space is expanded by constructing children of an existing node. Since the search space forms a tree with respect to child relation, we do not need to worry about the danger of cyclic constructions.

Below we give an algorithm based on these ideas. In this algorithm, a set S keeps track of current "frontier" sample points from which the search space can be expanded. In order to make the algorithm simpler, we introduce a hypothetical sample point of $\mathbf{R}^0$. This allows us to treat **Sample1** as a special case of **Lift**, namely lifting the hypothetical sample point of $\mathbf{R}^0$. We also assume that the projection operations have been already carried out before the following algorithm is called.

Algorithm Decide.
(1) [Initialize.] Initialize S to be the set of the hypothetical sample point of $\mathbf{R}^0$.
(2) [Choose.] Choose (and remove) a sample point s from S.
(3) [Generate.] Lift the sample point s, obtaining its children sample points.
(4) [Test.] For each child of s, test whether it satisfies Φ.
(5) [Satisfiable?] If there is at least one child which satisfies Φ, then Φ is satisfiable and we are done.
(6) [Expand.] Insert into S each child of s on which the satisfaction of Φ is not decided yet.
(7) [Un-satisfiable?] If S is empty, then Φ is un-satisfiable and we are done. Otherwise, goto Step (2). □

Now we make an important observation on the algorithm: Step (2) of the algorithm is *non-deterministic* in the sense that it does not specify which sample point is chosen from S in case S has more than one sample points. This motivates the following definition.

Definition 8 Choice Function. A *choice function* is a computable function which chooses a sample point from a given frontier set. Precisely, let Ω be a search space, and let $\bar{S}$ be the set of all frontier sets. Then a choice function Γ is a computable function from $\bar{S}$ to Ω such that $\Gamma(S) \in S$ for every $S \in \bar{S}$. □

Now a question naturally arises: *What choice function should be used by Step (2) of the algorithm?* First, we have the following basic fact which tells us that it is safe to use any choice function.

Theorem 9 Correctness. The algorithm gives a correct answer no matter what choice function it uses. □

However, the total computation time might heavily depend on the choice function used. One obviously wishes to find a *optimal* choice function which would guide the algorithm so that its total computing time is *minimum*. However it is extremely difficult (impossible) to find such an optimal choice function and prove that it is in fact optimal. Thus we are naturally led to *heuristics*, which is expected to give relatively good results most times. One branch of AI studies and analyzes various heuristic search strategies such as: *Breadth-First, Depth-First with Backtracking, Hill-climbing, Best-First*, and their diverse variations, specializations, and combinations For a survey on heuristic search, see Pearl's monograph [29].

The original algorithm can be viewed as adopting a *Breadth-First* strategy, where all the sample points in $\mathbf{R}^1$ is constructed first, then all the sample points in $\mathbf{R}^2$ are constructed next, and so on. The other extreme is to use a *Depth-First* strategy. The corresponding choice function is very cheap to compute, but it is a blind search which might often yield a bad search path. Thus we decided to use *Best-First* strategy. Best-First strategy is roughly a strategy where the most "promising" node among the candidate nodes is chosen. This motivates the following definition:

Definition 10 Heuristic Promise Function. A *heuristic promise function* f is a function from a search space Ω to the set of integers $\mathbf{Z}$.

Definition 11 Best-First Choice Function. The *Best-First Choice Function* Γ_f based on a heuristic promise function f is defined by the condition: $\Gamma_f(S)$ is the element s in S such that $f(s) \leq f(s')$ for every $s' \in S$. □

Now we only have to decide a heuristic promise function. For that purpose, we first take account of the computation time for lifting a sample point. Roughly speaking, we will assume that a sample point with cheaper lifting is more promising. In order to make this idea precise, we go into the details of the representation of sample points.

A sample point of $\mathbf{R}^n$ constructed by lifting operations is a tuple $(v_1, \ldots, v_n)$ where each v_i is a real algebraic number. More precisely, $v_i \in \mathbf{Q}(\alpha_1) \cdots (\alpha_i)$ where each α_j is a real root of a polynomial in $\mathbf{Q}(\alpha_1) \cdots (\alpha_{j-1})[x]$. However, the current implementation of the algorithm **Decide** carries out most computations only in simple extension fields of rational numbers. This is made possible by applying the following well known theorem from algebraic number theory:

Theorem 12 Primitive. Let α and β be real roots of polynomials respectively in $\mathbf{Q}[x]$ and $\mathbf{Q}(\alpha)[x]$. Then there exists (and we can find) γ such that γ is a real root of a polynomial in $\mathbf{Q}[x]$ and $\mathbf{Q}(\gamma) = \mathbf{Q}(\alpha)(\beta)$. □

We apply this theorem before each lifting, in order to carry out all the computation during the lifting in a simple extension field of $\mathbf{Q}$. We will call this operation *Conversion*.

Experiments show that conversion operation is in general extremely time-consuming except for certain trivial cases. One thing we can do is to quickly check whether a conversion is trivial one. Precisely, let a sample point be represented in $\mathbf{Q}(\alpha)(\beta)$. If β is an element of $\mathbf{Q}(\alpha)$, we have $\mathbf{Q}(\alpha)(\beta) = \mathbf{Q}(\alpha)$. Thus we can trivially set $\gamma = \alpha$. This kind of trivial conversion occurs often, so it is important to consider it while devising a heuristic promise function.

Another factor which influences the computing time of lifting is the *algebraic degree* of the sample point after conversion. By the degree of a sample point represented in $\mathbf{Q}(\alpha)$, we mean the degree of an irreducible polynomial M in $\mathbf{Q}[x]$ such that $M(\alpha) = 0$.

Finally we consider one more factor, namely the *level* of a sample point in the search tree. Roughly sample points on higher level are more likely to yield to successful tests, and thus we will assume that a sample point on higher level more promising.

Below we give a schema of a heuristic promise function f based on the three factors described above. We chose this one because of its simplicity. One might wish to try different schemas.

Definition 13 Heuristic Promise Function for "Decide".

$$f(s) = -w_c \cdot c(s) - w_d \cdot d(s) + w_l \cdot l(s)$$

where

$c(s)$	is 0 if the conversion of the sample point s is trivial, 1 otherwise.
$d(s)$	is the algebraic degree of the sample point s after conversion.
$l(s)$	is the level of the sample point s.
w_c, w_d, w_l	are non-negative integers (weights). $\square$

A "good" combination of the weights are to be found through experiments.

4   Experiments

In this section, we report the experimental results, comparing the original algorithm (Section 2) and the heuristic algorithm (Section 3) on several satisfaction problems.

4.1   Test Programs and Settings

We have implemented both the original algorithm and the heuristic algorithm in the C language on top of the C version of the computer algebra library SAC2 [12, 27]. All the experiments were carried out on a DECstation 5000/200 with 32 megabyte main memory, out of which 4 megabytes were used for dynamic memory (list cells). The timings were obtained by using the system cpu clock which is accurate upto 17 milli-seconds.

We have used the following weight combination for our experiments:

$$w_c = 100, \quad w_d = 10, \quad w_l = 1.$$

We do not claim that this combination is best, but the preliminary experiments suggest that this seems to work well most times. A rough justification for this combination is that the computing time for conversion can be enormously huge that we need to put most weight on it. Next the computation time for dealing with algebraic numbers of high degrees can be also great, so we put the middle weight on it.

4.2 Test Problems

The following test problems were randomly generated. The first four problems involve two variables, and the remaining six problems involve three variables.

$$P_1 \equiv \begin{cases} -12x^2y^2 - 4xy^2 + 2y^2 - 13x^2y - 3x^2 + 14x < 0 \\ 9x^2y^2 - 8xy^2 + 14y^2 + 11x^2y - 9xy + 7 < 0 \end{cases}$$

$$P_2 \equiv \begin{cases} -5xy - 3y + 15x^2 - 5 = 0 \\ -13x^2y^2 - 7xy^2 + xy + 6y + 7x^2 + 2 = 0 \end{cases}$$

$$P_3 \equiv \begin{cases} 2x^2y^2 + 7xy^2 - 8xy - 12y < 0 \\ 6x^2y^2 - 4x^2 + 11 < 0 \\ 9x^2y^2 + 14xy^2 - 3x^2 - 3 < 0 \end{cases}$$

$$P_4 \equiv \begin{cases} -7y^2 - 13xy - 4y + 5x - 9 = 0 \\ -11x^2y^2 - 5y^2 - x^2 - 11x = 0 \\ 6y^2 - 4xy + x^2 - x = 0 \end{cases}$$

$$P_5 \equiv \begin{cases} 12yz^2 - 8yz + 11x < 0 \\ -5x^2yz^2 - 14x^2z^2 - 11xyz < 0 \end{cases}$$

$$P_6 \equiv \begin{cases} -10y^2z^2 - 4x^2yz^2 - 6y^2z + 11xyz = 0 \\ 12y^2z^2 + 11yz^2 + 14xy^2z + 5xyz = 0 \end{cases}$$

$$P_7 \equiv \begin{cases} 14y^2z^2 + 12y^2z - 7xyz + 6yz - 8xz + 14y^2 + 11x < 0 \\ -2y^2z^2 - 8x^2z^2 - 5xyz - 14xz - 11y - x < 0 \end{cases}$$

$$P_8 \equiv \begin{cases} -8yz^2 - 11x^2y - 9x^2 = 0 \\ -10xz^2 - 4z^2 + 4xy^2z - 6yz + 11xz - 7xy^2 - 9xy - 15x^2 = 0 \end{cases}$$

$$P_9 \equiv \begin{cases} 12yz^2 - 8yz + 11x < 0 \\ -5x^2yz^2 - 14x^2z^2 - 11xyz < 0 \\ -13x^2z^2 - y^2z - 2z < 0 \end{cases}$$

$$P_{10} \equiv \begin{cases} -8y^2z^2 + 11x^2z^2 + 2x^2y = 0 \\ -10xy^2z^2 + 6x^2yz^2 - 2xz^2 = 0 \\ -13z + 11x^2y^2 - 12y^2 = 0 \end{cases}$$

4.3 Test results

The test results are shown in Table 1. All the timings are in milli-seconds. The following notations are used in the table:

Table 1. Experimental Data (Timings in milli-seconds)

Problem	Satisfiable?	Method	T_{total}	T_{lift}	T_{choice}	N_{lift}	N_{sample}	N_{front}
P_1	Yes	Original	37683			22	130	
		Heuristic	117	117	0	2	31	21
		Orig/Heur	322.0			11. 0	4.1	
P_2	Yes	Original	6000			16	106	
		Heuristic	1383	1383	0	13	92	15
		Orig/Heur	4.3			1.2	1.1	
P_3	Yes	Original	9183			30	310	
		Heuristic	83	83	0	2	43	29
		Orig/Heur	110.6			15.0	7.2	
P_4	No	Original	880			12	80	
		Heuristic	884	884	0	12	80	11
		Orig/Heur	1.0			1.0	1.0	
P_5	Yes	Original	7000			107	735	
		Heuristic	84	84	0	3	30	18
		Orig/Heur	83.3			35.6	24.5	
P_6	Yes	Original	75034			361	2415	
		Heuristic	116	116	0	2	43	39
		Orig/Heur	646.8			180.5	56.1	
P_7	Yes	Original	∞			$\gg$26	$\gg$211	
		Heuristic	1350	1350	0	5	124	103
		Orig/Heur	$\gg$444.4			$\gg$5.2	$\gg$1.7	
P_8	Yes	Original	∞			$\gg$ 14	$\gg$91	
		Heuristic	7783	7733	50	191	1206	117
		Orig/Heur	$\gg$77.0			$\gg$0.0	$\gg$ 0.0	
P_9	Yes	Original	∞			$\gg$ 24	$\gg$229	
		Heuristic	233	233	0	3	46	32
		Orig/Heur	$\gg$2575.1			$\gg$8.0	$\gg$4.9	
P_{10}	Yes	Original	∞			$\gg$5	$\gg$68	
		Heuristic	400	384	16	9	90	35
		Orig/Heur	$\gg$1500.0			$\gg$0.5	$\gg$0.7	

∞ : the program was aborted after 10 minutes (600,000 mill-seconds).

T_{total} The total time taken by the algorithms, except for the time for projections. We did not include the time for projection because we are only interested in the effect of heuristics during the lifting steps.

T_{lift} The time taken for lifting sample points. Actually it also includes the times for testing sample points for satisfaction. However their times are negligible compared to the times for lifting. Thus one can safely assume that T_{lift} is completely used up for lifting.

T_{choice} The time taken for choosing a sample point to be lifted. This only applies to the heuristic algorithm, since the original algorithm does not use any heuristics.

N_{lift} The number of sample points lifted.

N_{sample} The number of sample points generated (through lifting).

N_{front} The maximum number of sample points stored in the frontier set. It is the set S in the algorithm **Decide** in Section 3. This applies only to the heuristic algorithm, since the original algorithm does use any heuristics.

On several problems, the original algorithm was aborted after about 10 minutes. In such cases, we listed the statistics gathered upto the point when it was aborted. The actual statistics would have been far greater if the program had not been aborted. In the table, this fact is indicated by $\gg$ symbols in front of the intermediate statistics.

4.4   Observations on the test data

The table (especially the column T_{total}) shows that the heuristic algorithm is overall superior to the original algorithm, at least for the ten problems that we have tested. Recalling that the inputs were generated randomly, we conjecture that this will be the case for most other inputs.

For the last four problems, the original algorithm was aborted (by the author) after 10 minutes. During these 10 minutes, it carried out small number of lifting operations. Especially on the problem P_{10}, only 5 liftings had been done. A more detailed statistical data (not included in this paper) show that most of 10 minutes were spent in lifting one sample point which either required an expensive conversion or had a high algebraic degree. In contrast, the heuristic algorithm "skillfully" avoided such expensive liftings.

Note that the times T_{choice} for evaluating choice function are negligible in these ten problems. It is partly due to the simplicity of choice function we used. But more importantly it is due to the short length N_{front} of the frontier set it had to manage. We conjecture that the value of T_{choice} will grow as the problem size grows. But then, the value of T_{lift} will also grow. So it might happen again that a new T_{choice} is negligible compared to a new T_{lift}.

Note that

$$\frac{\text{Original } T_{\text{total}}}{\text{Heuristic } T_{\text{total}}} \gg \frac{\text{Original } N_{\text{lift}}}{\text{Heuristic } N_{\text{lift}}}.$$

Since most computing time of both algorithms are used up for lifting, this implies:

$$\frac{\text{Original } T_{\text{lift}}}{\text{Original } N_{\text{lift}}} \gg \frac{\text{Heuristic } T_{\text{lift}}}{\text{Heuristic } N_{\text{lift}}}$$

which says that the average computing time of a lifting for the original algorithm was bigger than that for the heuristic algorithm. It again says that the heuristic algorithm skillfully avoided expensive liftings.

Note that there is no speedup for the problem P_4. It is because the system is un-satisfiable. Whenever an input system is un-satisfiable, the heuristic algorithm will eventually have to construct the whole search space just like the original algorithm, finding out that none of the sample points satisfies the input.

In this connection, it is interesting to note that out of 10 randomly generated systems, only one system was un-satisfiable. One might explain this by observing that there were only two "over-determined" systems: P_3 and P_4. By an over-determined system, we mean a system where the number of polynomials is greater than the number of variables.

5 Conclusion

In this paper, we investigated the possibility of applying AI techniques to algebraic algorithms developed in computer algebra. In particular, we applied the *Best-First* search strategy to the cylindrical algebraic decomposition algorithm for deciding the satisfaction of polynomial constraints. The preliminary experimental results show that such heuristic search technique can often speed up the whole computation time, sometime dramatically.

We believe that various other algebraic algorithms developed in computer algebra are also amenable to such AI techniques, and we plan to investigate such possibilities.

References

1. D. S. Arnon. Algorithms for the geometry of semi-algebraic sets. Technical Report 436, Computer Sciences Dept, Univ. of Wisconsin-Madison, 1981. Ph.D. Thesis.
2. D. S. Arnon. A bibliography of quantifier elimination for real closed fields. *Journal of Symbolic Computation*, 5(1,2):267–274, 1988.
3. D. S. Arnon. A cluster-based cylindrical algebraic decomposition algorithm. *Journal of Symbolic Computation*, 5(1,2):189–212, 1988.
4. D. S. Arnon, G. E. Collins, and S. McCallum. Cylindrical algebraic decomposition I: The basic algorithm. *SIAM J. Comp.*, 13:865–877, 1984.
5. M. Ben-Or, D. Kozen, and J. H. Reif. The complexity of elementary algebra and geometry. *J. Comput. System Sci.*, 32(2):251–264, 1986.
6. W. Böge. Decision procedures and quantifier elimination for elementary real algebra and parametric polynomial nonlinear optimization. Manuscript in preparation, 1980.
7. B. Buchberger and H. Hong. Speeding-up quantifier elimination by Groebner bases. Technical Report 91-06.0, Research Institute for Symbolic Computation, Johannes Kepler University A-4040 Linz, Austria, 1991.
8. J. Canny. Some algebraic and geometric computations in PSPACE. In *Proceedings of the 20th annual ACM symposium on the theory of computing*, pages 460–467, 1988.
9. P. J. Cohen. Decision procedures for real and p-adic fields. *Comm. Pure and Applied Math.*, 22:131–151, 1969.

10. G. E. Collins. Quantifier elimination for the elementary theory of real closed fields by cylindrical algebraic decomposition. In *Lecture Notes In Computer Science*, pages 134–183. Springer-Verlag, Berlin, 1975. Vol. 33.

11. G. E. Collins and H. Hong. Partial cylindrical algebraic decomposition for quantifier elimination. *Journal of Symbolic Computation*, 12(3):299–328, September 1991.

12. G. E. Collins and R. Loos. *The SAC-2 Computer Algebra System*. Research Institute for Symbolic Computation, Johannes Kepler University, Linz, Austria A-4040.

13. N. Fitchas, A. Galligo, and J. Morgenstern. Algorithmes repides en séquential et en parallele pour l'élimination de quantificateurs en géométrie élémentaire. Technical report, UER de Mathématiques Universite de Paris VII, 1987. To appear in: Séminaire Structures Algébriques Ordonnées.

14. D. Yu. Grigor'ev. The complexity of deciding Tarski algebra. *Journal of Symbolic Computation*, 5(1,2):65–108, 1988.

15. D. Yu. Grigor'ev and N. N. Vorobjov (Jr). Solving systems of polynomial inequalities in subexponential time. *Journal of Symbolic Computation*, 5(1,2):37–64, 1988.

16. J. Heintz, M-F. Roy, and P. Solernó. On the complexity of semialgebraic sets. In *Proc. IFIP*, pages 293–298, 1989.

17. C. Holthusen. *Vereinfachungen für Tarski's Entscheidungsverfahren der elementaren reellen Algebra*. PhD thesis, University of Heidelberg, January 1974.

18. H. Hong. An improvement of the projection operator in cylindrical algebraic decomposition. Technical Report OSU-CISRC-12/89 TR55, Computer Science Dept, The Ohio State University, 1989.

19. H. Hong. An improvement of the projection operator in cylindrical algebraic decomposition. In *International Symposium of Symbolic and Algebraic Computation*, pages 261–264, 1990.

20. H. Hong. *Improvements in CAD-based Quantifier Elimination*. PhD thesis, The Ohio State University, 1990.

21. H. Hong. Comparison of several decision algorithms for the existential theory of the reals. Technical Report 91-41.0, Research Institute for Symbolic Computation, Johannes Kepler University A-4040 Linz, Austria, 1991. Submitted to *Journal of Symbolic Computation*.

22. H. Hong. Parallelization of quantifier elimination on workstation network. Technical Report 91-55.0, Research Institute for Symbolic Computation, Johannes Kepler University A-4040 Linz, Austria, 1991.

23. H. Hong. Simple solution formula construction in cylindrical algebraic decomposition based quantifier elimination. Technical Report 92-02.0, Research Institute for Symbolic Computation, Johannes Kepler University A-4040 Linz, Austria, 1992. To appear in *International Conference on Symbolic and Algebraic Computation* ISSAC-92.

24. J. R. Johnson. *Algorithms for Polynomial Real Root Isolation*. PhD thesis, The Ohio State University, 1991.

25. L. Langemyr. The cylindrical algebraic decomposition algorithm and multiple algebraic extensions. In *Proc. 9th IMA Conference on the Mathematics of Surfaces*, September 1990.

26. D. Lazard. An improved projection for cylindrical algebraic decomposition. Unpublished manuscript, 1990.

27. R. G. K. Loos. The algorithm description language ALDES (Report). *ACM SIGSAM Bull.*, 10(1):15–39, 1976.

28. S. McCallum. *An Improved Projection Operator for Cylindrical Algebraic Decomposition*. PhD thesis, University of Wisconsin-Madison, 1984.

29. J. Pearl. *Hueristics (Intelligent Search Strategies for Computer Problem Solving)*. Addison-Wesley, 1984.

30. J. Renegar. On the computational complexity and geometry of the first-order theory of the reals (part I). Technical Report 853, Cornell University, Ithaca, New York 14853-7501 USA, July 1989.

31. J. Renegar. On the computational complexity and geometry of the first-order theory of the reals (part II). Technical Report 854, Cornell University, Ithaca, New York 14853-7501 USA, July 1989.

32. J. Renegar. On the computational complexity and geometry of the first-order theory of the reals (part III). Technical Report 856, Cornell University, Ithaca, New York 14853-7501 USA, August 1989.

33. A. Seidenberg. A new decision method for elementary algebra. *Ann. of Math*, 60:365-374, 1954.

34. A. Tarski. *A Decision Method for Elementary Algebra and Geometry*. Univ. of California Press, Berkeley, second edition, 1951.

Unified Domains and Abstract Computational Structures

J. CALMET, K. HOMANN and I.A. TJANDRA

Universität Karlsruhe
Institut für Algorithmen und Kognitive Systeme
Am Fasanengarten 5; Postfach 69 80; D-76128 Karlsruhe; Germany

Abstract. This paper introduces a formalism to specify abstract computational structures (ACS) of mathematical domains of computation. This is a basic step of a project aiming at designing an environment for symbolic computing based upon knowledge representation and relying, when needed, on AI methods.
We present a method for the specification of these ACS's which is embedded in the framework of algebraic specifications and of unified domains. The first part of this paper deals with the theoretical solution of this specification problem. The second part reports on the implementation in the hybrid knowledge representation system MANTRA.

Keywords: Abstract Computational Structures, Unified Domains, Hybrid Knowledge Representation System.

1   Introduction

Mathematical domains of computations, such as finite groups, polynomial rings or finite fields for instance, are inherently modular. Furthermore, there exist inter-relationships among these domains. It is thus meaningful to design an environment for symbolic computing within the framework of the theory of algebraic specification [Gu 75], [Gu], [Go,Bu]. An algebraic specification basically introduces constants and operators together with properties that have to be imposed on their intended interpretation, or class of interpretations. This has several pragmatic benefits, such as the re-use of subspecifications within a specification taking into account the dependency relationship between particular specification modules.

Because we are mainly concerned with executable specifications, it turns out that the framework of *unified domains*[1] [Mo86] developed from the framework of order-sorted domains, cf. section 2, is appropriate to handle mathematical domains of computation. Furthermore, when defining the semantics of vertical or *inclusion polymorphism* in a Computer Algebra System one routinely deals with operations that map sorts to sorts, for instance mapping two sorts to their union or to their subset. This is well supported by unified domains since there is

---

[1] originally called unified algebras

a unified treatment of the *elements* of an abstract computational structure and their classifications into *sorts*. Thus the operations of a unified domain may take sorts and/or elements as arguments, and give sorts or elements as results [Mo86]. This framework improves much our first approach to this problem [Ca-Tj91a].

This paper outlines how to specify abstract computational structures (ACS's) in the framework of unified domains. The specifications are represented internally using the language that is embedded in the knowledge representation system MANTRA. We have specifically designed MANTRA [Ca et al. 91b] as a suitable environment for symbolic mathematical computation. Many aspects must be considered when planning such an environment. Among them is the capability to handle inclusion polymorphism, which is fundamental for symbolic mathematical computations. Also, it is known that no universal algorithm does exist for problems such as type inference for mathematical domains when properties are considered or canonical forms for mathematical expressions. Possible solutions consist in relying on methods from AI. Such problems were among the many motivations to design our environment.

Basically, the procedure of performing a specification can be depicted as follows: Having recognized the syntax correctness of an input (a specification) according to the underlying specification language, a parse tree is given to a program transformer that generates code, possessing the intended semantics, for the language of MANTRA. The code will be executed by MANTRA by invoking its inference engine.

The paper is organized as follows. As the framework of unified domains is developed from the framework of *order-sorted domains* we introduce in section 2 some basic notions and notations of order-sorted domain [Ba-Wo][Hu,Op]. Section 3 introduces the notion and notations of unified domains which is a specialization of order-sorted domains. Section 4 deals with an introduction to MANTRA. We outline only those features of MANTRA that we need in the sequel. The internal representation of the specification of ACS's is described in section 5. Finally, some concluding remarks are given in the last section.

2 Basic Notions and Notations

$\Sigma = \langle S, \Omega \rangle$ is an order-sorted signature consisting of S, a finite set of order-sorted sorts, and Ω, the set of operator symbols. Ω is the union of pairwise disjoint subsets of $\Omega_{w,s}$, the sets of operator symbols with argument sorts $w \in S^*$, i.e. $w = s_1 \times \cdots \times s_n$ for $n \geq 0$ and $s_i \in S$, and range sort $s \in S$.

A domain $\mathcal{D}$ with the signature $\Sigma = \langle S, \Omega \rangle$, also called Σ-domain, is a pair $\mathcal{D} = \langle \mathcal{U}, \mathcal{F} \rangle$ such that $\mathcal{U}$ is an S-indexed family of sets and $\mathcal{F}$ is a Ω-indexed family of functions with: (i) $\mathcal{U}_s$ are sets for all $s \in S$, called the universe of sorts s in the domain. (ii) $\mathcal{F}_f \in \mathcal{U}_s$ if $f \in \Omega_{\lambda,s}$ and $s \in S$, called constants of $\mathcal{D}$ of sort s. (iii) $\mathcal{F}_f \in \mathcal{U}_{s_1} \times \cdots \times \mathcal{U}_{s_n} \to \mathcal{U}_s$, functions for all operator symbols f, if $f \in \Omega_{s_1 \times \cdots \times s_n, s}$ with $n \geq 1$, $s, s_i \in S$ and $i = 1, \cdots, n$.

Let $\Sigma = \langle S, \Omega \rangle$ be a signature and X_s for each $s \in S$ a set of variables of sort s. We assume that the variables of distinct sorts are distinct and that no variable is a member of Ω. The union $X = \cup_{s \in S} X_s$ is called the set of variables with respect to Ω.

The sets $T_{s,\Omega}(X)$ of terms of sort s are inductively defined as follows: (i) $X_s \cup \Omega_{\lambda,s} \subseteq T_{s,\Omega}(X)$ (*basic terms*), (ii) $f(t_1, \cdots, t_n) \in T_{s,\Omega}(X)$ (*composite terms*) for all operator symbols $f \in \Omega_{s_1 \times \cdots \times s_n, s}$ and all terms $t_i \in T_{s_i,\Omega}(X)$ $i = 1, \cdots, n$, (iii) there are no further terms of sort s in $T_{s,\Omega}(X)$. The set $T_{s,\Omega}$ of terms without variables of sort s, also called *ground terms* of sort s, is defined for the empty set $X = \emptyset$ of variables as follows: (iv) $T_{s,\Omega} = T_{s,\Omega}(\emptyset)$. The set of terms $T_{\Sigma}(X)$ and the set of terms without variables T_{Σ} are defined as follows: (v) $T_{\Sigma}(X) = \bigcup_{s \in S} T_{s,\Omega}(X)$ with $s_1, s_2 \in S$, $s_1 \sqsubseteq s_2$, $t \in T_{s_1,\Omega}(X) \Rightarrow t \in T_{s_2,\Omega}(X)$ and $T_{\Sigma} = \bigcup_{s \in S} T_{s,\Omega}$ with $s_1, s_2 \in S$, $s_1 \sqsubseteq s_2$, $t \in T_{s_1,\Omega} \Rightarrow t \in T_{s_2,\Omega}$. This also implies that each sort (type) inherits all terms possessed by its subsorts (types).

Let T_{Σ} be the set of terms of signature $\Sigma = \langle S, \Omega \rangle$ and $\mathcal{D}$ a Σ-domain. The *interpretation* $\varphi : T_{\Sigma} \to \mathcal{D}$ is recursively defined as follows: (i) $\varphi(f) = \mathcal{F}_f$ for all $f \in \Omega_{\lambda,s}$, $s \in S$, (ii) $\varphi(f(t_1, \cdots, t_n)) = \mathcal{F}_f(\varphi(t_1), \cdots, \varphi(t_n))$ for all $f(t_1, \cdots, t_n) \in T_{\Sigma}$.

Given a set of variables X for $\Sigma = \langle S, \Omega \rangle$, a Σ-domain $\mathcal{D}$ and an assignment $\psi : X \to \mathcal{D}$ with $\psi(x) \in U_s$ for $x \in X_s$ and $s \in S$, the *extension* $\xi : T_{\Sigma}(X) \to \mathcal{D}$ of the assignment $\psi : X \to \mathcal{D}$ is recursively defined as follows:

(i) $\xi(x) = \psi(x)$ for all variables $x \in X$.

$\xi(f) = \mathcal{F}_f$ for all $f \in \Omega_{\lambda,s}$, $s \in S$.

(ii) $\xi(f(t_1, \cdots, t_n)) = \mathcal{F}_f(\xi(t_1), \cdots, \xi(t_n))$ for all $f(t_1, \cdots, t_n) \in T_{\Sigma}(X)$

Given a signature $\Sigma = \langle S, \Omega \rangle$ and variables X with respect to Σ. A triple $p = \langle X, L, R \rangle$ with $L, R \in T_{s,\Omega}(X)$ for some $s \in S$ is called a property of sort s with respect to Σ. The property $p = \langle X, L, R \rangle$ is said to be *valid* in a Σ-domain $\mathcal{D}$ if for all assignment $\psi : X \to \mathcal{D}$ we have $\xi(L) = \xi(R)$. *Ground properties* are properties $p = \langle X, L, R \rangle$ with $X = \emptyset$. In this case L and R are ground terms.

An *abstract computational structure* $ACS = \langle \Sigma, \mathcal{P} \rangle$ consists of a signature $\Sigma = \langle S, \Omega \rangle$ and $\mathcal{P}$, a set of properties with respect to Σ. Based on other known ACS's we can construct a new ACS by integrating all operators and properties possessed by the known ACS's and by adding additional operators and properties into the new one. This implies that each ACS inherits all operators and properties possessed by the ACS on which it is based.

3 Unified Domains

Initially, we define signatures and properties for unified domains by specializing the notation for abstract computational structures and domains as described in section 2.

A unified signature Σ^u is a *homogeneous first-order signature*. In contrast to order-sorted signatures there is only one sort, S^u, instead of a set of order-sorted sorts[2]. Let Σ^u be a pair $\langle S^u, \Omega^u \rangle$, where $\Omega^u \supseteq \Omega^{u0} = \{\bot, |, \&\}$. $\bot, |$ and

---

[2] As we shall see below, we still can simulate an order-sorted signature using a unified domain, since the carrier of a unified domain is a distributive lattice.

& represent the bottom of a lattice, the joint and meet operations on lattices, respectively. As we deal with homogeneous first-order signatures we can write $\Omega^u = \{\Omega^u_n \mid n \geq 0\}$, where Ω^u_n is the set of operator symbols of arity n.

Let X be a set of variables, disjoint from Ω^u. A Σ^u-unified property is a universal Horn Clause involving equalities with variables from X, operator symbols form Ω^u and the binary predicate symbols **Identical-to, Subsumes** and **Element-of**.

A Σ^u-unified domain $\mathcal{D}$ is a homogeneous Σ^u-domain –cf section 2– such that:

(i) $|\mathcal{D}|$ is a distributive lattice with $\_|\_\mathcal{D}$ as join, $\_\&\_\mathcal{D}$ as meet and $\perp_\mathcal{D}$ as bottom. Let **Subsumes**$_\mathcal{D}$ be the partial order of the lattice.

(ii) There is a distinguished subset of incomparable values, $\mathcal{E}_\mathcal{D} \subseteq |\mathcal{D}|$.

(iii) For each $f \in \Omega^u$, the function $f_\mathcal{D}$ is monotone with respect to **Subsumes**$_\mathcal{D}$.

The intended interpretation of the binary predicate symbols, in a unified domain $\mathcal{D}$ is as follows:

- x **Identical-to** y :$\Leftrightarrow$ x is identical to y
- x **Subsumes** y :$\Leftrightarrow$ x **Subsumes**$_\mathcal{D}$ y
- x **Element-of** y :$\Leftrightarrow$ x $\in$ $\mathcal{E}_\mathcal{D}$ and x **Subsumes**$_\mathcal{D}$ y

4 MANTRA

The *MANTRA* system [Ca et al. 91b] is a hybrid system with the following characteristics: All features are semantically motivated and all the inference algorithms involved are decidable. The decidability requirement has been met with the adoption of a four-valued semantics based on the works of Patel-Schneider [PS85], Frisch [Fr] and Thomason et al. [Th et al.]. This semantics is used throughout the system and ensures that it is semantically consistent.

The language of MANTRA can be thought of as an abstract data type allowing the creation and manipulation of knowledge bases. The *knowledge bases* consist of a set of knowledge base partitions, each associated to an independent formalism. The division of the language into several formalisms has two advantages: The computability problems associated to each formalism can be solved independently and the integration of new formalisms to the system is facilitated.

Each formalism is characterized by a set of definitions and a set of questions. The *definitions* allow the storage of knowledge into the knowledge bases and the *questions* allow the interrogation of these knowledge bases. Definitions are used to store knowledge only into the partition associated to the formalism, but questions can be directed to this partition or to a combination of two or more partitions of the knowledge base. The language is based on a new architecture, figure 1, consisting of three levels: The epistemological level, the logical level and the heuristic level.

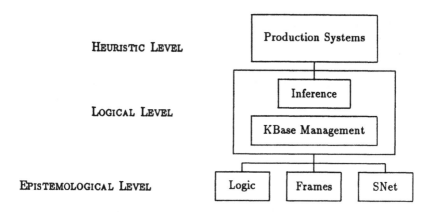

Fig. 1. The architecture of *MANTRA*

The first level consists of three modules: an assertional module, based on a decidable first-order logic language [PS85], a frame module, based on the terminological box of Krypton [Br et al.], and a semantic network module, providing inheritance with exceptions [Et]. These modules offer original features with respect to previously existing systems. The primitives of the modules of this first level define the epistemological primitives of the language. These primitives are not complete expressions of the language but are used as parameters for the Ask and Tell primitives of the logical level.

The primitives of the *assertional module* correspond to the usual operators of the first-order logic languages, but the meaning of these operators is based on a four-valued semantics.

This module is intended to be used to represent a terminology by means of concepts, the categories of objects, and relations, the properties of objects. These categories are described by restricting the values of the properties of the objects forming them. The notion of relations is an extension of the notion of roles, usually used in terminological languages. Roles are binary relations and relations are arbitrary n-place relations.

The terminological language embedded into the system has some additional characteristics usually not possessed by other terminological languages or hybrid systems: (i) It possesses a rich set of primitives, including disjunction and negation of both concepts and relations, (ii) It provides special symbols for the universal concept and for the bottom concept as well as for the universal relation and for the bottom relation and (iii) It includes tests for subsumption and for equality between concepts and between relations.

This module manipulates the notions of classes and hierarchies. The hierarchies can be explicitly created by defining links among classes. Two types of links are provided: *Default* links and *Exception* links. The hierarchies are used as inheritance paths between classes. The main inference procedure of this module calculates the *Subclasses* relation taking into account the explicit exception.

The second level introduces the notion of knowledge base. The language can be thought of as an abstract data type whose access functions are the primitives of this level. These primitives use the primitives of the first level as parameters. Two types of primitives are provided, **Tell** and **Ask**. These primitives are used, respectively, to store facts and to interrogate knowledge bases. The Ask primitives are defined in such a way that new facts can be inferred from evidence provided by the knowledge acquired only by one or by a combination of several of the first level modules.

Finally, the third level consists of primitives allowing the definition of production systems according to several configurations [Le-De]. These production systems are used to automatically manipulate knowledge bases according to heuristic rules. The rules used in these production systems are formed by queries to the knowledge bases defined in the logical level. Conflict resolution strategies and control strategies can be explicitly chosen by special primitives. The idea underlying the heuristic level is to allow the introduction of ad hoc rules in the inference process. These rules can directly introduce domain knowledge in the knowledge bases or they can specify strategies for the utilization of the logical level Ask and Tell primitives. This level enables thus to design expert systems.

This brief overview of *MANTRA* covers only those features which are used in the sequel.

5 Representation of ACS's using MANTRA

As properties are embodied by Horn Clauses, the class of Ω^u-unified domains that satisfy $\mathcal{P}$ has an initial domain [Go et. al.]. The proof of this theorem can be found in [Mo89].

If we impose initial constraints on Σ^u-unified domains then we are capable of specifying ACS's using an Σ^u-unified domain.

We describe the syntax of the language for algebraic specifications of abstract computational structures and its intended interpretation by means of some examples, cf. figure 2.

The specification of an ACS has two parts:

(a) A header, containing its name and

(b) A body, containing its signature that is divided into the following parts:

   (i) **Based-on:** This declaration is to indicate those abstract computational structures that are inherited by the ACS being specified.

   (ii) **Parameter:** This declaration is used to specify an ACS possessing particular ACS's as parameters, e.g. ACS Module has parameter ACS Rg and based on additive abelian group.

   (iii) **Sort:** To declare its new sorts. Subsort relationships are declared using the notations as introduced above, e.g. using | and &.

ACS Semi-Group {
      Based-on set;
      Sort SG;
      Operators _ f _ :: Elt − > Elt;
      Initial-Properties
      !V x,y,z : x Element-of Elt, y Element-of Elt, z Element-of Elt
                  => (x f y) f z = x f (y f z)}
ACS Monoid {
      Based-on Semi-Group;
      Sort Mo, ne Element-of Elt;
      Initial-Properties
      !V x: x Element-of Elt
            => ne f x = x}
ACS Group {
      Based-on Monoid;
      Sort Gr;
      Operators _ inv _ :: Elt − > Elt;
      Initial-Properties
      !V x: x Element-of Elt
            => inv(x) f x = ne}
ACS Abelian-Group {
      Based-on Group;
      Sort AG;
      Initial-Properties
      !V x, y: x Element-of Elt, y Element-of Elt
            => x f y = y f x}
ACS Rg {
      Based-on Semi-Group(**rename:** (f,×), (ne,1)),
            Abelian-Group(**rename:** (f,+), (ne,0), (inv, −));
      Sort R;
      Initial-Properties
      !V x, y, z: x Element-of Elt, y Element-of Elt, z Element-of Elt
      => x×(y+z) = (x × y) + (x × z)
      !V x, y, z: x Element-of Elt, y Element-of Elt, z Element-of Elt
            => (y+z)×x = (y × x) + (z × x)}

Fig. 2. Examples of some basic ACS

(iv) **Operator-Symbols:** In this part all new operator symbols are declared. :: and − > are special symbols used to define the functionality of an operator.

(v) **Initial-Properties:** To declare properties of the operators. !V represents the universal quantifier.

The use of **Based-on** declaration permits the re-use of parts of certain specifications in another specification, which is very convenient to design abstract computational structures for mathematical domains according to their inherently modular structure. The intended interpretation of this declaration is the inheritance of sorts – e.g. implicitly, in ACS Monoid the sort becomes Mo | SG where Mo and SG denote the sorts of Monoid and Semi Group, respectively, and so on– , operator symbols and initial properties possessed by the ACS embedded in this specification. Inherited symbols are not changed unless we rename explicitly old symbols using **rename**, e.g. in the ACS **Rg** below.

We use the mixfix notation for operator symbols according to the positions of place holders.

Recalling that classifications of elements into sorts are represented directly as values in the corresponding carriers, we define constants or nullary functions using **Element-of**.

We represent initial properties as rules, rather than merely writing them as comments, since we use them to simplify terms. Generally, a set of properties is not complete in the sense that terms can not canonically be simplified using these properties. Therefore we classify properties into two classes: Initial properties and learned properties. The method to complete a set of properties [Ca-Tj90] is beyond the scope of this paper. In this paper, we merely take into account initial properties.

The main principle of representing an ACS can be described as follows:

(i) The signature of an ACS, as shown in figure 2, is represented as frames. The relationships among ACS's are represented as semantic networks wherein each node corresponds to a particular frame, e.g. Semi-Group-Operators and Monoid-Operators. The link specifies which objects are inherited by which ones within a particular hierarchy, e.g. in the hierarchy **Based-on** there is a link from Semi-Group-Operators to Monoid-Operators meaning that, according to the examples, Semi-Group-Operators is inherited by Monoid-Operators.

(ii) The initial properties of an ACS are represented as rules and they are stored in a rule base. To identify the rule base we introduce a relation **rule-base**, possessed by an ACS frame, whose value is the identifier of the corresponding rule base.

Thus, an ACS is represented by a frame having a unique identifier, **id**. This frame possesses a relation **rule-base** and subsumes the frame *sort*, i.e. the frame identifier coincides with the sort identifier, and the frame **id**-Operators.

The frame *sort* subsumes another frames if the relation **Element-of** occurs in the sort declaration, e.g. in ACS Monoid: ne **Element-of** Mo. The frame **id**-Operators subsumes frames, representing each operator symbol, possessing the relation **domain** and the relation **codomain** specifying the domains of an operator and its codomain, respectively, cf. figure 3.

Having recognized the declaration **Based-on** within an ACS the program transformer build or modify the hierarchy (semantic network) *based-on* according to the specification. This is illustrated by figure 4.

The Interrogation of a knowledge base need a hybrid inference taking into account frames and semantic network simultaneously, e.g. using the primitive

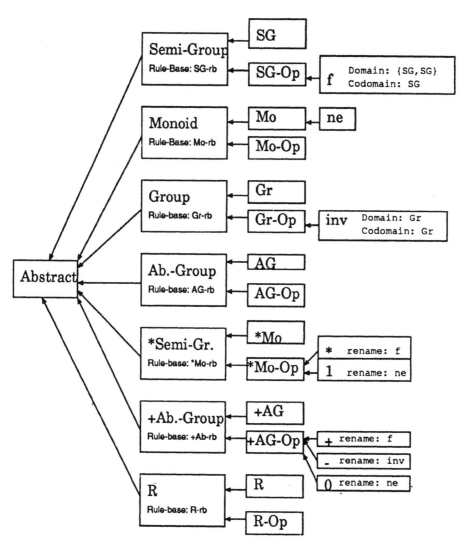

Fig. 3. Frame Structure

```
(ASK 'KB '(FROM-SNET-FRAME ........)).
```

Basically, mathematical domains of computations are models of ACS's. According to our approach they can be regarded as unified domains, since initial constraints are imposed on unified domains. We shall outline this part briefly.

The specification of a domain is quite similar to that of an ACS. The semantic specification of operations plays a crucial role and we adopt action semantics to specify the semantic functions.

Generally, the specification of a unified domain has two parts: (a) a header,

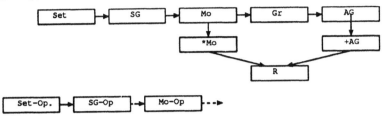

Fig. 4. The hierarchy *based-on*

containing its name and (b) a body as defined for ACS's but without initial properties and with two additional declaration: (i) **Model**, identifying the corresponding ACS and (ii) **Semantic-Equations**. The given specification will be transformed into the language of MANTRA where semantic equations will be represented accordingly by rules in such a way that they can be performed by the inference engine.

6 Conclusion

We have outlined the framework of Unified Domains and its application to specify abstract computational structures of mathematical domains of computation.

Unified domains make intensive use of nondeterministic operations, which correspond to union of classifications. We can use unified domains not solely to specify but, furthermore, it underlies the inclusion partial ordering which is preserved by all the other operations. This is the major advantage of using unified domains since it amounts to an easy implementation of vertical or inclusion polymorphism. We deal with this kind of polymorphism since we design an environment for symbolic computing.

In order to represent such domains we make use of the hybrid knowledge representation system MANTRA. The characteristic features of MANTRA consist in its decidable algorithms and its unified semantics. This ensures the security of its performance in the sense that each algorithm terminates. The hybrid inference mechanisms of MANTRA are used to process knowledge bases.

To design an executable or operational semantic function for mathematical domains of computation we make use of Action Semantics. In Action Semantics the semantic functions are defined inductively by *semantic equations*, called actions. These actions may be regarded as algebraic equations, but their *well-foundedness* is essential and they have a more operational nature than, for instance, the higher-order functions used as denotations in denotational semantics [St]. The details of this approach is out of the scope of this paper. We have merely depicted the major principle briefly.

Finally, it can be summarized that Unified Domains have made a significant contribution to specifying ACS's and a hybrid knowledge representation can be used to represent such domains.

References

[Ba-Wo] Bauer, F.L. and Woessner, H., *Algorithmische Sprache und Programmentwicklung(2nd Edition)*, Springer-Verlag Berlin Heidelberg New York Tokyo, 1984.

[Br et al.] Brachman, R.J., Gilbert, V.P., Levesque,H.J., *An Essential Hybrid Reasoning System: Knowledge and Symbol Level Accounts of KRYPTON*, Proceedings of IJCAI 9, pp. 532–539, 1985.

[Ca-Tj90] Calmet,J., Tjandra,I.A., *Learning Complete Computational Structures*, in Emrich et al (Eds), 5th International Symposium on Methodologies for Intelligent Systems (Selected Papers), Knoxville – USA, October 24 – 27, 1990, pp. 63 – 71, ICAIT.

[Ca-Tj91a] Calmet,J., Tjandra,I.A, *Representation of Mathematical Knowledge*, Z. W. Ras et al.. (Eds.) Proceedings of the 6th International Symposium on Methodologies for Intelligent Systems, Charlotte USA, October 16 – 19, 1991, Springer-Verlag.

[Ca et al. 91b] Calmet,J., Bittencourt,G., Tjandra, I.A., *MANTRA: A Shell for Hybrid Knowledge Representation*, Proceedings of the third International conference on Tools for Artificial Intelligence, San Jose USA, November 5 – 8, 1991, IEEE Computer Society Press.

[Et] Etherington,D.W., *On Inheritance Hierarchies with Exceptions*, Proceedings of AAAI-83, pp. 104–108, 1986.

[Fr] Frisch,A.M., *Knowledge Retrieval as Specialized Inference*, Report No.214, Department of Computer Science, University of Rochester, May 1987.

[Go,Bu] Goguen, J.A. and Burstall, R.M., *Introducing Institutions*, in Clarke, E. and Kozen, D. (Eds.), Proceedings Logics of Programming Workshop, Springer-Verlag, 1984.

[Go et. al.] Goguen, J.A., Thatcher, J.W. and Wagner, E., *An Initial Algebra Approach to the Specification Correctness and Implementation of Abstract Data Types*, in Yeh, R.T. (Ed.), Current Trends in Programming Methodology IV, Prentice Hall, 1978.

[Gu 75] Guttag, J. : *The Specification and Application to Programming of Abstract Data Types*, University of Toronto, Department of Computer Science, Ph. D. Thesis, Report CSRG-59 ; 1975.

[Gu] Guttag,J.V., *The Algebraic Specification of Abstract Data Types*, Acta Informatica 10, 27–52, Springer-Verlag, 1978.

[Hu,Op] Huet, G., Oppen D.: *Equations and Rewrite Rules: a survey*; in Book R., editor, Formal Language Theory: Perspective and Open Problems, Academic Press, 1980.

[Le-De] Lenat, D.B., McDermott,J., *Less Than General Production Systems Architectures*, Proceedings of IJCAI 5, pp. 928–932, 1977.

[Mo86] Mosses, P.D., *Unified Algebras and Action Semantics*, in Monien, B. and Cori, R. (Eds.), Proceedings of 6th Annual Symposium on Theoretical Aspects of Computet Science, Springer-Verlag, 1986.

[Mo89] Mosses, P.D., *Unified Algebras and Institutions*, in Proceedings IEEE-Logics in Computer Science, IEEE-Press, 1989.

[PS85] Patel-Schneider, P.F., *A Decidable First-Order Logic for Knowledge Representation*, Proceedings of IJCAI 9, pp. 455–458, 1985.

[St] Stoy, J.E., *Denotational Semantics: The Scott-Strachey Approach toProgramming Language Theory*, the MIT Press, 1977.

[Th et al.] Thomason, R.H, Horty, J.F. and Touretzky, D.S., *A Calculus for Inheritance in Monotonic Semantic Nets*, Technical Report CMU-CS-86-138, Computer Science Department, carnegie mellon University, Pittsburgh, PA, 1986.

Completion and Invariant Theory in Symbolic Computation and Artificial Intelligence

Eugen E. Ardeleanu*

Research Institute for Symbolic Computation
Johannes Kepler University, Linz, A-4040 Austria
eardelea@risc.uni-linz.ac.at

Abstract. An outline for the study of invariant theoretic (as structural) and completion (as syntactical) concepts in symbolic computation and artificial intelligence is presented on a level of abstraction which permits a unifying viewpoint on problems in symbolic computation and artificial intelligence. We refer to applications in computational polynomial ideal theory and in general problem-solving in the sense of AI research.

1 Introduction

In this article we discuss two methodological principles commonly used in mathematics and computer science.

The first principle we refer to, integrates *structural methods* mainly used to provide for an *external characterization* of relational (algebraic) structures.

The second envisioned principle embodies *syntactical methods* usually in force when dealing with an *internal characterization* of equational theories.

Correspondingly, we will concentrate on two instances of these principles:

1. the *search for invariants* method, and
2. the *completion* method.

It is our opinion that both areas of symbolic computation and artificial intelligence could benefit from a combined structural-syntactical approach to some of their problems.

In Section 2 we succinctly present some methodological considerations on the role the two principles play in mathematics and computer science and on their interrelation. In Section 3 we concentrate on the completion method and its most pertinent application to the construction of standard (Gröbner) bases by Buchberger's completion algorithm in polynomial ideal theory, which, in turn, represents an important segment of the new emerging border area (between mathematics and computer science) of symbolic computation. In Section 4 we consider general problem-solving (from AI viewpoint) and stress the parallelism with symbolic computation in applying the two methods. An example is also included. Section 5 concludes the article.

---

* Supported by the Austrian Ministry for Science and Research.

2 Methodology

2.1 The Methodology of Structural Theories: Invariants, Representations, Completeness

The relation between *Invariants, Representation Theorems*, and *Completeness Theorems* appears frequently in algebra and logic.

Let C be an arbitrary category of algebraic structures (i.e objects and morphisms obeying the usual axioms). Consider in C a fixed universal construct $\mathcal{U}$, i.e an object and a set of distinguished morphisms (e.g sums, products), assuming, of course, it exists. Then, one is interested in the following problems formulated on a general level:

- Determine the structure of the objects in C using $\mathcal{U}$.
- Determine the objects with the simplest structure which are preserved under (a certain class of) morphisms, i.e the invariants.
- Determine the complete set of invariants with which one is capable of characterizing any object in C.

Example 1. The theory of modules (with the well known instantiations, e.g the theory of finitely generated abelian groups, the theory of linear transformations for finite dimensional vector spaces), cf. [1].

The theory of quadratic forms, cf. [2].

The algebraic theory of dynamical systems, cf. [3].

The representation theorem for boolean functions, cf. [4].

The concept of indexed systems, e.g logical fiberings, cf. [5], and indexed categories and related topics of increasing interest for the foundations of computer science, logic and mathematics.

The typical gender of results one would expect by following this method is illustrated by the following theorem of Birkhoff in the context of universal algebra, cf. [6]:

> Every universal algebra is a subdirect product of subdirectly irreducible algebras (the *structural* characterization of universal algebras).

2.2 The Methodology of Syntactical Theories

It is often the case in computer science that the *syntactical* approach is emphasized cf. [7], mainly because of immediate possibilities for applications and implementations, e.g equational theories (varieties in universal algebras defined by a set of equations) in algebraic specification of programming languages, rewriting systems (with instantiations as grammars, Thue systems, etc), unification (with implementations as Prolog).

The main problems in syntactical theories are not to determine the structure of the objects, but to develop (constructive) methods for "simplifying" them.

The process is known as *reduction to normal forms* and the *termination* and *uniqueness* of the result of reduction are the crucial points.

To prove termination the common technique consists of defining a total well ordering on the set of objects compatible with the reduction relation between the "more" and "less" complex objects, cf. [8].

To prove uniqueness comes down to prove that the reduction relation has (some variant of) the Church-Rosser property, which, roughly, ensures that the reduction to normal forms does not dependent on the different ways it can be performed, i.e normal forms are unique, cf. [8].

There have been also attempts to introduce a categorical viewpoint in dealing with syntactical theories; the objects are the well-formed expressions in the theory and the morphisms are the reduction rules. In this way some fundamental properties of syntactical theories correspond to the existence of universal categorical constructs, cf. [9].

Finally we refer again to an illustrative result of Birkhoff, cf. [6].

A class **C** of universal algebras is a variety (or equational class) if and only if it has the following closure properties (the *internal* characterization of varieties):
If $A \in$ **C**, then every subalgebra of A is in **C**;
If $A \in$ **C**, then every homomorphic image of A is in **C**;
If $\{A_\alpha : \alpha \in I\} \subseteq$ **C**, then the product $\Pi A_\alpha \in$ **C**.

3   The Completion Principle

The general method of designing algorithms by completion (for a survey see [10]) is of ancient vintage in algebra. It is used nowadays in symbolic computation as a basic tool. The most famous instance is Buchberger's algorithm for "completion" to standard (Gröbner) bases in polynomial ideal theory. We will first briefly describe the problem and the idea of the method in this context (for the exact definitions and proofs see [11]).

Definition 1 *The ideal membership problem:*
   Given a polynomial, say $f \in \mathbb{R}[x_1, ..., x_n]$ and a finite set of polynomials $S = \{f_1, ..., f_m\} \subset \mathbb{R}[x_1, ..., x_m]$, decide whether f belongs to the ideal generated by S, $f \in I[S]$.

The natural idea to solve this problem is to start to divide f by polynomials f_i, $f = p_i f_i + r$ (we say that f *reduces* to r by f_i, in symbols $f \rightarrow_{f_i} r$) and continue like in Euclid's algorithm (in the case of univariate polynomials it *is* the Euclidean algorithm), with r as f. We can view this as a reduction process and we have to ensure termination and uniqueness. Termination is ensured by introducing a total ordering $<$ on $\mathbb{R}[x_1, ..., x_n]$, (e.g a graded lexicographical ordering) compatible with the reduction relation. Thus, if f reduces to r, then $r < f$. It turns out that this ordering is noetherian, by Hilbert's basis theorem.

Definition 2 *We say that f reduces to f' with respect to S, $f \to_S f'$ if and only if there exist $f_i, ..., f_j \in S$ such that*

$$f \to_{f_i} \cdots \to_{f_j} f'.$$

It is clear that, if $f \to_S 0$, then $f \in I[S]$. However, the converse is in general not true. For the converse to hold, we have to "complete" the set S to a so-called *standard (Gröbner) base*, first introduced by Buchberger, cf. [11]. If no polynomial in this set can be reduced via some other polynomial in the set, we have a *reduced* standard base. Motivated by the following Theorem 1, we will take the liberty of introducing some new terminology: we will call standard (Gröbner) bases *complete generating sets* and reduced standard (Gröbner) bases, *complete bases.*

There are many (syntactical) characterizations for the complete sets, cf. [12]; among others, the one stating that S is a complete set if and only if every polynomial f reduces to a unique normal form (i.e a polynomial f' which cannot be further reduced) with respect to S.

Definition 3 *A set of polynomials S is a complete set if one of the following equivalent conditions is fulfilled:*
The relation $\to_S$ has the Church-Rosser property.
For all $f_i, f_j \in S$, $spoly(f_i, f_j) \to_S 0$.

(Here $spoly(f_i, f_j)$ is the *substraction polynomial* of f_i and f_j defined as the difference between the polynomials obtained by reducing the least common multiple of the leading power products of the two polynomials with respect to f_i and f_j.) The equivalence of the conditions in Definition 3. is known as the Buchberger's theorem. This result makes it possible to design an algorithm which accepts as input S' a finite set of polynomials and produces as output a complete set S such that $I[S] = I[S']$, cf. [11]. Hence the ideal membership problem is decidable.

We can now restate the ideal membership problem (cf. Definition 1) as:

Definition 4 *The ideal membership problem is decidable by reduction with respect to S if and only if it can be formulated in the following way:*

$$f \in I[S] \quad \text{if and only if} \quad f \to_S 0.$$

We note that if $f \to_S f'$ then the coset $\overline{f}_{I[S]}$ of f modulo $I[S]$ is an *invariant* with respect to the reduction relation $\to_S$, i.e

$$\overline{f}_{I[S]} = \overline{f'}_{I[S]}.$$

The following theorem states a necessary and sufficient condition for the solvability of the ideal membership problem.

Theorem 1 *The set S is a complete set for $I[S]$ if and only if the ideal membership problem is decidable by reduction with respect to S.*

Proof. Cf. [11].

4   Problem-Solving

We consider "problem-solving" in the sense used in artificial intelligence, as the discipline concerned with the development of general applicable intelligent strategies which implemented on computers can (help to) provide solutions by imitating human reasoning.

For defining what we mean by a problem we employ a similar to "state-operators" description, cf. [14]. Given a relational structure **A**, we first define a *state space* having as elements *states*. A state is defined by a finite set of formulas in **A** (e.g., in the example at the end of this section a state is an enumeration of formulas $\bigwedge_{ij} \phi_{ij}$ where $\phi_{ij} = 0$ if and only if $cover(i,j) = true$); sets of formulas defining distinct states differ only in the values of the constants appearing in the formulas. A *state construction problem* expressed in **A** consists of a couple $StConP = (StSp, ConR)$ where $StSp$ represents the state space, and $ConR$ the *set of construction rules*. The states in $StSp$ and the construction rules in $ConR$ are characterized by formulas involving only constants, relations and functions from **A**. Let $s, s' \in StSp$ and $cr \in ConR$. If rule cr applied to state s transforms it into state s' we say that s' is *reachable* from s. In order to establish an analogy with the reduction technique in symbolic computation described in Section 3, we will adopt the hypothesis that every rule is reversible in the sense that, if s' can be reached from s by cr then s can be reached from s' by applying the corresponding reversed rule r. We say that s' *reduces* to s via r, in symbols $s' \rightarrow_r s$, and call r a *reduction rule*. We will use $RedR$ to denote the set of reduction rules. A state s' can be reached from a state s if there exist a finite set of construction rules $\{cr_i\} \subseteq ConR$ which successively applied to s transforms it into s'. If this is the case, we will say that s' is in the *reachable space* of s, $s' \in ReSp[s]$.

In this context we will not be concerned with finding a solution (i.e a way to transform s in s' by specifying the construction rules $\{cr_i\}$ and the order in which these are applied, cf. [13]) but merely to state necessary and sufficient conditions for the existence of such a solution. Thus, we assume that a necessary and sufficient condition for state s' to be reached from state s is that s' can be reduced to s by successively applying reduction rules from $RedR$.

"Solving problems by heuristically guided, trial-and-error search in a space of possible solutions is a dominant theme in artificial-intelligence research." (cf. [14])

We propose a new approach to the subject by considering the study of necessary and sufficient conditions for a problem to be solvable, i.e:

Finding necessary and sufficient conditions for the solvability of the problems by heuristically guided, trial-and-error search for invariants corresponding to the reduction rules R.

One could name such an attempt: *Invariant Driven Reasoning.*

Definition 1' *The state reachability problem consists of a triple $StReP = (InSt, R, FinSt)$, where $InSt \subseteq StSp$ is the set of initial states, $FinSt \subseteq StSp$ is the set of final states, and $R \subseteq RedR$ is the set of admissible reduction rules.*

(For simplicity reasons we will consider the case in which $InSt = \{is\}$ and $FinS = \{fs\}$ are singletons and $R = ConR \cup RedR$; which corresponds to considering the transitive symmetric closure of the reduction (or construction relation), cf. [12].)

The problem can now be stated as: Decide if the final state fs can be reduced to the initial state is by successively applying reduction rules from R, i.e $fs \in ReSp[is]$.

We will introduce now the basic concept for the proposed approach. A formula ϕ expressed in **A** which holds in state s' is said to be an *invariant formula* for the set of reduction rules R if for all $s \in StSp$ and $r \in R$ such that $s' \rightarrow_r s$ the formula ϕ holds also in the state s, i.e $\phi(s) = \phi(s')$, where $\phi(s)$ denotes the *value* of the invariant formula ϕ in state s.

Let $\Phi = \{\phi_i\}$ be a set of invariants for problem $StReP$.

Definition 2' *A set of invariants Φ is compatible with the problem $StReP = (is, R, fs)$ if the values of the invariants in the initial state are respectively equal to the values of the invariants in the final state*

$$\phi_i(is) = \phi_i(fs) \quad \text{for all} \quad i.$$

Definition 3' *The set Φ is a complete set of invariants (or an invariant system) for the problem $StReP$ if every invariant formula of the problem $StReP$ can be expressed as a formula involving only invariants from Φ.*

Conceivably, the invariant system of a problem is not uniquely defined. Let Φ be an invariant system for the problem $StReP$.

Definition 4' *The state reachability problem $StReP$ is decidable by reduction with respect to Φ if and only if it can be formulated in the following way:*

The state reachability problem has a solution if and only if Φ is a compatible invariant system for the problem.

The analog of Theorem 1 is the following statement.

Theorem 1' *The set Φ is a complete set of invariants for the problem $StReP$ if and only if the state reachability problem $StReP$ is decidable by reduction.*

Proof. Suppose Φ is complete with respect to R and compatible with respect to $StReP = (is, R, fc)$, and that $StReP$ has no solution. Let $s \in StSp$ be an arbitrary state and consider the formula $\phi(s)$: $\exists R'$ ($R' \subseteq R$ finite; $fc \rightarrow_{R'} s$). Then, if $\phi(s)$ holds then $\phi(s')$ also holds for all states s' and rules r such that $s \rightarrow_r s'$, and if $\phi(s)$ does not hold then also $\phi(s')$ does not hold (recall that we have considered $\rightarrow_R$ as the symmetric transitive closure of the reduction relation).

This shows that ϕ is an invariant formula for R. But now $\phi(fs)$ holds (trivially) while $\phi(is)$ does not, since $StReP$ has no solution. This is a contradiction with the fact that Φ is a complete, compatible invariant system.

Suppose now that Φ is complete and $StReP$ is solvable. Then clearly Φ is compatible.

For the reverse implication of the theorem suppose $StReP$ is decidable by reduction with respect to R and Φ is not complete. Then we can construct an invariant function ϕ_0 such that $\phi_0(fc) \neq \phi_0(ic)$ (because Φ is not complete) and this would contradict the assumption that $StReP$ is solvable.

Example 2. Let $\mathbf{A} = \mathbb{N}$, the set of natural numbers. Consider a rectangular board with dimensions $a, b \in \mathbb{N}$, completely divided into squares of dimension 1. Suppose we have at our disposal at least $\lfloor a \cdot b/2 \rfloor$ domino tiles of dimensions 2×1. We can take away squares from the board without changing the position of the remaining squares obtaining in this way different configurations. The problem is:

Given a certain configuration, decide whether it can be completely covered with non-overlapping domino tiles such that only the squares present in the configuration are covered.

We will first restate the problem as a state construction problem $StConP$. For this, let $StSp$ be the state space corresponding to a given configuration, having as elements (states) different coverings of the given configuration (board) with domino tiles, and $ConR$ the set of construction rules containing just one rule cr: *put a domino tile on two adjacent not covered squares in the configuration* (adjacent means here that they have a common edge).

The state construction problem $StConP$ can now be formulated as:

Given a board configuration and an initial state (assume, for simplicity, it is the one in which no tile has been placed, thus it actually coincides with the board configuration) and a final state (assume it is the one in which the given configuration has been completely covered with tiles) decide whether, using the construction rule cr, the initial state can be transformed into the final one, and, if so, show how can this be performed.

The reduction rule corresponding to cr is the rule r: *remove a tile from two adjacent squares in the configuration.*

The corresponding state reachability problem $StReP$ can now be formulated as:

Given a board configuration, an initial and a final state (as in $StConP$) decide whether the final state can be reduced to the initial one.

Consider a coordinate system $\mathbb{N} \times \mathbb{N}$ with the origin in the lower left corner of the board. Then, each square is represented by a pair of natural numbers $(x, y) \in \mathbb{N} \times \mathbb{N}, 0 \leq x \leq a, 0 \leq y \leq b$. For each state $s \in StSp$, i.e partial covering with tiles of the initial configuration, let us define the following predicates:

$cover(x, y) = true$ if and only if the square (x, y) is covered by a tile in state s; $cover$ characterizes the state;

$sing(x, y) = true$ if and only if the square (x, y) is in a *singular position* in state s, i.e all adjacent squares are covered with tiles;

$even(x, y) = true$ if and only if $x + y \equiv 0 \pmod{2}$.

Let us denote by $N_{cover}, N_{sing}, N_{even}$ the number of positions (x, y) for which the corresponding predicates take the value *true* (in state s). These all are functions defined on $StSp$ with values in $\mathbb{N}$.

For the initial state is, let us define the predicate:

$in - config(x, y) = true$ if and only if the square at position (x, y) is in the board configuration; thus $in - config$ characterizes the configuration of the board and, since we have considered by hypothesis that the board configuration coincides with the initial state, it also characterizes the initial state. Clearly, $N_{in-config}$ represents then the number of positions (squares) in the initial state (board configuration). The solvability of $StReP$ is equivalent to the conditions: $N_{cover}(is) = 0$ and $N_{cover}(fs) = N_{in-config}$.

Let us now put into evidence some of the invariants. It is easy to see that $N_d = N_{even} - N_{\neg even}$ is an invariant formula for the reduction rule r because a tile can be removed only from two adjacent squares. For the same reason N_{sing} is also an invariant. Clearly these two functions are independent, in the sense that they can take any value (in their codomains) independently of one another. It is also clear that N_0 in $N_{cov} \equiv N_0 \pmod{2}$ is an invariant, but, since $N_{cover} = N_{even} + N_{\neg even} = 2N_{even} - N_d \equiv N_d \pmod{2}$, it is not an independent (invariant) function. Now, since $N_{cover}(is) = 0 \equiv 0 \pmod{2}$ it follows $N_0 = 0$, hence $N_{in-config} \equiv 0 \pmod{2}$ is a necessary condition. Further, since $N_d(fs) = 0$ and $N_{sing}(fs) = 0$ it follows that, for the set of invariants $\{N_d, N_{sing}\}$ to be compatible, $N_d(is) = N_{sing}(is) = 0$. However the set $\{N_d, N_{sing}\}$ is not a complete invariant set. We were not able to find such a set for the problem.

We should also point out here that, in general we cannot reasonably hope for the existence of an algorithm which, performing on $StRePs$ as input, will give an invariant system as output (even in the simple problem considered in the example, the relational structure we had to consider for expressing the invariants, contains as subsystem the number-theoretic formal system where incompleteness theorems hold, cf. [4]). This situation contrasts with the situation we had when we considered Buchberger's completion algorithm for polynomials. However, it is not difficult construct simple $StRePs$ where a complete set of invariants can be exhibited.

5   Conclusions and Further Research

We have presented the basic idea of a new approach to problem solving from a certain unifying general perspective as invariant driven reasoning. While there is no way to systematically search for invariants, (i.e computer programs that "produce" such invariant formulas) we can envision programs that could test for invariant properties. In general, no completion algorithm exist, but since we know examples (in symbolic computation) were such algorithms do exist, it seems worthy to undertake such a research also in other particular fields.

An attempt for describing Robinson's resolution procedure as an instance (model) of the completion principle was made in [10]. It would be of interest to restate the resolution method using the notions introduced in this article.

Another research direction could be, given a relational structure **A** in which we express a problem $StReP$, to embed **A** into a richer algebraic structure where the invariants are solutions (fixed points) of certain equations (operators). Such a program was realized by Engeler [15], by constructing a combinatory model for programmable functions defined on a relational structure.

We are aware of the fact that our sketch for a new approach to problem-solving does not (yet) stand a comparison with the more developed techniques as resolution, rule-based strategies, etc., and (still) lacks convincing examples. However, viewed as a natural extrapolation of already proven successful methods in symbolic computation, there is a reasonable hope that analogies could perform well.

References

[1] Lang, S., (1984). *Algebra*. Addison-Wesley.

[2] Rota G.-C., Sturmfels B., (1988). Introduction to Invariant Theory in Superalgebras. *Invariant Theory and Tableaux*, ed., Stanton, D., Springer.

[3] Kalman, R., Arbib, M., Falb, P., (1965). *Topics in Mathematical System Theory*, McGraw-Hill.

[4] Kleene, S., (1967). *Introduction to Metamathematics*. North-Holland.

[5] Pfalzgraf, J., (1991). Logical Fiberings and Polycontextural Systems. In: *Fundamentals of Artificial Intelligence Research*, eds., Jorrand, Ph., Kelemen, J., LNCS **535**, Springer.

[6] Jacobson, N., (1980). *Basic Algebra II*. Freeman.

[7] Jouannaud, J-P., (1990). Syntactic Theories. In: *Mathematical Foundations of Computer Science*, ed., Rovan, B., LNCS **452**, Springer.

[8] Dershowitz, N., Jouannaud, J.-P. (1990). Rewrite Systems. In: *Handbook of Theoretical Computer Science. Vol B: Formal Methods and Semantics*, ed., van Leeuwen, J., North-Holland.

[9] Goguen, J., (1986). What Is Unification? In: *Resolution of Equations in Algebraic Structures. Vol. 1: Algebraic Techniques*, eds., Nivat, M., Ait-Kaci, H., Academic Press.

[10] Buchberger, B., (1987). History and Basic Features of the Critical-pair/completion Procedure. *J. Symbolic Computation*, **3**.

[11] Buchberger, B., (1976). A Theoretical Basis for the Reduction of Polynomials to Canonical Form. *ACM SIGSAM Bull.*, **10**.

A Type-Coercion Problem in Computer Algebra

Andreas Weber

Wilhelm-Schickard-Institut
Universität Tübingen
W-7400 Tübingen, Germany
⟨weber@informatik.uni-tuebingen.de⟩

Abstract. An important feature of modern computer algebra systems is the support of a rich type system with the possibility of type inference.

Basic features of such a type system are polymorphism and coercion between types. Recently the use of order-sorted rewrite systems was proposed as a general framework.

We will give a quite simple example of a family of types arising in computer algebra whose coercion relations cannot be captured by a finite set of first-order rewrite rules.

Keywords: Computer algebra, type systems, subtyping, type coercion, type inference, order-sorted rewriting, universal algebra.

1 Introduction

Early computer algebra systems did not have a sophisticated type system. This is mainly due to the fact that the types occurring in traditional programming languages are not fully appropriate for algebraic computations.

With the progress in computer algebra the number of applications and so of computational domains grew. As a result of not having language constructs that aid in organization the larger systems are considered to be at or near the ceiling of their extendability. Therefore modern languages for symbolic computation come with a type concept [1], [2], and [3].

An important feature that has to be accomplished by a type system is the possibility of an automatic *type inference*. As an example, the user of a system would like to write down

$$x^2 + 3x + \frac{1}{2}$$

and the system should infer that this is a polynomial over the rationals.

Since the problem of type inference in computer algebra is largely predefined by mathematical practice, finding a general mechanism for doing type inference in computer algebra has turned out to be difficult. Some suggestions are given in [4], [5], [6], and [7]. The suggestions given in [4], [6], and [7] are also an attempt to give a safe theoretical foundation for the type inference facilities of an existing system as AXIOM[1] [1].

---

[1] AXIOM is a trademark of *The Numerical Algorithms Group Ltd.*

[12] Winkler, F., (1984). The Church-Rosser Property in Computer Algebra and Special Theorem Proving. *PhD Thesis*, University. of Linz.

[13] Huet, G., (1987). Deduction and Computation. In: *Logic of Programming and Calculi of Discrete Design*, ed., Broy, M., Springer.

[14] Nilsson, N., (1971). *Problem Solving Methods in Artificial Intelligence*. McGraw-Hill.

[15] Engeler, E., (1990). Combinatory Differential Fields, *Theoretical Computer Science* **72**.

A notion that is common to all these approaches is that of *coercion*. It should be possible that the system automatically coerces one type to another, e. g. an integer into a rational number or into an integral polynomial.

In [6] the coercion rules are interpreted as rewrite rules over an order-sorted algebra. Types are terms in an order-sorted algebra and term rewriting techniques are used for type inference. This approach seems to be very promising. By the use of type variables many coercion problems arising in a computer algebra system can be handled in a uniform way. Moreover, the use of rules guarantees an easy extendability for future applications. Since term rewriting techniques are well established, a transfer of useful results should be possible to gain practically applicable systems.

However, in a certain sense the approach in [6] is too general to be applicable to an important range of problems. If arbitrary (first-order) terms are allowed in the rewrite rules, the general type inference problem becomes undecidable. Therefore we study typical examples from computer algebra in order to find restrictions on the rewrite rules under which the type inference problem becomes decidable. Unfortunately there are not only examples which would suggest such restrictions. There is also a quite simple example of a family of types arising in computer algebra whose coercion relations cannot be captured by a finite set of first-order rewrite rules at all!

We will present this example in Sect. 4. A technical result which is needed to set up the example will be proved in Sect. 3.

2 Preliminaries

We will assume that the reader is familiar with the basic notions of rewrite systems as can be found in [8]. We will also need some concepts of universal algebra. A comprehensive reference is [9].

The set of non-negative integers will be denoted by $\mathbf{N}$. We will write $\aleph_0$ for the first infinite cardinal, the cardinality of $\mathbf{N}$. The set of strings over an alphabet L will be L^*, where ε is the empty string.

We will use Σ to denote a first-order signature. $\mathcal{E}, \mathcal{E}_0$ will be sets of equations over Σ, in which we will use $t, x, y, v_0, v_1, \ldots$ as variables. The *size of a term* is the number of function symbols occurring in it.

If an algebra $\mathcal{A}$ is a *model* of a set of equations $\mathcal{E}$, we will write $\mathcal{A} \models \mathcal{E}$. A set of equations $\mathcal{E}$ over Σ is *axiomatized* by a set of equations $\mathcal{E}_0$ iff for any algebra $\mathcal{A}$ of the type given by Σ, $\mathcal{A} \models \mathcal{E}$ implies $\mathcal{A} \models \mathcal{E}_0$. In this case we will write $\mathcal{E}_0 \models \mathcal{E}$. $\mathcal{E}$ is *finitely based* iff there is a finite set of equations $\mathcal{E}'$ which axiomatizes it.

Recall that for any cardinal α and any set of equations $\mathcal{E}$ the free algebra on α generators for the equational class defined by $\mathcal{E}$ exists (see [9, p. 167]). It can be constructed as the term algebra over the set of generators modulo the equivalence relation given by the equations and is unique up to isomorphism. We will call this algebra the *free model of α generators over $\mathcal{E}$.*

We will use the notation of [6]. Since we are basically dealing with terms of the same sort in the example (integral domains), the order-sorted framework collapses to a single sorted one. Thus we will just write down first-order terms to simplify notation.

3 A Technical Result

Definition 1. Let $f : \{P, F\}^* \longrightarrow \{P, F\}^*$ be the function, which is defined by the following algorithm:

If no F is occurring in the input string, then return the input string as output string. Otherwise, remove any F except the leftmost occurrence from the input string and return the result as output string.

Let $\equiv$ be the binary relation on $\{P, F\}^*$ which is defined by

$$\forall v, w \in \{P, F\}^* : v \equiv w \iff f(v) = f(w).$$

Obviously, the function f can be computed in linear time and the relation $\equiv$ is an equivalence relation on $\{P, F\}^*$.

Let Σ be the first-order signature consisting of the two unary function Symbols F and P. We will now lift the equivalence relation $\equiv$ to a set of equations over Σ.

Definition 2. Let $\mathcal{E}$ be the following set of equations:

$$\mathcal{E} = \{ \ S_1(S_2(\cdots S_k(x) \cdots)) = S_{k+1}(S_{k+2}(\cdots S_r(x) \cdots)) :$$
$$S_i \in \{F, P\} \ (1 \leq i \leq r) \text{ and } S_1 S_2 \cdots S_k \equiv S_{k+1} S_{k+2} \cdots S_r \ \}$$

Theorem 3. $\mathcal{E}$ is not finitely based, i.e. there is no finite set of axioms for $\mathcal{E}$.

Proof. Assume towards a contradiction that there is such a finite set $\mathcal{E}_0$. Let $\mathcal{M}$ be the free model of $\aleph_0$ generators over $\mathcal{E}$ and let $\mathcal{M}_0$ be the free model of one generator over $\mathcal{E}_0$.

Except for a possible renaming of the variable symbol x, $\mathcal{E}_0$ has to be a subset of $\mathcal{E}$. Otherwise, $\mathcal{E}_0$ would contain an equation of the form

$$S_1(S_2(\cdots S_k(x) \cdots)) = S_{k+1}(S_{k+2}(\cdots S_r(y) \cdots)),$$

or of the form

$$S_1(S_2(\cdots S_k(x) \cdots)) = S_{k+1}(S_{k+2}(\cdots S_r(x) \cdots)), \ S_1 S_2 \cdots S_k \not\equiv S_{k+1} S_{k+2} \cdots S_r.$$

However, none of these equations holds in $\mathcal{M}$.

Now let $n \in \mathbf{N}$ be the maximal size of a term in $\mathcal{E}_0$. Then the equation

$$F(\underbrace{P(P(\cdots(P(x))\cdots)))}_{n} = F(P(F(\underbrace{P(P(\cdots(P(x))\cdots)))}_{n-1}))$$

holds in $\mathcal{M}$, but it does not hold in $\mathcal{M}_0$. $\qquad\square$

4 A Type-Coercion Problem

If R is an integral domain, we can form the field of fractions $\mathrm{FF}(R)$. We can also built the ring of univariate polynomials in the indeterminate x which we will denote by $\mathrm{UP}(R, x)$ — the ring of polynomials $R[x]$ in the standard mathematical notation — which is again an integral domain by a Lemma of Gauß. Thus we can also built the field of fractions of $\mathrm{UP}(R, x)$, $\mathrm{FF}(\mathrm{UP}(R, x))$ — the field of rational functions $R(x)$.

Starting from an integral domain R we will always get an integral domain and can repeatedly built the field of fractions and the ring of polynomials in a "new" indeterminate.

Thus if a computer algebra system has a fixed integral domain R and names for symbols $x_0, x_1, x_2 \ldots$, it should also provide types of the form

1. R,
2. FF(R),
3. UP(R, x_0),
4. UP(FF(R), x_0),
5. FF(UP(R, x_0)),
6. UP(UP(R, x_0), x_1),
7. UP(FF(UP(R, x_0)), x_1),
8. FF(UP(UP(R, x_0), x_1)),
9. FF(UP(FF(UP(R, x_0)), x_1),
10. UP(UP(UP(R, x_0), x_1), x_2),

$\vdots$

It is convenient to use the same symbols for a mathematical object and the symbolic expression which denotes the object. In order to clarify things we will sometimes use additional $\langle\!\langle \cdot \rangle\!\rangle$ for the mathematical objects.

There are canonical embeddings from an integral domain into its field of fractions and into the ring of polynomials in one indeterminate (an element is mapped to the corresponding constant polynomial).

It is common mathematical practice to identify the integral domain with its image under these embeddings. Thus the type system should also provide a coercion between these types.

If t is a type variable which ranges over "integral domains" and x is a "symbol", this property can be expressed by the rules

$$t \rightarrow \mathrm{FF}(t)$$

and

$$t \rightarrow \mathrm{UP}(t, x)$$

in the framework of [6].

However, not all of the types built by the type constructors FF and UP should be regarded to be different. If the integral domain R happens to be a field, then R will be isomorphic to its field of fractions. Especially, for any integral domain R, $\langle\!\langle \mathrm{FF}(R) \rangle\!\rangle$ and $\langle\!\langle \mathrm{FF}(\mathrm{FF}(R)) \rangle\!\rangle$ are isomorphic.

The fact that also $\langle\langle FF(FF(R))\rangle\rangle$ can be embedded in $\langle\langle FF(R)\rangle\rangle$ can be expressed by a rule

$$FF(FF(t)) \rightarrow FF(t),$$

which is one of the examples given in [6, p. 354].

But there are more isomorphisms which govern the relations of this family of types.

If we assume that an application of the type constructor UP always uses a "new" indeterminate as its second argument, any application of the type constructor FF except the outermost one application is redundant.

This observation will be captured by the following formal treatment. In order to avoid the technical difficulty of introducing "new" indeterminates, we will use an unary type constructor up instead the binary UP. The intended meaning of $up(t)$ is $UP(t, x_n)$, where x_n is a new symbol, i.e. not occurring in t.

Definition 4. Define a function trans from $\{F, P\}^*$ into the set of types recursively by the following equations. For $w \in \{F, P\}^*$,

- $\text{trans}(\varepsilon) = R,$
- $\text{trans}(Fw) = FF(\text{trans}(w)),$
- $\text{trans}(Pw) = up(\text{trans}(w)).$

If we take $\langle\langle R\rangle\rangle$ to be the ring of integers, the following lemma will be an exercise in elementary calculus.[2]

Lemma 5. *Let $\langle\langle R\rangle\rangle$ be the ring of integers. For any $v, w \in \{F, P\}^*$, the integral domains $\langle\langle \text{trans}(v)\rangle\rangle$ and $\langle\langle \text{trans}(w)\rangle\rangle$ are isomorphic iff $v \equiv w$.*

Moreover, $\langle\langle \text{trans}(v)\rangle\rangle$ can be embedded in $\langle\langle \text{trans}(w)\rangle\rangle$ and $\langle\langle \text{trans}(w)\rangle\rangle$ can be embedded in $\langle\langle \text{trans}(v)\rangle\rangle$ iff $\langle\langle \text{trans}(v)\rangle\rangle$ and $\langle\langle \text{trans}(w)\rangle\rangle$ are isomorphic.

The "simplifications" of rational expressions are some of the most frequently used operations in many computer algebra systems. Very often, they are just concrete implementations of the embeddings of Lemma 5.

Thus it would be of practical interest if these coercions between types could also be captured within the type system. However, this cannot be done by the use of a term rewriting system over the corresponding signature, since the finite set of rules corresponds to a finite set of equations only.

Theorem 6. *Let Σ be the signature consisting of the unary function symbols FF and up and the constant R. Let $\langle\langle R\rangle\rangle$ be the ring of integers.*

Then there is no finite set of Equations $\mathcal{E}'$ over Σ, such that for ground terms t_1 and t_2 the following holds.

$$\mathcal{E}' \models \{t_1 = t_2\} \iff \langle\langle t_1\rangle\rangle \text{ and } \langle\langle t_2\rangle\rangle \text{ are isomorphic.}$$

Proof. If t_1 and t_2 are ground terms, then there are $v, w \in \{F, P\}^*$ such that $t_1 = \text{trans}(v)$ and $t_2 = \text{trans}(w)$. Now we are done by Lemma 5 and Theorem 3. $\square$

---

[2] If we started with the ring of polynomials in infinitely many indeterminates over some domain, then there would be additional isomorphisms.

The problem is that the equational theory which describes the coercion relations in the example we gave is not finitely based. Since this property of an equational theory is *equivalence-invariant* in the sense of [9, p. 382], the use of another signature for describing the types does not help.

5 Conclusion

The use of rewrite rules seems to be a promising way to describe the coercions between types which occur in a computer algebra system. Important examples can be nicely described in a very short way and it is possible to extend a system by simply adding new rules.

Unfortunately there are simple examples of families of types arising in computer algebra whose coercion relations cannot be captured by a finite set of first-order rewrite rules at all. In the present paper we have given a quite simple example of such a system, whose behavior on the object-level can be handled by many existing computer algebra systems.

It is of major practical importance to have a safe and reliable interaction between various parts of a computer algebra system. Since many parts of the type system of such a system can be nicely described by rewrite rules, it would be useful if the example that we have given could be incorporated in this framework. This seems to be possible. The relations of the types can be described by means of the equational theory $\mathcal{E}$ of Definition 2. Since equivalence under $\mathcal{E}$ is decidable, it is possible to use methods of *class-rewriting* (see [8, Sect. 2.5]).

For most practical purposes it will be sufficient to distinguish only between the integers, the rationals, polynomials (in arbitrary many indeterminates) over the integers, polynomials (in arbitrary many indeterminates) over the rationals, and rational functions (in arbitrary many indeterminates) over the integers. This approach is taken in the type system of AXIOM.

Acknowledgments. I am indebted to R. Loos, R. Bündgen, and F. Haug for helpful discussions. The suggestions of an anonymous referee helped to improve the paper.

References

1. R. S. Sutor and R. D. Jenks. The type inference and coercion facilities in the Scratchpad II interpreter. *ACM SIGPLAN Notices*, 22(7):56–63, 1987. SIGPLAN '87 Symposium on Interpreters and Interpretive Techniques.
2. J. H. Davenport and B. M. Trager. Scratchpad's view of algebra I: Basic commutative algebra. In Miola [10], pages 40–54.
3. S. K. Abdali, G. W. Cherry, and N. Soiffer. A Smalltak system for algebraic manipulation. *ACM SIGPLAN Notices*, 21(11):277–283, November 1986. OOPSLA '86 Conference Proceedings, Portland, Oregon.
4. D. L. Rector. Semantics in algebraic computation. In E. Kaltofen and S. M. Watt, editors, *Computers and Mathematics*, pages 299–307, Massachusetts Institute of Technology, June 1989. Springer-Verlag.

5. G. Baumgartner and R. Stansifer. A proposal to study type systems for computer algebra. Technical Report 90-07.0, Research Institute for Symbolic Computation Linz, A-4040 Linz, Austria, March 1990.

6. H. Comon, D. Lugiez, and Ph. Schnoebelen. A rewrite-based type discipline for a subset of computer algebra. *Journal of Symbolic Computation*, 11:349–368, 1991.

7. A. Fortenbacher. Efficient type inference and coercion in computer algebra. In Miola [10], pages 56–60.

8. N. Dershowitz and J.-P. Jouannaud. Rewrite systems. In J. van Leeuwen, editor, *Formal Models and Semantics*, volume B of *Handbook of Theoretical Computer Science*, chapter 6, pages 243–320. Elsevier, Amsterdam, 1990.

9. G. Grätzer. *Universal Algebra*. Springer-Verlag, New York-Heidelberg-Berlin, second edition, 1979.

10. A. Miola, editor. *Design and Implementation of Symbolic Computation Systems (DISCO '90)*, volume 429 of *Lecture Notes in Computer Science*, Capri, Italy, April 1990. Springer-Verlag.

Algorithmic Development of Power Series

Wolfram Koepf

Fachbereich Mathematik, Freie Universität Berlin
Arnimallee 3, W-1000 Berlin 33, Germany

Abstract. There is a one-to-one correspondence between formal power series (FPS) $\sum_{k=0}^{\infty} a_k x^k$ with positive radius of convergence and corresponding analytic functions. Since a goal of Computer Algebra is to work with formal objects and preserve such symbolic information, it should be possible to automate conversion between these forms in Computer Algebra Systems (CASs). However, only MACSYMA provides a rather limited procedure powerseries to calculate FPS from analytic expressions in certain special cases.

We present an algorithmic approach to compute an FPS, which has been implemented by the author and A. Rennoch in MATHEMATICA, and by D. Gruntz in MAPLE. Moreover, the same algorithm can be reversed to calculate a function that corresponds to a given FPS, in those cases when an initial value problem for a certain ordinary differential equation can be solved.

Further topics of application like infinite summation, and asymptotic expansion are presented.

1 Introduction

We consider formal power series (FPS) of the form

$$F := \sum_{k=0}^{\infty} a_k x^k$$

with coefficients $a_k \in \mathbb{C}$ ($k \in \mathbb{N}_0$). All the algebraic operations for FPS like addition, multiplication, division, and substitution, can be done by finite algorithms if one truncates the resulting FPS, i.e. only evaluates the first N coefficients of it (where N is an arbitrary fixed positive integer), which gives a truncated power series. These algorithms are implemented in certain Computer Algebra Systems (CAS), e.g. in AXIOM[1], MACSYMA[2], MAPLE[3], MATHEMATICA[4], and REDUCE[5] (see [1], [10], [11], [16], and [7], respectively).

---

[1] AXIOM is a trademark of the Numerical Algorithms Group Ltd.
[2] MACSYMA is a registered trademark of Symbolics, Inc.
[3] MAPLE is a registered trademark of Waterloo Maple Software.
[4] MATHEMATICA is a registered trademark of Wolfram Research, Inc.
[5] REDUCE is a trademark of the RAND Corp.

Moreover, all CAS provide a procedure to find a truncated power series expansion for a function f. By Taylor's Theorem the power series coefficients of a function f can be calculated by the formula

$$a_k := \frac{f^{(k)}(0)}{k!} \quad ,$$

which provides an algorithmic procedure to calculate a truncated power series of certain degree N. The Taylor algorithm, however, does not generate an explicit formula for a_k. Moreover it generally has exponential complexity in the order N: The differentiation of a product by the product rule may generate 2^N summands. Furthermore in the general case, e.g. for $f(x) := \frac{\sin x}{x}$, the evaluations $f^{(k)}(0)$ must be replaced by limits

$$a_k := \frac{\lim\limits_{x \to 0} f^{(k)}(x)}{k!}$$

whose evaluation are, in general, of exponential complexity as well. Moreover, the larger the number N, the larger the chance that the CAS even fails to evaluate these limits, when the expressions $f^{(k)}$ get more and more complicated.

For the following derivative free approach to generate the Taylor coefficients a_k of a function f recursively which is purely based on limits, similar restrictions apply. If

$$f_k(x) := \begin{cases} f(x) & \text{if } k = 0 \\ \dfrac{f_{k-1}(x) - \lim\limits_{x \to 0} f_{k-1}(x)}{x} & \text{if } k \in \mathbb{N} \end{cases} \quad ,$$

then the Taylor coefficients are given by

$$a_k := \lim_{x \to 0} f_k(x) \quad .$$

Most decisive, however, is the fact that none of these algorithms leads to a formula for a_k, i.e. the formal transformation "function expression of variable x towards coefficient expression of variable k" cannot be supported.

Thus in existing CAS the work with power series is restricted to truncated power series. Some of the systems, like AXIOM (previously SCRATCHPAD) and MAPLE, internally work with streams, and lazy evaluation, i.e. series objects are given by a finite number of initial terms, and an (internally used) formula to calculate further coefficients, see e.g. [12]. Infinite series representations, however, are not supported in these systems, either.

We shall give an outline of how to resolve these issues for FPS of some special types, including many special functions. Our procedure then produces the exact formal result, i.e. an explicit formula for the coefficients a_k. Obviously this solves the complexity problem, and as only a small number of limits have to be found, the chance to succeed is even larger than in the truncated case.

2 Laurent-Puiseux Series of Hypergeometric Type

We require an assumption that every FPS F has positive radius of convergence $r := 1/\limsup_{k\to\infty} |a_k|^{1/k}$. In this situation the FPS represents an analytic function $f(x) = \sum_{k=0}^{\infty} a_k x^k =: F$ in its disk of convergence $\mathbb{D}_r := \{x \in \mathbb{C} \mid |x| < r\}$, i.e. its sum converges locally uniformly in $\mathbb{D}_r$ to f. So there is a one-to-one correspondence between the functions f analytic at the origin and the FPS F with positive radius of convergence represented by their coefficient sequences $(a_k)_{k\in\mathbb{N}_0}$. We denote this correspondence by $f \leftrightarrow F$. As we are interested in the conversion $f \leftrightarrow F$ the restriction to FPS with positive radius of convergence makes sense even though algebraically this restriction is not necessary.

To deal with many special functions, it is a good idea to consider the *(generalized) hypergeometric series*

$$
{}_pF_q \left(\begin{matrix} a_1 \; a_2 \; \cdots \; a_p \\ b_1 \; b_2 \; \cdots \; b_q \end{matrix} \bigg| x \right) := \sum_{k=0}^{\infty} \frac{(a_1)_k \cdot (a_2)_k \cdots (a_p)_k}{(b_1)_k \cdot (b_2)_k \cdots (b_q)_k \, k!} x^k \tag{1}
$$

where $(a)_k$ denotes the *Pochhammer symbol* (or *shifted factorial*) defined by

$$
(a)_k := \begin{cases} 1 & \text{if } k = 0 \\ a \cdot (a+1) \cdots (a+k-1) & \text{if } k \in \mathbb{N} \end{cases}
$$

Note that $\frac{(a)_k}{k!} = \binom{a+k-1}{k}$, where $\binom{\alpha}{k}$ is the *binomial coefficient*

$$
\binom{\alpha}{k} := \begin{cases} 1 & \text{if } k = 0 \\ \frac{\alpha \cdot (\alpha-1) \cdots (\alpha-k+1)}{k!} & \text{if } k \in \mathbb{N} \end{cases} ,
$$

and $k!$ denotes the *factorial*

$$
k! := \begin{cases} 1 & \text{if } k = 0 \\ 1 \cdot 2 \cdots k & \text{if } k \in \mathbb{N} \end{cases} .
$$

The coefficients A_k of the hypergeometric series $\sum_{k=0}^{\infty} A_k x^k$ are the unique solution of the special recurrence equation (RE)

$$
A_{k+1} := \frac{(k+a_1) \cdot (k+a_2) \cdots (k+a_p)}{(k+b_1) \cdot (k+b_2) \cdots (k+b_q)(k+1)} \cdot A_k \quad (k \in \mathbb{N})
$$

with the initial condition

$$
A_0 := 1 .
$$

Note that $\frac{A_{k+1}}{A_k}$ is rational in k. Moreover if $\frac{A_{k+1}}{A_k}$ is a rational function $R(k)$ in the variable k then the corresponding function f is connected with a hypergeometric series; i.e., if $k = -1$ is a pole of R, then f corresponds to a hypergeometric series evaluated at some point Ax (where A is the quotient of the leading coefficients of the numerator and the denominator of R); whereas, if $k = -1$ is no pole of R, then f may be furthermore shifted by some factor x^s ($s \in \mathbb{Z}$).

We further mention that the function f corresponding to the hypergeometric series

$$f \leftrightarrow F := {}_pF_q \left(\begin{array}{cccc} a_1 & a_2 & \cdots & a_p \\ b_1 & b_2 & \cdots & b_q \end{array} \middle| x \right)$$

satisfies the differential equation (DE)

$$\theta(\theta + b_1 - 1) \cdots (\theta + b_q - 1)f = x(\theta + a_1) \cdots (\theta + a_p)f \qquad (2)$$

where θ is the differential operator $x\frac{d}{dx}$. An inspection of the hypergeometric DE (2) shows that it is of the form ($Q := \max\{p, q\} + 1$)

$$\sum_{j=0}^{Q} \sum_{l=0}^{Q} c_{lj} x^l f^{(j)} = 0 \qquad (3)$$

with certain constants $c_{lj} \in \mathbb{C}$ ($l, j = 0, \ldots, Q$). Because of their importance in our development, we call a DE of the form (3), i.e. a homogeneous linear DE with polynomial coefficients, *simple*. We will show the existence of a simple DE for a more general family of functions.

It is remarkable that many elementary functions can be represented by hypergeometric series.

$$(1+x)^\alpha \leftrightarrow {}_1F_0(-\alpha \mid x), \quad e^x \leftrightarrow {}_0F_0(x), \quad -\ln(1-x) \leftrightarrow x \cdot {}_2F_1 \left(\begin{array}{cc} 1 & 1 \\ & 2 \end{array} \middle| x \right),$$

$$\sin x \leftrightarrow x \cdot {}_0F_1 \left(3/2 \middle| -\frac{x^2}{4} \right), \qquad \cos x \leftrightarrow {}_0F_1 \left(1/2 \middle| -\frac{x^2}{4} \right),$$

$$\arcsin x \leftrightarrow x \cdot {}_2F_1 \left(\begin{array}{cc} 1/2 & 1/2 \\ & 3/2 \end{array} \middle| x^2 \right), \qquad \arctan x \leftrightarrow x \cdot {}_2F_1 \left(\begin{array}{cc} 1/2 & 1 \\ & 3/2 \end{array} \middle| -x^2 \right).$$

Note that an FPS of the form $F(x^m)$ is called *m-fold symmetric*. Even functions are 2-fold symmetric and odd functions are shifted 2-fold symmetric functions.

By the above examples one is led to the following more general definition. First we extend the considerations to *formal Laurent-Puiseux series* (LPS) with a representation

$$F := \sum_{k=k_0}^{\infty} a_k x^{k/n} \qquad (a_{k_0} \neq 0) \qquad (4)$$

for some $k_0 \in \mathbb{Z}$, and $n \in \mathbb{N}$. LPS are formal Laurent series, evaluated at $\sqrt[n]{x}$. A formal Laurent series ($n = 1$) is a shifted FPS, and corresponds to a meromorphic f with a pole of order $-k_0$ at the origin. The number n in development (4) is called the *Puiseux number* of (the given representation of) f.

Definition (Functions of hypergeometric type). An LPS F with representation (4) — as well as its corresponding function f — is called to be *of* *hypergeometric type* if it has a positive radius of convergence, and if its coefficients a_k satisfy a RE of the form

$$a_{k+m} = R(k)\, a_k \quad \text{for } k \geq k_0 \tag{5}$$
$$a_k = A_k \qquad \text{for } k = k_0, k_0 + 1, \ldots, k_0 + m - 1$$

for some $m \in \mathbb{N}$, $A_k \in \mathbb{C}$ ($k = k_0 + 1, k_0 + 2, \ldots, k_0 + m - 1$), $A_{k_0} \in \mathbb{C} \setminus \{0\}$, and some rational function R. The number m is then called the *symmetry number* of (the given representation of) F. A RE of type (5) is also called to be of hypergeometric type. $\triangle$

We want to emphasize that the above terminology of functions of hypergeometric type is definitely more general than the terminology of a generalized hypergeometric function. It covers e.g. the function $\sin x$ which is *not* a generalized hypergeometric function as obviously no RE of the type (5) holds for its series coefficients with $m = 1$. So $\sin x$ is not of hypergeometric type with symmetry number 1; it is, however, of hypergeometric type with symmetry number 2. A more difficult example of the same kind is the function $e^{\arcsin x}$ which is neither even nor odd, and nevertheless turns out to be of hypergeometric type with symmetry number 2, too (see [8], Sect. 9). Further functions like $\frac{\sin x}{x^5}$ are covered by the given approach. Moreover the terminology covers composite functions like $\sin \sqrt{x}$, which do not have a Laurent, but a Puiseux series development.

If F is m-fold symmetric, then there is a hypergeometric representation (5) with symmetry number m, whereas such a representation does not guarantee any symmetry. In fact, if F is of hypergeometric type with symmetry number j, then it is of hypergeometric type with each multiple mj ($m \in \mathbb{N}$) of m as symmetry number since by induction we get the RE

$$a_{k+jm} = R(k)\, R(k+j)\, R(k+2j) \cdots R(k+(m-1)j)\, a_k \ ,$$

and $R(k)\, R(k+j)\, R(k+2j) \cdots R(k+(m-1)j)$ is rational, too. In particular, each hypergeometric type function with symmetry number $j = 1$ is of hypergeometric type for arbitrary symmetry number m.

On the other hand, it is clear that each LPS with symmetry number m, and Puiseux number n, can be represented as the sum of nm shifted m-fold symmetric functions.

The derivative and the antiderivative of an LPS given by (4) are given by the rules

$$F' := \frac{1}{n} \sum_{k=k_0}^{\infty} k a_k x^{k/n-1} = \frac{1}{n} \sum_{k=k_0-1}^{\infty} (k+1) a_{k+1} x^{k/n} \ ,$$

$$\int F := \sum_{\substack{k=k_0 \\ k \neq -1}}^{\infty} \frac{a_k}{k/n+1} x^{k/n+1} + a_{-1} \ln x = n \sum_{\substack{k=k_0+1 \\ k \neq 0}}^{\infty} \frac{a_{k-1}}{k} x^{k/n} + a_{-1} \ln x \ .$$

Now we give a list of transformations on LPS that preserve the hypergeometric type.

Lemma *Let F be an LPS of hypergeometric type with representation* (4). *Then*

(a) $x^j F$ $(j \in \mathbb{N})$,   (b) F/x^j $(j \in \mathbb{N})$,   (c) $F(Ax)$ $(A \in \mathbb{C})$,

(d) $F\left(x^{\frac{p}{q}}\right)$ $(p, q \in \mathbb{N})$,   (e) $\dfrac{F(x) \pm F(-x)}{2}$,   (f) F',

are of hypergeometric type, too. If $a_{-1} = 0$, then also

(g) $\int F$ *is of hypergeometric type.*   □

For a proof of this Lemma we refer to ([8], Lemma 2.1, Theorem 8.1). We note that as $\cos x = \frac{e^{ix} + e^{-ix}}{2}$ and $\sin x = \frac{e^{ix} - e^{-ix}}{2i}$, a combination of (c) and (e) shows that the hypergeometric type of $\cos x$ and $\sin x$ follows from that of the exponential function.

We remark further that one can extend the definition of functions of hypergeometric type to include also the functions defined in (g) for arbitrary LPS (see [8], Sect. 8). Note that because of the logarithmic terms these functions, in general, do not represent LPS.

It is essential for our development that functions of hypergeometric type satisfy a simple DE.

Theorem *Each LPS of hypergeometric type satisfies a simple DE.*

Proof. Let F be given by

$$F := \sum_{k=k_0}^{\infty} a_k x^{k/n} \qquad (a_{k_0} \neq 0) \ .$$

Define the differential operator $\theta_n := nx\frac{d}{dx}$ working on a function of variable x. θ_n has the property

$$\theta_n F = \sum_{k=k_0}^{\infty} k a_k x^{k/n} \ ,$$

and by induction for all $j \in \mathbb{N}$

$$\theta_n^j F = \sum_{k=k_0}^{\infty} k^j a_k x^{k/n} \ .$$

This shows that moreover, if P is any polynomial, we may formally write

$$P(\theta) F = \sum_{k=k_0}^{\infty} P(k) a_k x^{k/n} \ . \tag{6}$$

This commuting property is the reason why the differential operator θ_n is much more appropriate for the current discussion than the usual differential operator $\frac{d}{dx}$.

If the representation of F has symmetry number m, we know that there is also a representation with symmetry number nm

$$Q(k) a_{k+nm} = P(k) a_k \tag{7}$$

with two polynomials P, and Q, and we may assume without loss of generality that the polynomials P and Q are chosen such that $Q(k_0-1) = Q(k_0-2) = \cdots = Q(k_0-nm) = 0$. This goal can be reached by multiplying both P and Q with the factors $(k-k_0+j)$ $(j = 1, \ldots, nm)$. From (6) and (7) we get

$$Q(\theta_n - nm) F = \sum_{k=k_0}^{\infty} Q(k-nm) a_k x^{k/n} \qquad \text{by (6), as } Q \text{ is a polynomial}$$

$$= \sum_{k=k_0+nm}^{\infty} Q(k-nm) a_k x^{k/n} \quad \text{as } Q(k_0-1) = \cdots = Q(k_0-nm) = 0$$

$$= \sum_{k=k_0}^{\infty} Q(k) a_{k+nm} x^{k/n+m} \qquad \text{by an index shift}$$

$$= x^m \sum_{k=k_0}^{\infty} P(k) a_k x^{k/n} \qquad \text{by (7)}$$

$$= x^m P(\theta) F \qquad \text{by (6) again.}$$

This represents a DE for F which turns out to be of form (3). Note that for $m = 1$, $n = 1$, and $k_0 = 0$ we have exactly (2). $\qquad\qquad\qquad\square$

Now assume, a function f representing an LPS is given. In order to find the coefficient formula, it is a reasonable approach to search for its DE, to transfer this DE into its equivalent RE, and you are done by an adaption of the coefficient formula for the hypergeometric function corresponding to the transformations given in the Lemma. It turns out that this procedure can be handled by an algebraic algorithm.

3   The First Conversion Procedure

There are two obvious transformation procedures: $f \mapsto F$ and $F \mapsto f$. At the moment we want to emphasize on the first situation. This transformation procedure $f \mapsto$ powerseries$(f, x, x0)$ is implemented in MACSYMA (see [10]). The implementation in MACSYMA, however, uses heuristic maneuvers, and is not based on an algorithm. It fails to convert many important functions with a simple development like e.g.

```
powerseries(1/(x^2+2*x+2),x,0),  powerseries(1/(x^2-2*x-2),x,0),
powerseries(atan(x),x,b),         powerseries(atan(x+a),x,0),
powerseries(exp(x)*exp(y),x,0),  powerseries(exp(x)*sin(x),x,0).
```

Here we present an algorithm corresponding to a function call of the form PowerSeries$[f,x,0]$ — we use MATHEMATICA syntax as our implementation is written in MATHEMATICA language (see [16]) — i.e. producing a Laurent-Puiseux series expansion of the function f with respect to the variable x.

Algorithm (for `PowerSeries[f,x,0]`)

(1) **Rational functions** (for details, see ([8], Sect. 4))

If f is rational in x, then use the *rational algorithm* by calculating a *complex* partial fraction decomposition of f (see e.g. [14], p. 171) which can be algorithmically done at least if the denominator has a rational factorization. Finally expand termwise by the binomial series.

(2) **Find a simple DE** (for details, see ([8], Sect. 5), where we prove that this procedure always succeeds in finding the simple DE of lowest degree for f)

(a) Fix a number $N_{\max} \in \mathbb{N}$, the maximal order of the DE searched for; a suitable value is $N_{\max} := 4$.

(b) Set $N := 1$.

(c) Calculate $f^{(N)}$; *either*, if the derivative $f^{(N)}$ is rational, apply the *rational algorithm*, and integrate;

(d) *or* find a simple DE for f of order N

$$\sum_{j=0}^{N} p_j f^{(j)} = 0$$

where p_j $(j = 0, \ldots, N)$ are polynomials in the variable x. Therefore decompose the expression

$$f^{(N)}(x) + A_{N-1} f^{(N-1)}(x) + \cdots + A_0 f(x)$$

in elementary summands (with respect to the constants $A_0, \ldots, A_{N-1}$). Test, if the summands contain exactly N rationally independent expressions. (Two expressions are called rationally independent if their ratio is not rational.) Just in that case there exists a solution as follows: Sort with respect to the rationally independent terms and create a system of linear equations by setting their coefficients to zero. Solve this system for the numbers $A_0, A_1, \ldots, A_{N-1}$. Those are rational functions in x, and there exists a unique solution. After multiplication by the common denominator of $A_0, A_1, \ldots, A_{N-1}$ you get the DE searched for.

(e) If (d) was not successful, then increase N by one, and go back to (c), until $N = N_{\max}$.

(3) **Find the corresponding RE** (for details, see ([8], Sect. 6))

Suppose you found a simple DE in step (2), then transfer it into a RE for the coefficients a_k. The RE is then of the special type

$$\sum_{j=0}^{M} P_j a_{k+j} = 0 , \tag{8}$$

where P_j $(j = 0, \ldots, M)$ are polynomials in k, and $M \in \mathbb{N}$. This is done by the substitution

$$x^l f^{(j)} \mapsto (k+1-l)_j \cdot a_{k+j-l} \tag{9}$$

into the DE.

(4) **Type of RE** (for details, see ([8], Sect. 7))

Determine the type of the RE according to the following list

(a) If the RE (8) contains only two summands then f is of hypergeometric type, and an explicit formula for the coefficients can be found by the hypergeometric coefficient formula (1), and some initial conditions.

(b) If the DE has constant coefficients $(c_j \in \mathbb{C} \ (j = 0, \ldots, N))$

$$\sum_{j=0}^{N} c_j f^{(j)} = 0 \ ,$$

then f is of *exp-like type*. In this case the substitution $b_k := k! \cdot a_k$ leads to the RE

$$\sum_{j=0}^{N} c_j b_{k+j} = 0 \ ,$$

which has the same constant coefficients as the DE, and can be solved by a known algebraic scheme using the first N initial coefficients.

(c) If the RE is none of the above types, try to solve it by other known RE solvers (a few of which are implemented in the MATHEMATICA package `DiscreteMath'RSolve'`, e.g., and in MAPLE's `rsolve`).

The details of the single parts of the algorithm are presented in [8]. Here we prefer to give some examples for the use of the algorithm.

1. **(Exp-like type case)** Suppose $f(x) = e^x \sin x$, so $f' = e^x \left(\sin x + \cos x \right)$ and $f'' = 2e^x \cos x$. The first step of the algorithm does not apply. In the second step for $N := 1$ the expression $A_0 = -\left(1 + \cot x \right)$ is not rational in x. For $N := 2$ we get the expression

$$f'' + A_1 f' + A_0 f = 2e^x \cos x + A_1 e^x \left(\sin x + \cos x \right) + A_0 e^x \sin x \ .$$

Under the summands

$$2e^x \cos x \ , \quad A_1 e^x \sin x \ , \quad A_1 e^x \cos x \ , \quad A_0 e^x \sin x$$

there are exactly the two rationally independent terms $e^x \cos x$ and $e^x \sin x$. We set the coefficient sums of these expressions to zero. The linear equations system

$$2 + A_1 = 0 \ , \qquad A_1 + A_0 = 0$$

has the solution $A_1 = -2$, $A_0 = 2$, and leads so to the DE for f

$$f'' - 2f' + 2f = 0 \ .$$

This DE has constant coefficients, so f is of exp-like type. For $b_k := k! \, a_k$ we have the RE

$$b_{k+2} - 2b_{k+1} + 2b_k = 0 \ .$$

The initial conditions are

$$b_0 = a_0 = f(0) = 0 , \quad \text{and} \quad b_1 = a_1 = \frac{f'(0)}{1!} = 1 .$$

For a RE with constant coefficients the setup $b_k := \lambda^k$ leads to a solution. Possible values for λ are solutions of the equation $\lambda^{k+2} - 2\lambda^{k+1} + 2\lambda^k = 0$, or solutions of the equivalent *characteristic equation*

$$\lambda^2 - 2\lambda + 2 = 0 ,$$

and so the values $\lambda_{1,2} := 1 \pm i$. Superposition (the RE is linear) leads to the general solution $b_k = A\lambda_1^k + B\lambda_2^k$ with constants $A, B \in \mathbb{C}$. The initial conditions lead then to the linear equations system for A and B

$$0 = b_0 = A + B , \qquad 1 = b_1 = A(1+i) + B(1-i) ,$$

whose solution is $A = \frac{1}{2i}, B = -\frac{1}{2i}$. So we have finally

$$a_k = \frac{b_k}{k!} = \frac{1}{k!} \frac{(1+i)^k - (1-i)^k}{2i} = \frac{1}{k!} \operatorname{Im} (1+i)^k = \frac{1}{k!} \operatorname{Im} \left(\sqrt{2} e^{i\frac{\pi}{4}} \right)^k = \frac{1}{k!} 2^{\frac{k}{2}} \sin \frac{k\pi}{4} ,$$

and

$$F = \sum_{k=0}^{\infty} \frac{1}{k!} 2^{\frac{k}{2}} \sin \frac{k\pi}{4} x^k .$$

2. **(Hypergeometric type Puiseux series case)** Suppose next $f(x) = \sin\sqrt{x}$. Then $f' = \frac{\cos\sqrt{x}}{2\sqrt{x}}$, and $f'' = -\frac{\sin\sqrt{x}}{4x} - \frac{\cos\sqrt{x}}{4\sqrt{x}^3}$. The second step of the algorithm leads to the DE for f

$$4xf'' + 2f' + f = 0 ,$$

and the transformation via the rule (9) generates then the RE

$$(2k+1)(2k+2)a_{k+1} + a_k = 0 ,$$

which is of hypergeometric type as only two summands occur. From the fact that the largest zero $k = -\frac{1}{2}$ of the polynomial in front of a_{k+1} is nonintegral, we realize that we must consider a Puiseux series with Puiseux number 2. So we make the transformation $b_k := a_{k/2}$, i.e. consider $g(x) := f(x^2) = \sin x = \sum b_k x^k$ rather than $\sin \sqrt{x} = \sum b_k x^{k/2}$, and get the hypergeometric type RE

$$(k+1)(k+2)b_{k+2} + b_k = 0$$

for b_k with symmetry number $m = 2$.

The fact that $g(0) = 0$ tells us that g is odd. We work with the FPS

$$h(x) := \sum_{k=0}^{\infty} c_k x^k$$

for which $g(x) = xh(x^2)$, and so $c_k = b_{2k+1}$, leading to the RE for c_k

$$c_{k+1} = -\frac{1}{4}\frac{1}{(k+\frac{3}{2})(k+1)}c_k \ .$$

The initial condition is $c_0 = b_1 = g'(0) = 1$, so that finally by (1)

$$c_k = \frac{(-1)^k}{4^k\left(\frac{3}{2}\right)_k k!} = \frac{(-1)^k}{(2k+1)!} \ ,$$

and

$$F = \sum_{k=0}^{\infty}\frac{(-1)^k}{(2k+1)!}x^{k+1/2} \ .$$

3. **(Hypergeometric type power series case)** Now we consider the function

$$f(x) := \int_0^x \frac{\operatorname{erf}(t)}{t}\,dt$$

where erf denotes the error function

$$\operatorname{erf}(x) := \frac{2}{\sqrt{\pi}}\int_0^x e^{-t^2}\,dt \ .$$

Note that f is not represented by means of elementary functions, however the algorithm still applies. We have $f' = \frac{\operatorname{erf}(x)}{x}$, $f'' = \frac{2}{\sqrt{\pi}}\frac{e^{-x^2}}{x} - \frac{\operatorname{erf}(x)}{x^2}$, and $f''' = -\frac{4}{\sqrt{\pi}}e^{-x^2} - \frac{4}{\sqrt{\pi}}\frac{e^{-x^2}}{x^2} + 2\frac{\operatorname{erf}(x)}{x^3}$, leading to the DE

$$xf''' + (2x^2+2)f'' + 2xf' = 0 \ ,$$

and thus the RE

$$(k+2)^2(k+1)a_{k+2} + 2k^2 a_k = 0 \ .$$

This is also of hypergeometric type, and the same argumentation as in Example 2 shows that f is odd. We use the same substitutions and get for the coefficients c_k of h

$$c_{k+1} = -\frac{(k+\frac{1}{2})^2}{(k+\frac{3}{2})^2(k+1)}c_k$$

with initial condition $c_0 = a_1 = f'(0) = \frac{2}{\sqrt{\pi}}$, so that finally

$$c_k = \frac{2}{\sqrt{\pi}}\frac{(-1)^k\left(\left(\frac{1}{2}\right)_k\right)^2}{\left(\left(\frac{3}{2}\right)_k\right)^2 k!} = \frac{2}{\sqrt{\pi}}\frac{(-1)^k}{(2k+1)^2 k!} \ ,$$

and

$$F = \sum_{k=0}^{\infty}\frac{2}{\sqrt{\pi}}\frac{(-1)^k}{(2k+1)^2 k!}x^{2k+1} \ .$$

4. **(Rational type case)** Let's now look at $f(x) = \arctan x$. Here $f' = \frac{1}{1+x^2}$ is rational. So we apply the rational algorithm (see ([8], Sect. 4)), and integrate. We find the complex partial fraction decomposition

$$f'(x) = \frac{1}{1+x^2} = \frac{1}{2}\left(\frac{1}{1+ix} + \frac{1}{1-ix}\right) \; ,$$

from which we deduce for the coefficients b_k of the derivative $F'(x) = \sum_{k=0}^{\infty} b_k x^k$ by using the binomial series that

$$b_k = \frac{1}{2}\left(i^k + (-i)^k\right) = \frac{i^k}{2}\left(1 + (-1)^k\right) \; .$$

By the calculation

$$b_{2k+1} = \frac{i^{2k+1}}{2}\left(1 + (-1)^{2k+1}\right) = 0$$

it follows that F' turns out to be even. Moreover

$$b_{2k} = \frac{i^{2k}}{2}\left(1 + (-1)^{2k}\right) = (-1)^k \; ,$$

so that by integration

$$F = \sum_{k=0}^{\infty} \frac{(-1)^k}{2k+1} x^{2k+1} \; .$$

We remark that the arctan function can also be handled by the hypergeometric procedure similarly as Example 2.

5. **(Hypergeometric type Laurent series case)** We consider $f(x) := \frac{\arcsin^2 x}{x^4}$. Here the algorithm produces the DE

$$(x^5 - x^3)f''' + (15x^4 - 12x^2)f'' + (61x^3 - 36x)f' + (64x^2 - 24)f = 0 \; ,$$

converting to the hypergeometric type RE

$$(k+6)(k+5)a_{k+2} - (k+4)^2 a_k = 0$$

with symmetry number 2. It follows that $a_{-6} = a_{-5} = 0$, and thus for all $k \le -5$ we have $a_k = 0$. Therefore we consider the shifted function $g(x) := x^4 f(x)$ with the coefficients $b_k = a_{k-4}$ for which the RE

$$b_{k+2} = \frac{k^2}{(k+2)(k+1)} b_k$$

holds. The hypergeometric coefficient formula finally leads to

$$F = \sum_{k=0}^{\infty} \frac{(k!)^2 4^k}{(2k+1)!(k+1)} x^{2k-2} \; .$$

4 Scope of the Algorithm

In CAS's problems can be solved by means of

- implemented data bases, or
- algorithmic calculations.

In the first case, the implemented data base is fixed for all times until the software is released, whereas in the second case the use of the CAS *generates* a data base. Having an implementation of an algebraic algorithm makes a CAS an expert system: An arbitrary user may produce results, which had been unknown before, and can be added to a data base. Concerning the conversion of functions and power series [6] probably is the most exhaustive existing data base. It is a collection of numerical series, power series, products, and other material. It is easy to observe that most of its power series entries are of hypergeometric type, and so it is not surprising how many of them can be treated by our algorithm. The algorithm covers results about integrals that cannot be represented by elementary functions like Fresnel integrals, Bessel functions, and many other functions. More examples are given in ([8], Sect. 9), and will be published elsewhere [9].

When using [6] to find a power series representation, the scope is restricted to its finite contents, and moreover the success depends on the user's ability to *find* the entry he's searching for: The problem is thus converted into a search problem. On the other hand, the use of the CAS implementation of our algorithm does not have this kind of limitations. It is only a question of time when new results are discovered by its use.

5 The Algorithm as a Simplifier

One of the main questions of Computer Algebra is to decide whether a given expression algebraically is equivalent to zero or not. A simple example of this kind is the rational expression

$$\frac{2}{1-x^2} - \frac{1}{1-x} - \frac{1}{1+x}$$

which after an expansion with common denominator algebraically simplifies to zero. Much more difficult are nonrational algebraic or transcendental expressions like

$$\arcsin\frac{1}{3} + 2\arctan\frac{\sqrt{2}}{2} - \frac{\pi}{2}, \qquad \sqrt{x+1} + 1 - \sqrt{2\sqrt{x+1} + x + 2},$$

$$\sqrt{\frac{1-\sqrt{1-x}}{x}} - \frac{\sqrt{1+\sqrt{x}} - \sqrt{1-\sqrt{x}}}{\sqrt{2x}},$$

$$\cos(4\arccos x) - (1 - 8x^2 + 8x^4), \qquad \text{or} \qquad \cos(\arcsin x) - \sqrt{1-x^2},$$

all of which turn out to equal zero. In general, it cannot be decided if a given transcendental expression is equivalent to zero. Our algorithm, however, may assist with this decision. Assume, an expression involving a variable x is given, and we apply our algorithm to it. In principle, all expressions which are equivalent to zero, are of hypergeometric type; indeed, every polynomial, especially the zero polynomial, is of hypergeometric type. It may happen, however, that we cannot decide this since the hypergeometric type DE may not be found because of the lack of algebraic simplifications. If we are lucky, however, the expression is identified to be of hypergeometric type, in which case its series coefficients can be calculated, and are quotients of Pochhammer symbols. For this kind of expressions, however, it can be decided whether or not they are equivalent to zero, and so we will get the desired result. As examples, all above mentioned expressions that depend on x are recognized by the algorithm to equal zero.

More results in this direction can be found in ([8], Sect. 10). We are convinced that the use of the algorithm will lead to new identities.

6   The Second Conversion Procedure

The algorithm presented has a natural inverse $F \mapsto f$, calculating the function f from its LPS F. We omit the details of that procedure Convert[F,x] that converts an LPS F into its equivalent function f with respect to the variable x. Given a formula for the general coefficient a_k, the first step consists of finding a RE of the type (8). This step is algebraically equivalent to the search for the DE, presented in ([8], Sect. 5).

The next step is then the back-substitution that produces the left-hand side of the DE from the left-hand side of the RE, see ([8], Sect. 11).

The main part of the procedure Convert is to solve the finally generated simple DE together with some initial conditions. At the moment, with MATHEMATICA Version 2.0, this is, at least in the case of DE's of order greater than 2, in general beyond its capabilities. On the other hand, all but very few examples that we tested were solved by MACSYMA's ode procedure (version 417). By a theorem of Singer ([13], see e.g. [2], p. 192) there is an algorithm to decide whether the corresponding function has a representation in terms of elementary functions, in which case this representation is produced. It turns out, however, that the initial value problem may involve rather complicated nonlinear equations for the occuring integration constants.

This procedure Convert moreover is able to produce a closed form representation for (convergent) infinite sums

$$\sum_{k=0}^{\infty} a_k$$

whenever the numbers a_k satisfy a homogeneous, linear RE with polynomial coefficients P_j $(j = 0, \ldots, M)$

$$\sum_{j=0}^{M} P_j(k)a_{k+j} = 0 \ ,$$

and the generating function

$$f(x) := \sum_{k=0}^{\infty} a_k x^k$$

of the sequence (a_k) has a representation by means of elementary functions. **Convert** produces f, and finally

$$\sum_{k=0}^{\infty} a_k = \lim_{x \uparrow 1} f(x)$$

by Abel's continuity theorem for power series (see e.g. [14], p. 149).

This algorithm should be compared with the Gosper algorithm ([3], see e.g. [5], § 5.7) which finds a closed form representation for a sum $A_n = \sum_{k=0}^{n} a_k$ with variable upper bound, in the special case that A_n is the nth term of a hypergeometric function. In this case the generating function of the sequence (a_k) turns out to be hypergeometric, too.

Our procedure does not generate closed forms for sums with variable upper bound, except in the special case that the given sequence (a_k) satisfies $a_k = 0$ for $k > n$. This case, however, occurs e.g. when considering sums of products of binomial coefficients. Moreover, for infinite sums our algorithm has a much wider range of applicability.

We consider two examples.

1. **(Indefinite summation)** We search for a closed formula for $(n \in \mathbb{N})$

$$A_n := \sum_{k=0}^{n} \binom{n}{k} .$$

As $n \in \mathbb{N}$, the coefficients $a_k := \binom{n}{k} = 0$ for $k > n$ so that we can consider A_n as an infinite sum

$$A_n = \sum_{k=0}^{\infty} \binom{n}{k} ,$$

and the method applies. The first part of the procedure **Convert** produces the RE

$$(k+1)a_{k+1} - (n-k)a_k = 0$$

for the coefficients of the generating function

$$f(x) := \sum_{k=0}^{\infty} \binom{n}{k} x^k$$

which is transferred by the back-substitution into the DE

$$(1+x)f' - nf = 0$$

for f. The initial value problem

$$(1+x)f' - nf = 0 , \qquad f(0) = a_0 = 1$$

has the solution

$$f(x) = (1+x)^n ,$$

and so we have finally

$$A_n = \lim_{x \uparrow 1} f(x) = (1+1)^n = 2^n ,$$

the desired result.

2. **(Infinite summation)** As another example we consider the infinite sum which cannot be treated by Gosper's algorithm

$$\sum_{k=0}^{\infty} a_k = \sum_{k=0}^{\infty} \frac{(-1)^k}{2k+1}$$

with generating function

$$f(x) := \sum_{k=0}^{\infty} a_k x^k = \sum_{k=0}^{\infty} \frac{(-1)^k}{2k+1} x^k .$$

An application of the first part of procedure `Convert` produces the RE for the coefficients a_k

$$(2k+3)a_{k+1} + (2k+1)a_k = 0 ,$$

which holds for $k \geq 0$. If we multiply this by the factor $(k+1)$, the resulting RE

$$(k+1)(2k+3)a_{k+1} + (k+1)(2k+1)a_k = 0$$

holds for all $k \in \mathbb{Z}$. The back-substitution yields the initial value problem

$$(2x+2x^2)f'' + (3+5x)f' + f = 0 , \qquad f(0) = a_0 = 1 , \quad f'(0) = a_1 = -\frac{1}{3} ,$$

for f, which has the solution

$$f(x) = \frac{\arctan \sqrt{x}}{\sqrt{x}} .$$

Thus the original sum has the value

$$\sum_{k=0}^{\infty} \frac{(-1)^k}{2k+1} = \lim_{x \uparrow 1} f(x) = \arctan 1 = \frac{\pi}{4} .$$

We mention that the ability to deal with the conversion of series and generating functions algorithmically leads to the ability to produce binomial identities (see e.g. [15], § 4.3), to solve recurrence equations (see e.g. [5], § 7.3), and to solve problems in probability theory (see e.g. [5], § 8.3).

7 Asymptotic Series

One more field where the method described can be used, are asymptotics (see e.g. [5], Chap. 9). Assume, a function f on the real axis is given, and we are interested in an asymptotic expansion, i.e. a function g for which

$$\lim_{x \to \infty} \frac{f(x)}{g(x)} = 1 \ .$$

Asymptotic expansions by no means are unique; indeed each function is its own asymptotic expansion, but there may exist much simpler asymptotic expansions as well. It is a special property of a function f if one of its asymptotic expansions forms a Laurent-Puiseux series

$$F := \sum_{k=k_0}^{\infty} a_k \left(\frac{1}{x}\right)^{k/n} \qquad (a_{k_0} \neq 0) \ ,$$

and if f has such a Laurent-Puiseux asymptotic expansion (LPA), then it is unique. Note that as we consider only positive values x (or complex values x that lie in a certain sector) the LPA is only *one-sided*.

It is now easy to see that a slight modification of the algorithm PowerSeries can be used to produce the LPA of a function f by the following procedure:

1. Consider $h(x) := f\left(\frac{1}{x}\right)$.
2. Find a DE for h, and a RE for its one-sided series coefficients

$$h(x) = \sum_{k=k_0}^{\infty} a_k x^{k/n} \qquad (x \geq 0)$$

  by the PowerSeries algorithm.
3. Find the corresponding initial values by taking one-sided limits, e.g.

$$a_0 := \lim_{x \downarrow 0} h(x) \ ; ,$$

  and solve the RE.
4. Finally you have

$$F = \sum_{k=k_0}^{\infty} a_k \left(\frac{1}{x}\right)^{k/n}$$

We mention that the one-sided limits considered are complex limits if and only if the function $h(x) := x^{k_0} f(x^n)$ is analytic at ∞, and so the corresponding LPA of h equals its power series representation there. Moreover the radius of convergence equals zero, and so F is a divergent series, if one of the one-sided limits considered is *not* a complex limit.

Since MATHEMATICA only supports the calculation of complex limits, there is no direct way for an implementation of this procedure. On the other hand, MAPLE has some capabilities to calculate one-sided limits, and thus the MAPLE implementation covers asymptotic series expansions.

We give some examples of this procedure.

1. Suppose $f(x) = e^x(1 - \operatorname{erf}\sqrt{x})$. The algorithm leads to the DE

$$2x^3 h'' + (2x + 3x^2)h' - h = 0$$

for $h(x) = f\left(\frac{1}{x}\right) = e^{1/x}\left(1 - \operatorname{erf}\frac{1}{\sqrt{x}}\right)$, and the transformation via the rule (9) generates then the RE

$$a_{k+1} + k a_k = 0 \ ,$$

which is of hypergeometric type. The initial values lead then to the asymptotic series representation

$$F = \frac{1}{\sqrt{\pi}} \sum_{k=0}^{\infty} \frac{(-1)^k (2k)!}{4^k k!} \left(\frac{1}{x}\right)^{k+1/2}$$

2. If $f(x) := x e^{-x} \operatorname{Ei}(x)$ where Ei denotes the exponential integral function

$$\operatorname{Ei}(x) := \int_{-\infty}^{x} \frac{e^t}{t}$$

with a Cauchy principal value taken. We get the DE

$$x^2 h'' + (3x - 1)h' + h = 0$$

for $h(x) = f\left(\frac{1}{x}\right)$, and the RE

$$a_{k+1} = (k+1)a_k$$

for its coefficients, leading to the asymptotic expansion

$$F = \sum_{k=0}^{\infty} k! \left(\frac{1}{x}\right)^k \ .$$

References

1. AXIOM: *User Guide.* The Numerical Algorithms Group Ltd., 1991.
2. Davenport, J.H., Siret, Y., Tournier, E.: *Computer-Algebra: Systems and algorithms for algebraic computation.* Academic Press, 1988.
3. Gosper Jr., R.W.: Decision procedure for indefinite hypergeometric summation. Proc. Natl. Acad. Sci. USA **75** (1978) 40 – 42.
4. Gradshteyn, I.S., Ryzhik, I.M.: *Table of integrals, series, and products – corrected and enlarged version.* Academic Press, New York-London, 1980.
5. Graham, R.L., Knuth, D.E., Patashnik, O.: *Concrete Mathematics: A Foundation for Computer Science.* Addison-Wesley Publ. Co., Reading, Massachusetts, 1988.
6. Hansen, E.R.: *A table of series and products.* Prentice-Hall, Englewood Cliffs, NJ, 1975.
7. Hearn, A.: REDUCE *User's manual, Version 3.4.* The RAND Corp., Santa Monica, CA, 1987.

8. Koepf, W.: Power series in Computer Algebra. J. Symb. Comp. **13** (1992) 581–603.
9. Koepf, W.: Examples for the algorithmic calculation of formal Puiseux, Laurent and power series (to appear).
10. MACSYMA: *Reference Manual, Version 13.* Symbolics, USA, 1988.
11. MAPLE: *Reference Manual, fifth edition.* Watcom publications, Waterloo, 1988.
12. Norman, A. C.: Computing with formal power series. Transactions on mathematical software 1, ACM Press, New York (1975) 346–356.
13. Singer, M. F.: Liouvillian solutions of n-th order homogeneous linear differential equations. Amer. J. Math. **103** (1981) 661–682.
14. Walter, W.: *Analysis I.* Springer Grundwissen Mathematik 3, Springer-Verlag, Berlin-Heidelberg-New York-Tokyo, 1985.
15. Wilf, H.S.: *Generatingfunctionology.* Academic Press, Boston, 1990.
16. Wolfram, St.: MATHEMATICA. *A system for doing mathematics by Computer.* Addison-Wesley Publ. Comp., Redwood City, CA, 1991.

A Cooperative Approach to Query Processing: Integrating Historical, Structural, and Behavioral Knowledge Sources

Larry Kerschberg and Anthony Waisanen
Department of Information and Software System Engineering
George Mason University
Fairfax, Virginia 22030-4444
kersch@gmuvax2.gmu.edu
mxpya1.macmhs@mhs.safb.af.mil

Abstract. Query processing strategies are frequently considered as sequential processes of successive decompositions and evaluations. This paper describes an approach that exploits opportunities to bypass steps in this sequential process. Initially, descriptions of the database are used in responding to queries since no other knowledge about the database exists (no queries have been previously encountered by the query processor). As queries are processed, descriptions of the individual queries and query fragments are retained along with the resulting access strategy, a description of the response, and costs to retrieve the desired information. This knowledge is then used to develop future access strategies.

1 Overview

1.1 Position statement

Intuitively, the goal of query processing (and specifically query optimization) is to retrieve the answer to an input query while using the fewest resources possible. We define "query optimization" to be the transformation of a descriptive (or high-level) query into one or more prescriptive (or low-level) queries or access paths which use a minimal amount of specified resources. In "resources" we include system resources such as I/O, CPU, and communication costs, as well as user resources such as overall response time and quality of answer (explained later).

One aspect of query optimization is timing; when the optimization process is performed. There are three general approaches to query optimization based on timing [14]; *static, dynamic,* and *hybrid.* In static query optimization, the high-level query is transformed into one or more low-level queries and equivalent access methods. The optimal access method or path is determined by calculating the expected amount of resources that will be required to perform the access path's operations. In dynamic query optimization, the choice of operation to be performed is based on the results of the preceding operation(s) which, since they are *post priori*, are highly accurate descriptions of the database. Hybrid query optimization is a combination of static and dynamic optimization. This latter approach is central to the position presented in this paper, *an approach to query optimization that uses performance results of previously executed queries, is capable of better overall system efficiency than traditional approaches.*

1.2 Scope and limitations

In this paper, we limit our discussion to query processing in distributed database management systems with architecture models like that depicted in Figure 1 [14]. In other words, each site has a local internal schema (LIS) and local conceptual schema (LCS), and the mappings between the conceptual schema and the internal representations are supported by the respective local directory/dictionary. We further presume that each local database has a management system (DBMS) capable of optimizing local queries (i.e., efficiently accessing data at the physical level) and that each DBMS can report the cost of executing a query. User access and applications on the entire system are supported by external schemas (ESs) which are above the enterprise view of the distributed data, described by the global conceptual schema (GCS). Mappings between the global and external database schema processors and the various conceptual database schemas are performed by a global directory/dictionary. For the sake of discussion, we use descriptive and prescriptive queries based on the relational calculus and algebra.

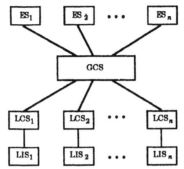

Fig 1. Distributed Database Reference Architecture

2 Previous efforts

Query processing and query optimization in distributed database systems has been extensively studied and reported by several researchers (see [5] and [14] for bibliographies and discussions). Query processing is typically considered to occur in stages or layers, in order to make the problem tractable. According to [14], a generic view is to consider four functional layers as in Figure 2. Processes in these layers convert the distributed query into a set of operations to be performed on the local databases. Basically, this process is to decompose the distributed query into its algebraic equivalent, to determine the disjoint subsets of relations that are referenced by the query, to find a near-optimal execution strategy based on the statistics available on the query fragments, and finally to have the local databases optimize their respective fragments. Thus, the problem of processing the distributed query is divided into several subproblems. Notice that "optimization" is really a misnomer. It has been formally shown in [8] that it is computationally intractable to find the optimal execution strategy for a multiple-join query. Thus, finding *the* optimal execution strategy is not as important as finding a *good* strategy while avoiding a *bad* strategy.

In the first layer of this generic view, query decomposition, the calculus query on distributed relations is decomposed into an algebraic query on global relations. The knowledge used in this layer is provided by the global schema which describes the global relations. Four sequential steps are used in this process. First, the calculus expression is converted to a normalized, or *canonical* form using logical transformations. This normalized query is then analyzed semantically for inconsistencies. The third step is to simplify the normalized and correct query to remove redundant predicates. In the last step, the query is restructured as an algebraic query. Since multiple, equally correct algebraic queries can be derived from the same calculus query, heuristics and transformation rules are typically used to select the "best" algebraic query.

The second layer determines which fragments of the relations are involved in the query and transforms the query into a *fragment* query, an algebraic query over the fragments of the relation. This is accomplished in two steps. First, each operator on a fragmented relation is transformed into operations on the fragments of the relation using the fragmentation schema. The fragment query is then simplified and reconstructed using the same rules as in the first layer. Though it is possible to execute this fragment query, it is likely to be far from a "good" strategy since no knowledge of the fragment statistics have been considered.

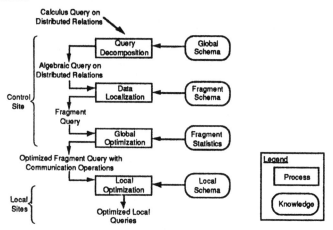

Fig 2. Generic Layering Scheme for Distributed Query Processing

The purpose of the third layer, global optimization, is to optimize the fragment query in terms of total execution cost, minimal execution time, or both. To do this, the expected cost of executing the operations in the query must be estimated. Since it is possible to vary the order of the operations as well as the method of executing them, several alternative fragment queries are possible. Specifically, it is necessary to find a "good" sequence of join, semi-join, and union operations. It is also necessary to determine how to best evaluate joins. The final selection of operation sequence and method is made using knowledge of how the attribute-values are distributed in the relation fragments and the limitations and capabilities of the system supporting the database. Communication operations to control data transfer between sites are included in the final, optimized fragment query.

The optimized fragment query (with communication operations) is then submitted to the local databases for execution. The individual database management systems then independently develop optimal execution paths or sequences based on local knowledge and using optimization algorithms developed for centralized systems.

2.1 Use of heuristics

Heuristics often play a large role in distributed query processing. One of the early references to the use of heuristic rules can be found in [2]. A particularly important use is in the Global Optimization layer in determining an optimal sequence of JOINs and SEMI-JOINs. This is because there is a combinatorial explosion of the number of options possible when the number of fragments becomes large. Furthermore, the general formulation of the problem is NP-complete, and only a limited class of queries admits a polynomial solution.

2.2 Use of models

Model-based analysis, analysis which uses a depiction of the distributed database, has been used in designing distributed databases [4 and 13], particularly in fragment allocation [6], and for developing precise optimization models and algorithms [9]. A disadvantage to directly using results from these analyses is that highly accurate cost estimations are only possible with highly specified models. When these models differ from the actual distributed databases, the results may be misleading. Thus, the results from these analyses are useful for forming general approaches, but must be used judiciously.

3 Knowledge existing in a database

Knowledge exists in several places in a distributed database. First, there is *available* knowledge; stored knowledge about the database that describes its contents or structure. Second, there is *derivable* knowledge; knowledge that can be generated by interrogating or analyzing the database. These will be discussed separately.

3.1 Available Knowledge

One source of available knowledge in a database system is the data dictionary, specifically, the schema, mappings, constraints, and rules. This knowledge is in the global data dictionary and also the data dictionaries and schemas of the local databases.

Another source of knowledge is the different views allowed the users by the database administrator (DBA). As shown by Motro in [11], views may be incorporated explicitly into the database by using predicates. If we extend this further, we can use these predicates to circumvent portions of the query processing. For example, by quickly responding to a query with "You are not authorized that information" rather than after the information had been retrieved and transferred to the sending site.

3.2 Derivable knowledge

Unlike available knowledge, derivable knowledge is not retrieved from the data dictionary or views of the database. It can only be determined from processing the database, either statistically or logically. Two sources of this knowledge are statistical profiles and intensional information.

Statistical profiles. As mentioned above, the Global Optimization layer uses statistical information about the fragments to develop the expected execution cost values for the various fragment queries. For example, the cardinality of each relation fragment, the size of each attribute in each fragment, and the number of distinct values for each attribute in each fragment is typically required for a quantitative analysis [5]. These values may supplied by the DBA, calculated by statistical sampling (in very large databases), or exhaustively determined.

Early analytical approaches to query optimization in centralized and distributed database systems were based on a number of assumptions: 1) attribute values are uniformly distributed over their domain; 2) attribute-values are independent of each other; 3) queries refer to attribute-values with equal frequency; 4) blocks of a file contain the same number of records; 5) the placement of record in a file does not effect that record's likelihood of qualifying in a query. Christodoulakis has shown in [3] that if these assumptions are incorrect, the expected execution costs will be over-estimated. And as the database is updated or changed, this knowledge may become obsolete.

Intensional information. In the first three layers of distributed query processing, significant processing time can be saved, and subsequent queries simplified, when contradictions can be detected between the selection criteria of the queries and the qualifications of fragments. When intensional information is part of the database, either as rules or as constraints, it is possible to transform it into predicates which are relevant to specific relations. For example, in [1], semantically constrained axioms are developed which can transform a query in Horn-clause form into one or more simpler queries. Similarly, Pirotte *et al* [15] describes an approach for developing *residuals* for each relation and using them for developing intensional answers.

When this information is not an explicit part of the database, it may be manually derived from the application programs or through exhaustive analysis of the database extensions. In the former case, machine-learning techniques that have been developed for discovering knowledge in legacy systems are applicable. In the latter, statistical analyses, particularly factorial analysis and correlation analysis, would be useful for discovering possible dependencies.

4 Applying cooperative knowledge sources

The traditional approaches to query processing have used multiple reasoning paradigms; statistics, heuristics, and rules. But what has been lacking in general is the active or dynamic use of the performance results. One exception to this assertion is the query optimization algorithm of Distributed INGRES [7]. However, this algorithm does not keep the results of the detachments from one query to another. Thus, the optimization costs for a query do not ameliorate the costs of subsequent queries. Our proposed approach supports this amelioration by retaining and using meta-information about the queries.

4.1 Approach

We see that there are two groups of information which can be used for query processing. One is the information that has typically been relied upon in the approaches described above; logical, statistical, heuristic, and (recently) static intensional. The other group concerns performance. That is, the actual (versus expected) cost of executing the query and a description of the result of the query ("Was NULL returned? A single tuple retrieved?"). It certainly is possible to update the statistical, heuristic, and static intensional information used in the traditional approaches, but this would not necessarily result in an overall savings. This is because most current algorithms precompile query optimization knowledge, expecting to ameliorate the cost of these calculations over the subsequent processing. Thus, while the cost/benefit and result information is current, it is not immediately useful.

Furthermore, traditional approaches are designed to determine all possible consequences of a change in this meta-information. And since this information is dependent on the current state of the database (the extensions, the fragmentation, current work-load at the various sites, and so forth) which may be highly dynamic, calculating all possible ramifications of a change in such knowledge would require a work-load which may not be justified. For example, a null response to a query that results because a tuple does not exist at execution time should not be considered sufficient justification for adding a rule to that effect. The very next update to the database could invalidate it. What is needed is an approach that detects when a query result is potentially useful for processing future queries, saves this knowledge in a form that can be readily retrieved and applied, and maintains this knowledge at an acceptable cost. Our approach is to enhance the traditional query process layers with a component capable of reasoning from previous instances, a case-based reasoning component.

4.2 Case-based reasoning component

The underlying notion in case-based reasoning is that solutions to previously encountered problems can be adapted to solve new problems [16]. Thus, the problem solving process for a case-based planner is:

1. Select features of the current problem which can be used for referencing past problems;
2. Retrieve past problems, and their solutions, from the database of cases that have those same features;
3. Select the subset of cases which most closely matches the current problem;

4. If a selected past problem exactly matches the current problem, reuse the solution to that problem, otherwise adapt the solution(s) in the subset of retrieved cases to match the current problem.

In the domain of query optimization, the last step depends on domain knowledge. Since we are focusing on a generic approach, we will not investigate the area of adaptation at this time.

The crux of the problem with using a case-based reasoning system is in defining what is meant by a *case*. Intuitively, a case is an example of an instance where a situation was encountered and a solution was, or was not, achieved. In the context of query processing, a case is a representation of a query or fragment together with the result of executing that query or fragment. A design feature of database management systems is to allow a high degree of flexibility in referencing the information contained in the database. Thus, it is highly unlikely that saving every input query and its result will be beneficial. In order for this to be a feasible strategy, users would have to frequently submit a small set of queries and the size of the results of these queries would have to be small. Furthermore, the relevant cases would no longer be valid should the database change. A more manageable representation of the encountered situation and result must be used.

Case knowledge. It is clear from the voluminous literature on query processing that processing performance can be significantly improved if there is *a priori* knowledge that a query operation will generate a small or known set of tuples. An extreme example is when a conflicting statement results as the input query is decomposed during the first layer of processing. In other words, a potentially useful case is when a query fragment is equivalent to a conflict. Notice that while logical transformations may detect these conflicts during decomposition and localization, null answers are detected *after* the optimized local queries have executed.

In our approach, the calculus query that eventually led to the null response is not very useful because of the several possible intermediate fragment queries that could have been, or were, derived from it. Rather, we propose that the null answers be associated with the optimized fragments from the global optimization layer. The resulting cases would then be referenced by the query fragment.

Format of cases. The principal constituent of a case in our approach is a description of the problem. In our problem domain, this means a description of the query or query fragment: 1) the attributes and relations involved in the query or query fragment and 2) the predicates which were in effect for that query or query fragment. As previously mentioned, the query fragment is needed to reference the case. Our cases also retain the type of query (SELECT, PROJECT, UNION, or JOIN). We do not include the other SQL algebraic operators since, as described by Ceri and Pelagatti in [5], these are the four critical operations for forming equivalence transformations.

Also, a description of the implemented solution is needed; the access strategy that was executed; a record of the conditions under which this case may be applied. We do not require for these conditions to be necessary, only sufficient. That is, there may be other conditions under which the particular query fragment would generate a NULL or a predicate, but they need not be specified. An exception is when a query fragment matches a query fragment in a case, the preconditions are not satisfied (so the case is not applied), but the query fragment results in a NULL or the same predicate. In this instance, the case that was not previously applied would have the precondition slot updated to include the new conditions. In other words, the case for the query fragment would have the new preconditions OR'd with the old preconditions.

The results of implementing the solution are a semantic representation of the set of retrieved tuples as well as the expected cost and the actual cost of executing this access strategy. By *semantic representation of the set of retrieved tuples*, we mean one of three possible responses: NULL, "more than one value," or a unary value. This is similar to the intensional answers suggested by Pirotte *et al* in [15] and constraint residuals suggested by Chakravarthy *et al* in [1]. However our predicates are a consequence of the current state of the distributed database (to possibly include the fragmentation of the database) while their approaches presume that integrity constraints and intensional predicates are explicitly defined. In this way, our approach is similar to one suggest by Motro in [12]. But here, too, the intensional answers are generated and used, not updated. In addition to retaining NULLs or predicates, we have considered retaining a two-dimension array when a single tuple has been retrieved. However, we have not seen where this complication offers benefits beyond simply retaining what we have already specified.

The cost of executing this query fragment is either calculated from data provided by the local database that executed the fragment, or it is assigned by the Control Site when the results are returned by the local databases. The idea behind associating a cost value with a query fragment is that during Global Optimization, statistics on the data fragments are used for optimizing the order and method of performing JOIN operations. One approach, if the local databases will supply the required statistics for calculating tuple densities, tuple distribution densities, and join ratios, is to use a Detailed Database Statistics Model, a DDSM [10]. A limitation of this approach is that these statistics must be precompiled. Thus, while accurate estimates are possible, if an inaccuracy has been introduced or has developed as the updates occur, it will continue to adversely influence the optimization process. With our method, the optimization process considers the past performance of query fragments before deciding on the specific join operation. The resulting format is depicted by Figure 3.

Slot:	Description:
Case:	Case identifier
Attribute(s):	List of attribute names
Relation(s):	Relation(s):
Predicate(s):	*Relation.attribute*=attribute.value or NULL
Access strategy:	Statement in relational algebra
Expected cost:	Initially estimated by query optimizer; then based on actual costs
Actual cost:	Resulting cost as a function of database accesses (reads), time to retrieve the information, and CPU load
Result:	either NULL, "more than one value," or a unary value
# of recalls:	0 initially, incremented with each access

Fig 3. Format of Case for Query Processing

Role of cases in query processing. Now that the notion of a case has been developed, three issues remain to be answered. First, when to use the cases; second, how to use the cases; and third, how to update the case-base.

When to use the cases is a primarily a function of how cases are derived. Recall that cases are formed when an "interesting" result is generated. At all levels, this is when a NULL is retrieved. At the Query Decomposition level, cases are formed when a simpler, equivalent query is discovered based on information in the lower levels. And at the Global Optimization level, a case is formed when the Control Site decides that the cost of executing a specific query fragment is excessive. Thus, the case base should be checked after Query Decomposition and during Data Localization and Global Optimization.

Cases are used for a similar purpose, to "skip" transformations or unifications required in traditional query processing. At the Query Decomposition level, cases reflect information about the materialized relations (a "meta-fragment"). Thus, we use them to determine when an answer is impossible because constraints or intensions are violated or when the answer is NULL, i.e., we are certain there is no answer. Thus, if the algebraic query derived from the calculus query can be answered from cases, the remaining processes are skipped. At the Data Localization level, fragments may unify with cases and either result in a NULL answer or help simplify the fragment query. Finally, cases are used during Global Optimization to not only simplify fragment formation but also to avoid accessing high cost attribute-value combinations.

The third issue, how to update the case-base is relatively straightforward. Recall that a feature of the case is a set of preconditions which must be satisfied for the case to be valid. While the database remains unchanged, the preconditions will remain valid. If we require the case-based component be notified whenever a database update occurs, it is possible to ensure validity by removing those cases whose preconditions can no longer be satisfied or altering cases when possible. For example, suppose we have the following case:

Query fragment: "Names of employees stored in database at site x"
Equivalent fragment: "Not Brown and not Smith"
Precondition: 1) "Not Brown"; 2) "Not Smith"

Obviously, inserting a tuple for Brown into the database at site x would result in this case being modified; the equivalent fragment would change to "Not Brown." And a subsequent inserted tuple for Smith would completely negate it. Notice though, that this does not result in a requirement that the entire case-base be recompiled or validated, only the relevant cases. And the relevance is readily determined by referring to the fragments specified in the query fragment.

5 Conclusion

By way of concluding remarks, we discuss potential side-benefits of this approach. We also briefly touch on how lessons learned in this research could be applied in another domain, fault diagnosis.

5.1 Advantages

There are several potential advantages of using this approach of cooperating knowledge sources, mostly in discovering knowledge and meta-knowledge in the database.

Two major issues in designing a distributed database are in determining an efficient fragmentation scheme and in allocating the fragments to the various sites. A top-down design approach would accomplish this by evaluating the requirements of frequently executed or critically required application programs. Notice that in the bottom-up approach, commonly encountered when merging existing database systems, this information may not be available until after the database is used. For example, suppose a shipping corporation buys out another company involved with shipping. The subsidiary will be interested in some fragments of information in the parent company's database such as stock-on-hand. It is conceivable that the stock-on-hand data had been fragmented in order to optimize its use by the original subsidiary companies. Now, suppose a fragment of the stock-on-hand data is more frequently accessed by this newly incorporated entity than by any other entity. The fragmentation should change. But how is this knowledge manifested? According to Ceri and Pelagatti [5], selection of predicates (i.e., knowing that a predicate is useful for describing fragmentation) relies on the intuition of the designer who understands the application programs. In other words, by a DBA with insight into marketing practices. But this presupposes a close and active cooperation between representatives in Marketing and technologists in Information Services. An alternative is to allow the Control Site to process queries from the new subsidiary. Since the knowledge about the corporation's database exists in the Fragment Schema and Global Schema, and the new subsidiary will be submitting queries to the Control Site, it will be possible to collect statistics which would indicate that an alternative fragmentation would be beneficial.

Periodic statistical analysis of the case-base contents could indicate a significant shift in fragment access. An added benefit with this approach is that the statistics could be used to quantify the potential cost savings.

Another area of discovery is directly related to the initial motivation for using knowledge cooperatively, discovering inconsistencies and deficiencies. Null answers result for three reasons: 1) there *is* no tuple in the requested relation (but there might be, someday); 2) there *may not* be a tuple (i.e., the relation is known to be complete and the query cannot be instantiated); 3) there *cannot* be a tuple (an integrity constraint or similar predicate is violated). Notice that only the last reason can be determined by logical transformations (an operation in the Query Decomposition layer). Notice too, that null answers generated for the first reason are not normally used in static query optimization and are only temporarily useful to dynamic optimizers (like the optimizer for INGRES). The second reason requires semantic knowledge which is not likely to be discovered but might be inferred if closure axioms are incorporated into the database (but this leads to a discussion of how non-monotonic logics could be applied to generating intensional answers, which is not the intent of this paper).

5.2   Applicability to fault diagnosis

The approach to query processing that we have described above where performance results are retained with preconditions and referenced by the stimulus conditions, and processing is via a cooperative use of information rather than competitive, is applicable to several other domains, particularly diagnosis. For example, diagnosis of telecommunications network faults. In this domain, the "tuples" are alarm messages in an alarm message log. The database schema and architecture are analogous to the topology and alarm models of the network. Finally, queries are analogous to inquiries into the cause of the alarms.

This is currently being investigated in parallel with further developing the aforementioned approach to query processing.

References

1. U.S. Chakravarthy, J. Grant, and J. Minker; "Logic-Based Approach to Semantic Query Optimization", *ACM Transactions on Database Systems*, vol 15, no. 2, (June 1990).

2. J.M. Chang; "A Heuristic Approach to Distributed Query Processing," *Proc. of Eighth International Conference on Very Large Data Bases*, (1982).

3. S. Christodoulakis; "Implications of Certain Assumptions in Database Performance Evaluation," *ACM Transactions on Database Systems*, vol. 9, no. 2, (1984).

4. S.B. Navathe, S. Ceri, and G. Wiederhold; "Distribution Design of Logical Database Schemas," *IEEE Transactions on Software Engineering*, vol. 9, no. 4, (1983).

5. S. Ceri and G. Pelagatti; *Distributed Databases; Principles and Systems*. McGraw-Hill, New York, NY (1984).

6. L.W. Dowdy and D.V. Foster; "Comparative Models of the File Assignment Problem," *ACM Computing Surveys*, vol. 14, no. 2, (1982).

7. R. Epstein, M. Stonebraker, and E. Wong; "Query Processing in a Distributed Relational Database System," *Proc. of the ACM SIGMOD International Conference on Management of Data*, Austin, TX (1978).

8. T. Ibaraki and T. Kameda "On the Optimal Nesting Order for Computing N-Relation Joins," *ACM Transactions on Database Systems*, vol. 9, no. 3, (1984).

9. L. Kerschberg, P.D. Ting, and S.B. Yao, "Query Optimization in Star Computer Networks," *ACM Transactions on Database Systems*, vol. 7, no. 4, (1982).

10. B. Muthuswamy and L. Kerschberg; "A DDSM for Relational Query Optimization," *Procs. of ACM Annual Conference*, (1985).

11. A. Motro; "An Access Authorization Model for Relational Databases Based on Algebraic Manipulation of View Definitions," *Proc. IEEE Computer Society 5th International Conference on Data Engineering*, Los Angeles, CA (1989).

12. A. Motro; "Using Integrity Constraints to Provide Intensional Answers to Relational Queries," *Proc. 15th International Conference on Very Large Databases*, Amsterdam (1989).

13. S. Mahmoud and J.S. Riordon, "Optimal Allocation of Resources in Distributed Information Networks," *ACM Transactions on Database Systems*, vol. 1, no. 1, (1976)

14. M.T. Özsu and P. Valduriez; *Principles of Distributed Database Systems*. Prentice Hall, Englewood Cliffs, NJ (1991).

15. A. Pirotte, D. Roelants, E. Zimanyi; "Controlled Generation of Intensional Answers," to appear in *IEEE Transactions on Knowledge and Data Engineering*, (1991).

16. C.K. Riesbeck and R.C. Schank; *Inside Case-based Reasoning*, Lawrence Erlbaum Associates, Hillsdale, NJ (1991).

A Desk-Top Sequent Calculus Machine

G. Cioni[1], A. Colagrossi[1,2], A. Miola[3]

[1] Istituto di Analisi dei Sistemi ed Informatica del C.N.R., Viale Manzoni 30, 00185 Roma, Italy; Phone/Fax +39-6-77161; E-mail cioni@iasi.rm.cnr.it

[2] Presidenza del Consiglio dei Ministri, Dipartimento di Informatica e Statistica, Via della Stamperia 7, 00187 Roma, Italy

[3] Dipartimento di Informatica e Sistemistica, Università degli Studi di Roma "La Sapienza", Via Salaria 113, 00198 Roma, Italy

Abstract. This paper presents an implementation of a sequent calculus to be used as a flexible tool for automated deduction. The proposed implementation represents a desk-top machine used in interactive mode to solve verification as well as generation and abduction problems. The object oriented design is described and some implementation remarks are given. Examples of working sessions are presented.

1 Introduction

A particular sequent calculus for automated reasoning able to solve validity problem as well as generation and abduction problems in first order logic has been recently proposed in [1]. In this paper the implementation of this inference method for automated extended deduction is presented. The proposed software system operates in a totally interactive mode, as a desk top machine, based on a single automated deduction mechanism.

In literature the usefulness of deduction methods for mathematical problem solving, embedded in an imperative or functional programming environment, is widely recognized. In particular, the lack of semantic correctness recognized in the existing symbolic computation systems can be overridden by addind reasoning capabilities [2].

In this paper we are mainly interested in mathematical problems solving where the manipulation of properties of mathematical objects is a central aspect of the computation.

From the methodological point of view problems can be characterized in three different classes:

This work has been partially supported by "Progetto Finalizzato Sistemi Informatici e Calcolo Parallelo" of CNR.

(a) *verificative problems*, also called "validity problems", which require the verification of a specific property for a given object or for a well defined class of objects;

(b) *generative problems*, which require the generation of a new property for a well defined class of objects;

(c) *abductive problems*, also called "causes for events" problems, which require the generation of new properties for a class of objects which a given property can be derived from.

Problems of type (a) are generally solved with a *refutation mechanism*; while different deductive methods are usually applied for problems of type (b) and (c).

Our aim is the definition of a single method to support the solution of problems of the three classes and, therefore, implementation requirements are crucial in order to obtain an effectively useful tool. The proposed automated deduction method has to be very natural, without requiring any overplus for express properties in some predefined form, and it must be usable in an interactive mode, allowing the user to stop or force the development of the deduction steps.

The method has been implemented and embedded in an object-oriented software system. In fact, the hierarchical structure obtained by the object oriented paradigm exactly matches with that of a logical calculus whose kernel is the propositional calculus [3].

In this paper the methodological basic approach to be used to solve different inference problems is summarized in the next section. For all the technical details and theorems about the proposed sequent calculus we refer to [1]. In the successive section the main characteristics of an object oriented design and of the used programming language are presented. Then, some remarks on the implementation are given and three examples of working sessions are described. In the concluding section, future developments are briefly presented.

2 A Sequent Calculus

The sequent calculus is a proof system characterized by three fundamental components: sequents, inference rules, inference procedures.

Every kind of proof on a given sequent (i.e. verificative, generative or abductive) can be performed by applying the inference rules to the sequent. From an operative point of view, a proving process is either directly driven by the user, in a full interactive mode, or obtained by executing two inference procedures (i.e. the Proving and the Reasoning procedures of the sequent calculus provided in [1]) when operating in automatic mode.

A *sequent* is a string:

$$\alpha_1, \cdots, \alpha_{k-1}, \alpha_k, \alpha_{k+1}, \cdots, \alpha_n \Rightarrow \beta_1, \cdots, \beta_{h-1}, \beta_h, \beta_{h+1}, \cdots, \beta_m \qquad (1)$$

where α_i and β_j, for i=1,$\cdots$,n; j=1,$\cdots$,m, denote wffs in first order logic.

In *sequent* (1) $\alpha_1, \cdots, \alpha_n$ is called *Antecedent* and $\beta_1, \cdots, \beta_m$ is called *Succedent*. The Antecedent or Succedent in a sequent may also be empty. If both are empty, the (1) denotes an empty sequent.

The Antecedent (Succedent) of a sequent, as in (1), is either valid, if $\alpha_1 \wedge \alpha_2 \wedge \cdots \wedge \alpha_n$ ($\beta_1 \vee \beta_2 \vee \cdots \vee \beta_m$) is valid, or a contradiction, if $\alpha_1 \wedge \alpha_2 \wedge \cdots \wedge \alpha_n$ ($\beta_1 \vee \beta_2 \vee \cdots \vee \beta_m$) is a contradiction, or falsifiable, otherwise.

A sequent is valid if either the Antecedent is a contradiction or the Succedent is valid; a sequent is a contradiction if the Antecedent is valid and the Succedent is a contradiction; a sequent is falsifiable if it is neither valid nor a contradiction.

In the following capital greek letters denote strings of formulas, while lower-case greek letters denote formulas. Then, with these notations, the set of inference rules can be given with the usual schema, where the lower element is the input sequent and the upper element(s) is the result of the application.

Each rule performs the decomposition of a sequent in one or two sequents by eliminating a connective from the input sequent. Thus the rules are also called *decomposition* rules and successive applications of such rules derive sequents without connectives.

Rule Set D^8

$\wedge A$
$$\frac{\Gamma, \alpha_1, \alpha_2, \Delta \Rightarrow \Lambda}{\Gamma, \alpha_1 \wedge \alpha_2, \Delta \Rightarrow \Lambda}$$

$\wedge S$
$$\frac{\Gamma \Rightarrow \Delta, \alpha_1, \Lambda \qquad \Gamma \Rightarrow \Delta, \alpha_2, \Lambda}{\Gamma \Rightarrow \Delta, \alpha_1 \wedge \alpha_2, \Lambda}$$

$\neg A$
$$\frac{\Gamma, \Delta \Rightarrow \alpha_1, \Lambda}{\Gamma, \neg\alpha_1, \Delta \Rightarrow \Lambda}$$

$\neg S$
$$\frac{\alpha_1, \Gamma \Rightarrow \Delta, \Lambda}{\Gamma \Rightarrow \Delta, \neg\alpha_1, \Lambda}$$

$\vee A$
$$\frac{\Gamma, \alpha_1, \Delta \Rightarrow \Lambda \qquad \Gamma, \alpha_2, \Delta \Rightarrow \Lambda}{\Gamma, \alpha_1 \vee \alpha_2, \Delta \Rightarrow \Lambda}$$

$\vee S$
$$\frac{\Gamma \Rightarrow \Delta, \alpha_1, \alpha_2, \Lambda}{\Gamma \Rightarrow \Delta, \alpha_1 \vee \alpha_2, \Lambda}$$

$\supset A$
$$\frac{\Gamma, \Delta \Rightarrow \alpha_1, \Lambda \qquad \alpha_2, \Gamma, \Delta \Rightarrow \Lambda}{\Gamma, \alpha_1 \supset \alpha_2, \Delta \Rightarrow \Lambda}$$

$\supset S$
$$\frac{\alpha_1, \Gamma \Rightarrow \alpha_2, \Delta, \Lambda}{\Gamma \Rightarrow \Delta, \alpha_1 \supset \alpha_2, \Lambda}$$

Other decomposition rules are defined to operate on sequents with quantified formulas. The computational definition of these rules also specifies the effects of their application on the following three auxiliary data structures:

$\Phi = \{\Phi_0, \Phi_1, \cdots, \Phi_k\}$, a family of sets $\Phi_i = \{\varphi_{i_0}, \varphi_{i_1}, \cdots, \varphi_{i_{r_i}}\}$ of quantified formulas, where φ_{i_j} is either $\forall x\beta$ or $\exists x\beta$ and β is a wff;

$\Theta = \{(y_0, \Phi_0), (y_1, \Phi_1), \cdots, (y_k, \Phi_k)\}$, a set of pairs where the Φ_i's belong to the family Φ and the y_i's are free variables or constants in the sequent under consideration;

$Y = \{y_{k+1}, y_{k+2}, \cdots\}$, the set of all the variables not included in any pair of Θ.

These auxiliary data structures define the global environment in which the application of the decomposition rules takes place and are manipulated in the different steps of the calculus. Namely, the application of the rules $\exists A$ and $\forall S$ removes the element y from Y and includes the pair $(y, \emptyset)$ into Θ, while the application of the rules $\exists S$ and $\forall A$ includes the quantified formula α into every set Φ_i for which the corresponding term t_i in Θ has been substituted for x in α (such substitution doesn't occur if α belongs to Φ_i).

The set $\mathbf{D}^4$ of decomposition rules for quantified formulas is the following:

$$\exists A \qquad \frac{\Gamma, \alpha[y/x], \Delta \Rightarrow \Lambda}{\Gamma, \exists x\alpha, \Delta \Rightarrow \Lambda}$$

$$\forall S \qquad \frac{\Gamma \Rightarrow \Delta, \alpha[y/x], \Lambda}{\Gamma \Rightarrow \Delta, \forall x\alpha, \Lambda}$$

$$\exists S \qquad \frac{\Gamma \Rightarrow \Delta, \alpha[t_1/x], \ldots, \alpha[t_k/x], \exists x\alpha, \Lambda}{\Gamma \Rightarrow \Delta, \exists x\alpha, \Lambda}$$

$$\forall A \qquad \frac{\Gamma, \alpha[t_1/x], \ldots, \alpha[t_k/x], \forall x\alpha, \Delta \Rightarrow \Lambda}{\Gamma, \forall x\alpha, \Delta \Rightarrow \Lambda}$$

The complete set of decomposition rules is $\mathbf{D}^{12} = \mathbf{D}^8 \cup \mathbf{D}^4$.

The following properties can be quoted as particularly useful for the inference process.

(a) The order of the component formulas of a sequent is immaterial.

(b) A sequent with a contradiction in the Antecedent is valid.

(c) A sequent with a tautology in the Succedent is valid.

(d) A sequent with two equivalent components one in the Antecedent and one in the Succedent is valid.

(e) The application of any rule of $\mathbf{D}^{12}$ to a valid sequent produces valid sequents.

(f) A sequent from which valid sequents are derived by applying any rule of $\mathbf{D}^{12}$ is valid.

(g) The order of application of the rules of $\mathbf{D}^{12}$ to a sequent is immaterial.

The proving procedure P, to verify the validity of a given sequent, takes a sequent as input and returns a set of sequents. If the input sequent is valid, Procedure P always terminates with an empty output set. If the input sequent is not valid, then Procedure P returns (when terminating) a not empty set of sequents.

Procedure P applies the decomposition rules to the input sequent: at each step the rule to be applied is automatically selected on the basis of a given priority criterium. Different priority criteria produce different degrees of efficiency.

Starting from a given sequent, Procedure P applies the decomposition rules D^{12} and discards all the sequents recognized as valid according to the basic properties of the sequent calculus. When Procedure P terminates, it returns either an empty set, if all the sequents have been discarded, or a set of *atomic sequents* (i.e. undecomposable sequents), if no more rules can be applied. In the first case the input sequent is valid.

A reasoning procedure R is defined to reason on a not valid sequent in order to generate/abduce new formulas, that, when added to the not valid sequent, make it valid.

Other inference rules are necessary to compose sequents from atomic sequents, i.e. by introducing connectives and quantifiers. These rules are:

$\underline{S\neg}$
$$\frac{\Gamma, \neg\alpha_1, \Delta \Rightarrow \Lambda}{\Gamma, \Delta \Rightarrow \alpha_1, \Lambda}$$

$\underline{A\wedge}$
$$\frac{\Gamma, \alpha_1 \wedge \alpha_2, \Delta \Rightarrow \Lambda}{\Gamma, \alpha_1, \alpha_2, \Delta \Rightarrow \Lambda}$$

$\underline{A\neg}$
$$\frac{\Gamma, \Delta \Rightarrow \neg\alpha_1, \Lambda}{\Gamma, \alpha_1, \Delta \Rightarrow \Lambda}$$

$\underline{S\vee}$
$$\frac{\Gamma, \Delta \Rightarrow \alpha_1 \vee \alpha_2, \Lambda}{\Gamma, \Delta \Rightarrow \alpha_1, \alpha_2, \Lambda}$$

The complete sets of rules for the generation and abduction processes are:

$\mathbf{G} = \mathbf{D}^{12} \cup \{\underline{S\neg}, \underline{A\wedge}\}$,

$\mathbf{A} = \mathbf{D}^{12} \cup \{\underline{A\neg}, \underline{S\vee}\}$.

Let $X \Rightarrow Y$ be a not valid sequent input of Procedure P. Let S be the output of Procedure P. Procedure R operates on the set S by applying the rules of the sets $\mathbf{G}$ and $\mathbf{A}$ and, by a combinatorial process, constructs two sets of formulas with connectives and quantifiers according to the decomposition process done by Procedure P and stored in the auxiliary data structures. Given a formula γ of the first set and a formula α of the second set generated by R, the following sequents:

$$X \Rightarrow Y, \gamma \qquad \text{and} \qquad X, \alpha \Rightarrow Y$$

are valid, on the basis of a theorem stated in [1].

The proposed approach always provides a solution to problems in propositional logic. In first order logic one of the characteristics of the proposed machine, namely the use in interactive mode, turns out to be very useful in order to stop the computation or to change proving strategy.

3 The Object Oriented Design

The proposed deduction tool is presently implemented using the language LOGLAN [4]. This is an object oriented language, with a PASCAL-like syntax and strictly influenced by SIMULA. It has been designed and implemented at the Institute of Informatics of Warsaw University and it has a good degree of portability.

A LOGLAN program is a set of separately compilable units, each one constituted by two parts: a declarative part and an executive part. Units as blocks, functions and procedures are the usual ones; units as classes, coroutines and processes are addressable units and have a different lifetime. Those units can, indeed, be allocated, remote referred and deallocated by a programmer's decision in the framework of a memory model with a virtual addressing system which is able of avoiding the dangling reference problem.

The implementation of our machine benefits from some fundamental features of this language: the *type class* units, the *multilevel prefixing*, the *virtual definition* of functions and procedures, the *type coroutine* units, the *parameterized types*, and the *exception handling*.

A class unit with prefixing allows one to build objects following a hierarchical structured specification at compile as well as at run time. On the basis of this feature, together with the virtual definition of functions and procedures and with the parameterized types, a flexible object concatenation is made possible, thus obtaining all the fundamental advantages of the object oriented paradigm, (e.g. redefinition and specialization of attributes and methods, parameterized units). Exception handling, embedded in the object oriented environment, is the basic feature for driving the execution after some errors have occurred. Coroutines turn to be very useful in order to define active entities, cooperating one with the other in a quasi-parallel fashion.

The object oriented methodology has proved to be a good approach in order to build an environment devoted to the treatment of mathematical objects. In fact the intrinsical structure of these objects fits perfectly with the modular design of an object oriented programming language [5]. In this context the aim of adding deductive capabilities can be pursued with the crucial benefit of the uniform semantics of the entire computation. Then, the basic idea of embedding an automated extended deduction tool in an object oriented environment implies to consider the mathematical objects both as computable entities and as objects owning properties on which different inference steps can be carried on. Moreover, the sequent calculus is itself a structured method and can be easily represented following an object oriented approach. As it is shown in the following figure, the propositional logic calculus is the ground level of the entire calculus and the decomposition rules for the connectives are its basic methods.

On top of this level, inheriting all the already defined methods and adding those concerning the quantified formulas, a predicative calculus level is defined.

The third level is the sequent calculus level which owns the specific methods of that calculus both for proving and for reasoning on the given sequents.

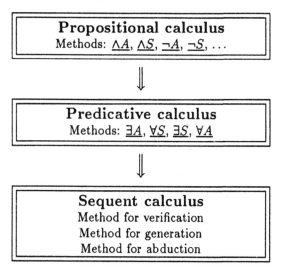

The design of the deductive tool turns to be consistent with the modularity of the general approach also from another point of view: a strictly hierarchical organization has been chosen in order to distinguish the parts which are problem dependent (i.e. the selected and implemented rules which, on the basis of this choice, can be easily changed) from the strategy under which the computation is carried out.

The flexibility of the strategy has been considered as a very crucial point, allowing one to change it at any time of the execution process. This objective is pursued by mantaining the possibility of changing the priority of the application of the transformation rules, with respect to the standard priority specified in the proof procedure, always preserving the soundness of the method. Moreover, also a total change of the searching process is possible, if necessary.

An important characteristic of the tool is the possibility of monitoring the system and eventually of stopping its execution, for example when the computation becomes too long. Then, a totally interactive (step by step) way of operating is offered and this last feature makes the deduction tool similar to a desk top sequent calculus machine which, on the basis of an input command, operates on a given sequent by applying a specified transformation rule, and furnishes as output some sequents partially decomposed or recomposed.

The interactivity of the method turns out to be very useful when steps of recomposition in the generative and abductive approaches have to be performed.

Actually the choice of the formulas to be considered is up to the user, but different automatic solutions can be proposed, defining intelligent strategies or user friendly interface components, such as automatic masks, in order to give relevance to some special elements of the formulas.

4 Implementation Remarks

In the design of the sequent calculus machine all the program units are organized using the multilevel prefixing feature offered by LOGLAN, in order to avoid any software duplication and to increase the software correctness.

The system is set up with two classes: the AND-OR-TREE class devoted to build the computation trees (i.e., the proving and the reasoning trees) and the KNOW-HOW class that embodies the rules of the sequent calculus.

A node of a computation tree is represented by the coroutine NODE that owns (as attributes) its level in the tree structure, the number of possible successors, a status flag to indicate the possibility of further expansions and pointers to other nodes of the tree. Together with these variable attributes, the coroutine also owns some virtual attributes, redefined in the problem oriented part, which allow one to operate on the node, to generate another node and to verify important computational conditions (e.g., the termination of the decomposition process).

Let us note that a node has been represented by a coroutine in order to allow further developments of the program, possibly, requiring an active role of the unit node (e.g., the definition of an intelligent searching strategy).

In class KNOW-HOW the main element is the class SEQUENT, which turns out to be an extension (using the prefixing) of coroutine NODE. In the same class inference rules are defined as subclasses following a hierarchical structure according to their computational characteristics.

The execution starts from procedure STARTING of class KNOW-HOW. An input sequent is read, its syntax is verified and the redundant quantifiers are removed. Then the auxiliary (global) data structures are initialized on the basis of the variables and constants of the input sequent. It is up to the user to choose between an automatic or an interactive execution.

5 Example of Working Sessions

In this section, we present three examples of interactive sessions for solving verificative, generative and abductive problems, respectively. The three problems derive from a single model of first order logic problem proposed in [6].

During the execution a deduction tree is built, whose root is the input sequent. In this tree nodes are labeled by sequents and numbered. At each step of the interactive session the sequent calculus machine provides the user with sufficient information to allow him to select the successive step of computation.

For instance, the machine provides the list of applicable rules for each (sequent) node together with the effect of their application. In this case, the user can select the appropriate rule on the basis of a precise knowledge of its effect.

5.1 Verificative case

In this working session the input sequent is

$$(1) \quad \forall x(U(x) \supset \neg B(x)), \exists x(B(x) \wedge D(x)) \Rightarrow$$

and it is associated to the root of the deduction tree as NODE(1).

The session starts by providing the user with information on the possible next steps. In this case, two rules are applicable to the NODE(1), the first rule to the second component formula and the second rule to the first component formula of the sequent (1), respectively.

$NODE(1)$. Applicable rules:
 1) Rule $\underline{\exists A}$ (Formula 2):
    $\forall x(U(x) \supset \neg B(x)), B(y_1) \wedge D(y_1) \Rightarrow$
 2) Rule $\underline{\forall A}$ (Formula 1):
    $\forall x(U(x) \supset \neg B(x)), U(x_1) \supset \neg B(x_1), \exists x(B(x) \wedge D(x)) \Rightarrow$

If the user chooses rule one, by using the command SELECT_RULE, the resulting sequent(s) labelling the current leaf node(s) of the tree are the following:
LEAF NODE(S):
    $NODE(2) : \forall x(U(x) \supset \neg B(x)), B(y_1) \wedge D(y_1) \Rightarrow$

The successive steps of the interactive session proceed analogously.

$NODE(2)$. Applicable rules:
 1) Rule $\underline{\wedge A}$ (Formula 2):
    $\forall x(U(x) \supset \neg B(x)), B(y_1), D(y_1) \Rightarrow$
 2) Rule $\underline{\forall A}$ (Formula 1):
    $\forall x(U(x) \supset \neg B(x)), U(x_1) \supset \neg B(x_1), U(y_1) \supset \neg B(y_1), B(y_1) \wedge D(y_1)) \Rightarrow$

SELECT_RULE 2

LEAF NODE(S):
    $NODE(3) : \forall x(U(x) \supset \neg B(x)), U(x_1) \supset \neg B(x_1), U(y_1) \supset \neg B(y_1), B(y_1) \wedge D(y_1) \Rightarrow$

$NODE(3)$. Applicable rules:
 1) Rule $\underline{\wedge A}$ (Formula 4):
    $\forall x(U(x) \supset \neg B(x)), U(x_1) \supset \neg B(x_1), U(y_1) \supset \neg B(y_1), B(y_1), D(y_1) \Rightarrow$
 2) Rule $\underline{\supset A}$ (Formula 2):
    $\forall x(U(x) \supset \neg B(x)), U(y_1) \supset \neg B(y_1), B(y_1) \wedge D(y_1)) \Rightarrow U(x_1)$
    and:
    $\forall x(U(x) \supset \neg B(x), \neg B(x_1), U(y_1) \supset \neg B(y_1), B(y_1) \wedge D(y_1) \Rightarrow$

3) Rule $\supset A$ (Formula 3):
   $$\forall x(U(x) \supset \neg B(x)), U(x_1) \supset \neg B(x_1), B(y_1) \land D(y_1)) \Rightarrow U(y_1)$$
   and:
   $$\forall x(U(x) \supset \neg B(x)), U(x_1) \supset \neg B(x_1), \neg B(y_1), B(y_1) \land D(y_1) \Rightarrow$$

Note that the rule $\forall A$ is no more applicable to the (sequent) NODE(3).
SELECT_RULE 1

LEAF NODE(S):
   $NODE(4)$: $\forall x(U(x) \supset \neg B(x)), U(x_1) \supset \neg B(x_1), U(y_1) \supset \neg B(y_1), B(y_1), D(y_1) \Rightarrow$

$NODE(4)$. Applicable rules:
   1) Rule $\supset A$ (Formula 2):
      $$\forall x(U(x) \supset \neg B(x)), U(y_1) \supset \neg B(y_1), B(y_1), D(y_1) \Rightarrow U(x_1)$$
      and:
      $$\forall x(U(x) \supset \neg B(x)), \neg B(x_1), U(y_1) \supset \neg B(y_1), B(y_1), D(y_1) \Rightarrow$$
   2) Rule $\supset A$ (Formula 3):
      $$\forall x(U(x) \supset \neg B(x)), U(x_1) \supset \neg B(x_1), B(y_1), D(y_1)) \Rightarrow U(y_1)$$
      and:
      $$\forall x(U(x) \supset \neg B(x)), U(x_1) \supset \neg B(x_1), \neg B(y_1), B(y_1), D(y_1) \Rightarrow$$

SELECT_RULE 2

LEAF NODES:
   $NODE(5) : \forall x(U(x) \supset \neg B(x)), U(x_1) \supset \neg B(x_1), B(y_1), D(y_1) \Rightarrow U(y_1)$
   $NODE(6) : \forall x(U(x) \supset \neg B(x)), U(x_1) \supset \neg B(x_1), \neg B(y_1), B(y_1), D(y_1) \Rightarrow$
(axiom)

Note that NODE(6) is an *axiom* (i.e., a valid sequent), and, as such, can be discarded.

$NODE(5)$. Applicable rules:
   1) Rule $\supset A$ (Formula 2):
      $$\forall x(U(x) \supset \neg B(x)), B(y_1), D(y_1) \Rightarrow U(y_1), U(x_1)$$
      and:
      $$\forall x(U(x) \supset \neg B(x)), \neg B(x_1), B(y_1), D(y_1) \Rightarrow U(y_1)$$

SELECT_RULE 1

LEAF NODE(S):
   $NODE(7) : \forall x(U(x) \supset \neg B(x)), B(y_1), D(y_1) \Rightarrow U(y_1), U(x_1)$
   $NODE(8) : \forall x(U(x) \supset \neg B(x)), \neg B(x_1), B(y_1), D(y_1) \Rightarrow U(y_1)$

Note that no rules are applicable to the (sequent) NODE(7) that remains as a leaf node, and the computation proceeds only on NODE(8).

$NODE(8)$. Applicable rules:
   1) Rule $\neg A$ (Formula 2):
      $$\forall x(U(x) \supset \neg B(x)), B(y_1), D(y_1) \Rightarrow U(y_1), B(x_1)$$

SELECT_RULE 1

LEAF NODE(S):

$NODE(7): \forall x(U(x) \supset \neg B(x)), B(y_1), D(y_1) \Rightarrow U(y_1), U(x_1)$

$NODE(9): \forall x(U(x) \supset \neg B(x)), B(y_1), D(y_1) \Rightarrow U(y_1), B(y_1)$

At this point, there are no more sequents to be decomposed, and the computation terminates. Since leaf nodes are not axioms the input sequent is not valid.

It has to be noted that the same deduction tree (and therefeore the same leaves) would have been obtained by an automatic execution.

5.2 Generative case

Let us now present a second working session to solve a generative problem. The problem consists in finding formula(s) such that, once added to the Succedent of the sequent (1), as in the previous example, the resulting sequent is valid.

Starting from the leaf nodes obtained at the end of the previous session, the machine provides the user with the following two sequents

(7) : $\quad B(y_1), D(y_1), \neg U(y_1), \neg U(x_1) \Rightarrow$

(9) : $\quad B(y_1), D(y_1), \neg U(y_1), \neg B(x_1) \Rightarrow$

from which some literals (at least one) have to be chosen, interactively. If the user chooses $D(y_1)$ and $\neg U(y_1)$, the sequent calculus machine builds the following wff:

$$\exists x(D(x) \wedge \neg U(x))$$

which can be added to the Succedent of the input sequent, thus obtaining the following valid sequent:

(2) $\quad \forall x(U(x) \supset \neg B(x)), \exists x(B(x) \wedge D(x)) \Rightarrow \exists x(D(x) \wedge \neg U(x))$

5.3 Abductive case

In this third session, devoted to solve an abductive problem, the input is the following sequent (3), derived from valid sequent (2), by eliminating one of the formulas from its Antecedent:

(3) $\quad \forall x(U(x) \supset \neg B(x)) \Rightarrow \exists x(D(x) \wedge \neg U(x))$

The machine decomposes this sequent in its elementary parts (atomic sequents), by building the proof tree in automatic mode, thus obtaining three not valid sequents as leaf nodes. Then the following sequents are provided and displayed to the user:

(6): $\Rightarrow D(x_1), U(x_1)$

(8): $\Rightarrow D(x_1), B(x_1)$

(12): $\Rightarrow B(x_1), \neg U(x_1)$

If the user chooses $D(x_1)$ from sequents (6) and (8), and $B(x_1)$ from sequent (12), the machine builds the following wff:

$$\exists x(D(x) \wedge B(x))$$

The addition of such a wff in the Antecedent of the input sequent (3) produces a valid sequent, namely the sequent (2).

6 Conclusions and Further Developments

In this paper a tool for proving and reasoning by a sequent calculus is presented both from the methodological and implementation points of view, and its use as a desk top logic machine is stressed.

The system architecture has been designed with the propositional calculus as a kernel. In fact, when adding the rules for the manipulation of quantifiers, the general philosophy does not change. The architecture is open to the insertion of a flexible strategy block (as proposed in [7]). On the basis of various experiments, it appears appropriate to provide the user with several strategies to choose the most useful sequents among those derived by the machine. Actually the machine may construct, in automatic mode, all the possible formulas obtained by combining the not valid sequents produced by the proving process, or it may ask the user (as in the given examples) to select appropriate component formulas, in an interactive mode.

Intelligent strategies may drive the user in this selection process. Those strategies are strictly related with the semantical context of the domain considered for the application and, therefore, cannot be defined at a general level.

For instance, in the context of mathematical problems, by interpreting logic predicates as known predefined properties, it is possible to provide the user with a strategy which builds formulas including only some of those predicates. Analogously, an alternative strategy can build formulas with a predefined syntactic form, when, with respect to a given interpretation, the aim is that of validating a formula by using some predefined premises.

References

1. Cioni G., Colagrossi A., Miola A.:   A sequent Calculus for Automated Reasoning, submitted for the pubblication, (1991).
2. Limongelli C., Miola A., Temperini M.: Design and Implementation of Symbolic Computation Systems, Proc. IFIP W.G. 2.5 Conference, North Holland, (1991).
3. Bonamico S., Cioni G., Colagrossi A.:   A Gentzen Based Deduction Method for mathematical problems solving, in Computer Systems and Applications, E. Balagurusamy and B. Sushila (eds.), Tata McGraw-Hill, (1990).
4. Kreczmar A., Salwicki A., Warpechowski M.: LOGLAN'88 - Report on the Programming Language, LNCS 414, Springer Verlag, (1990).

5. Regio M., Temperini M.: Object Oriented Methodology for the Specification and Treatment of Mathematical Objects, in Computer Systems and Applications, E. Balagurusamy and B. Sushila (eds.), Tata McGraw-Hill, (1990).

6. Chang C., Lee R.C.: Logica simbolica, Tecniche Nuove, (1988).

7. Bonamico S., Cioni G.: Embedding flexible control strategies into object oriented languages, Proc. AAECC 6, LNCS 357, Springer Verlag, (1988).

Gentzen-style Characterizations of Negation as Failure

Jan A. Plaza

University of Miami
Department of Mathematics and Computer Science
P.O. Box 249 085
Coral Gables, Florida 33124
U.S.A.
janplaza@math.miami.edu

Abstract. We investigate Negation as Failure as incorporated in (nondeterministic) SLDNF-resolution and in (deterministic) PROLOG's resolution. We formulate three Gentzen-style systems which characterize propositional SLDNF-resolution and propositional PROLOG's resolution in a sound and complete way (without assuming stratification or any other restrictions on programs.) Our analysis employs certain three-valued logics. The results of this paper can be of interest for rule-based expert systems which represent knowledge in propositional logic, and for other AI-systems which use Negation as Failure for non-monotonic reasoning.

Keywords: negation as failure, SLDNF-resolution, PROLOG, logic programming, Gentzen systems, soundness and completeness, many-valued logics, nonmonotonic reasoning, rule-based expert systems, knowledge representation in logic.

1 Introduction

Negation as Failure became a predominant method for handling negative information or lack of information in AI-systems. It constitutes a part of (nondeterministic) SLDNF-resolution and (deterministic) PROLOG's resolution. Unlike the classical negation, Negation as Failure allows nonmonotonic effects, unlike Closed World Assumption it is computationally feasible. Despite its widespread use of Negation as Failure its logical properties are not yet fully understood.

The problem of finding nonprocedural characterizations of SLDNF-resolution has been studied from the early days of logic programming. Characterizations of SLD-resolution and programs without negation have been found in early 1970's. Finding characterizations for programs with negation (interpreted as Negation as Failure) proved more difficult.

Here, we will restrict our attention to propositional programs. Given program P and a literal L as a query, SLDNF-resolution returns the answer *yes*, or returns the

---

* This research has been supported from the University of Miami's Summer '92 Award in Natural Sciences and Engineering

answer *no*, or loops infinitely. So, one would like to have three general conditions depending on P and L which are necessary and sufficient for getting answers *yes*, *no*, or no answer from the SLDNF. First such necessary conditions were formulated by K.L. Clark [6] in his soundness theorems; however they were not sufficient conditions. Further work of many researchers aimed at finding sufficient conditions, or in other words, at obtaining completeness. This has been done by restricting severely the class of programs and goals (by requiring various kinds of stratification) so that Clark's condition could be also sufficient. Other attempts, in which many-valued logics were used, resulted only in soundness theorems. In References we list some research papers related to both directions. In yet another approach D. Gabbay proved soundness and completeness of for the unrestricted class of propositional programs using a complex notion of program completion in the modal logic of arithmetical provability [14]. The author of this paper proved soundness and completeness for the unrestricted class of propositional programs using simple and intuitive notions of program completions and employing either classical, intuitionistic, intermediate or modal logics [30], [31].

In this way Negation as Failure has been characterized in the environment of SLDNF-resolution but not in PROLOG. Let us recall that SLDNF is considered an abstract (and approximate) model of PROLOG. At the level of propositional programs the two systems differ in the following: derivations of SLDNF involve non-deterministic choices, while PROLOG employs a height first left-to-right search in the tree of all possible derivations. We got convinced that mentioned above characterizations of SLDNF-resolution do not generalize to the situation of PROLOG, and if such a characterization is sought, another system is needed. As demonstrated in this paper Gentzen-style systems are suitable for this task.

The paper is organized as follows. Section 2 contains definitions of which will be used throughout the paper. Section 3 recalls some three-valued logics. In Section 4 we define Gentzen-style system **G1**, and formulate a completeness theorem for SLDNF-resolution. In Section 5 we show how to extend results of the previous section to obtain characterizations of PROLOG's resolution. In Section 6 we formulate two systems alternative to **G1**. Section 7 contains a conclusion. Section 8 contains proofs.

2 Propositional Programs

In this section we extend SLDNF and PROLOG's resolution to handle non-normal programs. In general we assume the terminology, definitions and basic results from [1] or from chapters 1-3 of [26] but some additions are needed.

Let us recall two well known transformations of propositional formulas. We will be concerned only with those formulas B which are built from propositional letters, truth constant $\top$ and falsehood constant $\bot$ by means of conjunction $\land$, disjunction $\lor$ and negation $\neg$.

The *deMorgan operation* transforms formula B into formula B^{deM} by using de-Morgan laws to push negations down to the atomic level and replacing $\neg\neg A$ by A, $\neg\top$ by $\bot$, and $\neg\bot$ by $\top$. A formula B is said to be in *deMorgan form* if $B^{deM} = B$. For example: $(\neg(p \land \neg(q \lor \neg\bot)) \land \neg\neg p)^{deM} = (\neg p \lor (q \lor \top)) \land p$.

The *dnf operation* transforms formula B into B^{dnf} by using deMorgan oration first, and then by using distributivity laws $B_1 \land (B_2 \lor B_3) \equiv (B_1 \land B_2) \lor (B_1 \land B_3)$

and $(B_1 \lor B_2) \land B_3 \equiv (B_1 \land B_3) \lor (B_2 \land B_3)$ to obtain disjunctive normal form. (It will be important that our dnf algorithm does not use any other laws, such as commutativity laws or simplifications $B \lor \neg B \equiv \top$, $B \land \neg B \equiv \bot$, etc.)

The following definition generalizes the notion of a program clause.

Definition 1.

By a *propositional program statement* we understand any formula $A \leftarrow B$ where A, called a *head*, is a propositional letter, and B, called a *body*, is built from propositional letters, $\top$, $\bot$, by means of $\land$, $\lor$, and $\neg$.

By a *propositional program* we understand any finite sequence of propositional statements. (The order of the statements is not essential for SLDNF-resolution but it is essential for PROLOG's resolution.)

By a *propositional goal* we understand any formula $\leftarrow B$, where B, called a *body*, is built from propositional letters, $\top$ and $\bot$, by means of $\land$, $\lor$ and $\neg$. For any propositional goal, we consider the formula $\bot$ as its *head*.

By a *propositional statement* we understand either a propositional program statement or a propositional goal.

Two sequences Γ_1, Γ_2 of propositional statements are considered *equal*, $\Gamma_1 = \Gamma_2$, if for any A (being a propositional letter or $\bot$) the relative order of the statements which have A as their head is identical in Γ_1 and Γ_2.

IF Γ is a sequence of propositional statements and γ is a propositional statement, we write Γ, γ to denote the concatenation of Γ and $\langle \gamma \rangle$.

In propositional programs $\neg$ will be interpreted not as classical negation but as Negation as Failure.

The following transformations of propositional statements will be used in the sequel.

Definition 2. Let Γ be a sequence of propositional statements.

1. *Clark's transformation* Γ^C is obtained as follows. For every propositional letter A in Γ, list in their relative order all the statements which have A in their heads: $A \leftarrow B_1, \ldots, A \leftarrow B_n$, and form $A \leftarrow B_1 \lor \ldots \lor B_n$. (If there are no statements with head A, this construction gives $A \leftarrow \bot$.) Also, if there are any propositional goals in Γ, list them: $\leftarrow B_1, \ldots, \leftarrow B_n$, and form $\leftarrow B_1 \lor \ldots \lor B_n$. Sequence Γ^C consists of all the statements obtained in this way (in any order.)
2. Γ is said to be in *Clark's form* if $\Gamma^C = \Gamma$.
3. *dnf transformation* Γ^{dnf} is obtained by replacing any statement's body B by B^{dnf}.
4. Γ is said to be in *dnf* if $\Gamma^{dnf} = \Gamma$.
5. Γ is said to be in *Clark's dnf* if $(\Gamma^C)^{dnf} = \Gamma$
6. If Γ in dnf, Γ^n is obtained by replacing each statement $A \leftarrow B_1 \lor \ldots \lor B_n$, where $B_1, \ldots, B_n$ are conjunctions of literals, by a sequence $A \leftarrow B_1, \ldots, A \leftarrow B_n$. (In this way a normal program results.)
7. Γ is *normal* if $(\Gamma^{dnf})^n = \Gamma$.
8. Γ^e is obtained by repeatedly replacing any statement $A \leftarrow B$ (with A being a propositional letter or $\bot$), where B is not a conjunction of literals, according to the following rules. If B is $B_1 \land B_2$, introduce new propositional letters A_1, A_2,

and replace $A \leftarrow B$ by the sequence $A \leftarrow A_1 \wedge A_2$, $A_1 \leftarrow B_1$, $A_2 \leftarrow B_2$. If B is $B_1 \vee B_2$, replace $A \leftarrow B$ by the sequence $A \leftarrow B_1$, $A \leftarrow B_2$. If B is $\neg B_1$, introduce a new propositional letter A_1, and replace $A \leftarrow B$ by the sequence $A \leftarrow \neg A_1$, $A_1 \leftarrow B$.

SLDNF-resolution was defined originally for a normal program and single normal goal, cf. [1], [26]. A generalization to a situation of a normal program and multiple normal goals is straightforward. The next definition introduces also a convenient notation.

Definition 3. We define function $SLDNF$ which takes an argument a sequence of propositional statements containing a propositional goal and returns one of the values Y, or N, or U. Let P be a normal propositional program and let $\leftarrow B_1, \ldots, \leftarrow B_n$ be a nonempty sequence of normal propositional goals.

$SLDNF(P, \leftarrow B_1, \ldots, \leftarrow B_n) = Y$ if or some i there is an SLDNF-refutation of $P; \leftarrow B_i$. (This can be read: "SLDNF-computation for $P, \leftarrow B_1, \ldots, \leftarrow B_n$ returns answer YES".)

$SLDNF(P, \leftarrow B_1, \ldots, \leftarrow B_n) = N$ if for each i there is a finitely failed SLDNF-tree for $P; \leftarrow B_i$. (This can be read: "SLDNF-computation for $P, \leftarrow B_1, \ldots, \leftarrow B_n$ returns answer NO".)

$SLDNF(P, \leftarrow B_1, \ldots, \leftarrow B_n) = U$ if neither of the cases above holds. (This can be read: "SLDNF-computation for $P, \leftarrow B_1, \ldots, \leftarrow B_n$ loops infinitely".)

Proposition 4. *Let Γ be a sequence of propositional statements containing a propositional goal. Then: $SLDNF(\Gamma^e) = SLDNF((\Gamma^{dnf})^n)$.*

Due to this proposition we can accept the following generalization of SLDNF-resolution to (possibly non-normal) propositional programs as reasonable.

Definition 5. Let Γ be a sequence of propositional statements containing a propositional goal. We define $SLDNF(\Gamma)$ as $SLDNF(\Gamma^e)$.

By PROLOG's resolution (for a normal program and a single normal goal) we understand deterministic version of SLDNF-resolution which searches the tree of all possible derivations in a depth-first, left-to-right manner. If an infinite branch is entered, PROLOG does not return any answer. If a branch fails finitely, backtracking occurs. So, by a *PROLOG-tree* for $P, \leftarrow B$ we understand an SLDNF-tree with the subgoals selected according to the computation rule "select the leftmost". By a *PROLOG-refutation* of $P, \leftarrow B$ we understand a successful branch in the PROLOG-tree such that all the branches to the left of it fail finitely. (If this explanation is not sufficient, the reader can look at Definition 34.)

As explained in the following definition, PROLOG's resolution can be extended to (possibly non-normal) propositional programs in a way similar SLDNF but with a modification to accommodate its deterministic character.

Definition 6. We define function $PROLOG$ which takes an argument a sequence of propositional statements containing a propositional goal and returns one of the values Y, or N, or U. Let P be a normal propositional program and let $\leftarrow B_1, \ldots, \leftarrow B_n$ be a nonempty sequence of normal propositional goals.

$PROLOG(P, \leftarrow B_1, \ldots, \leftarrow B_n) = Y$ if there exists i such that there is a PROLOG-refutation of $P; \leftarrow B_i$, and for every $j < i$ there exists a finitely failed PROLOG-tree for $P; \leftarrow B_j$. (This can be read: "PROLOG-computation for $P, \leftarrow B_1, \ldots, \leftarrow B_n$ returns answer YES".)

$PROLOG(P, \leftarrow B_1, \ldots, \leftarrow B_n) = N$ if for each $i = 1, \ldots, n$ there is a finitely failed PROLOG-tree for $P; \leftarrow B_i$. (This can be read: "PROLOG-computation for $P, \leftarrow B_1, \ldots, \leftarrow B_n$ returns answer NO".)

$PROLOG(P, \leftarrow B_1, \ldots, \leftarrow B_n) = U$ if neither of the cases above holds. (This can be read: "PROLOG-computation for $P, \leftarrow B_1, \ldots, \leftarrow B_n$ loops infinitely".)

We extend this function to possibly non-normal programs.

Definition 7. Let Γ be a sequence of propositional statements containing a propositional goal. We define $PROLOG(\Gamma)$ as $PROLOG(\Gamma^e)$.

Similarly as for SLDNF we have a proposition which shows that this definition is reasonable.

Proposition 8. *Let Γ be a sequence of propositional statements containing a propositional goal. Then: $PROLOG(\Gamma^e) = PROLOG((\Gamma^{dnf})^n)$.*

3 Many-valued logics

In this section we recall two important non-classical logics. In both logics the formulas are built from propositional letters, $\top$ and $\bot$, by means of $\wedge, \vee$ and $\neg$. In the semantics the following truth-values are used: Y (yes/true), N (no/false), and U (nonterminating/undefined).

In the *three-valued logic of lazy left-to-right evaluation* the operations of conjunction, disjunction and negation on the truth-values are defined as in the following tables:

$\wedge$	Y	N	U
Y	Y	N	U
N	N	N	N
U	U	U	U

$\vee$	Y	N	U
Y	Y	Y	Y
N	Y	N	U
U	U	U	U

$\neg$	
Y	N
N	Y
U	U

where the first argument of an operation is specified on the left border of the table, and the second argument (of $\wedge$ or $\vee$) is specified at the top border. Notice that the operations are not symmetrical: $U \wedge N \neq N \wedge U$, also $U \vee Y \neq Y \vee U$. Let us remark that this logic is used in LISP which denotes N by nil, Y by any atom different from nil, and in which U corresponds to the situation when the computation of a truth value does not terminate.

In the *three-valued logic of lazy parallel evaluation* (called also Kleene's strong three-valued logic) the operations of conjunction, disjunction and negation on the truth-values are defined as in the following tables:

$\wedge$	Y	N	U
Y	Y	N	U
N	N	N	N
U	U	N	U

$\vee$	Y	N	U
Y	Y	Y	Y
N	Y	N	U
U	Y	U	U

$\neg$	
Y	N
N	Y
U	U

Now the operations are symmetrical.

Given a valuation $\nu : Prop \rightarrow \{Y, N, U\}$ of propositional letters, for any formula B, its truth-value $\nu(B)$ is computed according to the tables above (with $\top$ interpreted as Y, and $\bot$ interpreted as N). Two formulas B_1, B_2 are called *equivalent*, if for every valuation ν, $\nu(B_1) = \nu(B_2)$.

In both logics defined above the laws of associativity, distributivity and double negation hold. The laws of commutativity hold in the logic of lazy parallel evaluation, but not in the logic of lazy left-to right evaluation. Finally, in either of the two logics $B \vee \neg B$ is not equivalent to $\top$ because $U \vee \neg U \neq N$.

The logics mentioned above are closely related to SLDNF-resolution and to PROLOG.

Proposition 9. *Let P be a propositional program and $\leftarrow B_1$ and $\leftarrow B_2$ be propositional goals.*

1. *If $\wedge, \vee$ and $\neg$ are interpreted as in the logic of lazy parallel evaluation then:*
   $SLDNF(P, \leftarrow \neg B_1) = \neg SLDNF(P, \leftarrow B_1)$
   $SLDNF(P, \leftarrow B_1 \wedge B_2) = SLDNF(P, \leftarrow B_1) \wedge SLDNF(P, \leftarrow B_2)$
   $SLDNF(P, \leftarrow B_1 \vee B_2) = SLDNF(P, \leftarrow B_1) \vee SLDNF(P, \leftarrow B_2)$
2. *If $\wedge, \vee$ and $\neg$ are interpreted as in the logic of lazy left-to-right evaluation then:*
   $PROLOG(P, \leftarrow \neg B_1) = \neg PROLOG(P, \leftarrow B_1)$
   $PROLOG(P, \leftarrow B_1 \wedge B_2) = PROLOG(P, \leftarrow B_1) \wedge PROLOG(P, \leftarrow B_2)$
   $PROLOG(P, \leftarrow B_1 \vee B_2) = PROLOG(P, \leftarrow B_1) \vee PROLOG(P, \leftarrow B_2)$

Definition 10.

1. Two propositional goals $\leftarrow B_1$ and $\leftarrow B_2$ are called *SLDNF-equivalent* if for every propositional program P, $SLDNF(P, \leftarrow B_1) = SLDNF(P, \leftarrow B_2)$.
2. Two propositional goals $\leftarrow B_1$ and $\leftarrow B_2$ are called *PROLOG-equivalent* if for every propositional program P, $PROLOG(P, \leftarrow B_1) = PROLOG(P, \leftarrow B_2)$.

For instance $\leftarrow \neg\neg B$ and $\leftarrow B$ are SLDNF-equivalent and PROLOG-equivalent. $\leftarrow B_1 \wedge B_2$ and $\leftarrow B_2 \wedge B_1$ are SLDNF-equivalent but not PROLOG-equivalent. $\leftarrow B \vee \neg B$ and $\leftarrow \top$ are neither SLDNF-equivalent nor PROLOG-equivalent. We leave to the reader to determine whether $\leftarrow B_1 \wedge (B_2 \vee B_1)$ and $\leftarrow B_1$ are SLDNF-equivalent and whether they are PROLOG-equivalent.

From Proposition 9 we can obtain the following characterization of equivalence of goals.

Corollary 11. *Let $\leftarrow B_1, \leftarrow B_2$ be propositional goals.*

1. *$\leftarrow B_1$ and $\leftarrow B_2$ are SLDNF-equivalent if and only if B_1 and B_2 are equivalent in the three-valued logic of lazy parallel evaluation.*
2. *$\leftarrow B_1$ and $\leftarrow B_2$ are PROLOG-equivalent if and only if B_1 and B_2 are equivalent in the three-valued logic of lazy left-to-right evaluation.*

4 Gentzen-style Characterization of SLDNF-resolution

The Gentzen-style system which will be defined in this section deals with *sequents* such as $P \vdash B$ where P is a propositional program in Clark's form, and $\leftarrow B$ is a propositional goal in the language of P. The intuitive meaning of $P \vdash B$ is "SLDNF-resolution with program P and goal $\leftarrow B$ returns answer Y". Later we will use this system to characterize also PROLOG's resolution.

Below, we write A to denote arbitrary propositional letter. If we write $P, A \leftarrow B$ we assume that the program $P \cup \{A \leftarrow B\}$ is in Clark's form, which implies that P doesn't contain any statement with the head A.

Definition 12. The system G1 consists of the following rules:

$$
\begin{array}{llll}
P \vdash B_1 \lor B_2 & \Longleftrightarrow P \vdash B_1 & \text{or} & P \vdash B_2 \\
P \vdash \neg(B_1 \lor B_2) & \Longleftrightarrow P \vdash \neg B_1 & \text{and} & P \vdash \neg B_2 \\
P \vdash B_1 \land B_2 & \Longleftrightarrow P \vdash B_1 & \text{and} & P \vdash B_2 \\
P \vdash \neg(B_1 \land B_2) & \Longleftrightarrow P \vdash \neg B_1 & \text{or} & P \vdash \neg B_2 \\
P \vdash \neg\neg B & \Longleftrightarrow P \vdash B & & \\
P, A \leftarrow B \vdash A & \Longleftrightarrow P, A \leftarrow B \vdash B & & \\
P, A \leftarrow B \vdash \neg A & \Longleftrightarrow P, A \leftarrow B \vdash \neg B & &
\end{array}
$$

and of the following axioms:

$$
P \vdash \top
$$
$$
P \vdash \neg\bot
$$

(In this system, the second, fourth and fifth rule, and the second axiom could be replaced by a more general rule $P \vdash B \iff P \vdash B^{deM}$.) Notice that in each rule of G1 the goals change as we pass from the left to the right side, but the program stays the same.

Definition 13. By a *G1-derivation* of sequent $P \vdash B$ we understand a tree whose nodes are sequents, such that the following conditions hold:

1. The tree is finite.
2. The root of the tree is $P \vdash B$.
3. The leaves are axioms of G1.
4. If a node is $P \vdash B_1 \lor B_2$ then it has one child: either $P \vdash B_1$ or $P \vdash B_2$.
5. If a node is $P \vdash \neg(B_1 \lor B_2)$ then it has two children: $P \vdash \neg B_1$, and $P \vdash \neg B_2$.
6. If a node is $P \vdash B_1 \land B_2$ then it has two children: $P \vdash B_1$, and $P \vdash B_2$.
7. If a node is $P \vdash \neg(B_1 \land B_2)$ then it has one child: either $P \vdash \neg B_1$ or $P \vdash \neg B_2$.
8. If a node is $P \vdash \neg\neg B_1$ then it has one child: $P \vdash B_1$.
9. If a node is $P \vdash A$ where A is a propositional letter and $A \leftarrow B_1$ is the unique statement in P with the head A, then it has one child: $P \vdash B_1$.
10. If a node $P \vdash \neg A$ where A is a propositional letter and $A \leftarrow B_1$ is the unique statement in P with the head A, then it has one child: $P \vdash \neg B_1$.

Example 1. Consider the following program P:

$$
p \leftarrow \neg(\neg p \land q)
$$
$$
q \leftarrow \bot
$$

As $SLDNF(P, \leftarrow p) = Y$, according to the intuition mentioned at the beginning of this section, we expect that the sequent $P \vdash p$ is G1-derivable. Indeed:

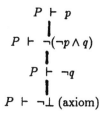

$$P \vdash p$$
$$\mid$$
$$P \vdash \neg(\neg p \wedge q)$$
$$\mid$$
$$P \vdash \neg q$$
$$\mid$$
$$P \vdash \neg\bot \text{ (axiom)}$$

Theorem 14 on soundness and completeness. *Let P be a propositional program in Clark's form, and let $\leftarrow B$ be a goal in the language of P. Then:*
1. *$SLDNF(P, \leftarrow B) = Y$*   *iff*   *$P \vdash B$ is G1-derivable.*
2. *$SLDNF(P, \leftarrow B) = N$*   *iff*   *$P \vdash \neg B$ is G1-derivable.*
3. *$SLDNF(P, \leftarrow B) = U$*   *iff*   *neither $P \vdash B$ nor $P \vdash \neg B$ is G1-derivable.*

Notice that in this theorem we do not assume stratification or any other restrictions on programs.

5   Characterizations of PROLOG's Resolution

Before we can characterize PROLOG's resolution, we have to look more closely at derivations in our Gentzen-style systems. The set of all possibly infinite derivations starting with a sequent $P \vdash B$ can be arranged into an and/or tree.

Definition 15. We define the *full G1-derivation tree* for a sequent $P \vdash B$. This tree is ordered (i.e. for each node, the order of its children is essential.) Each node has 0, 1, or 2 children. The nodes which have 2 children are classified either as *and-nodes* or as *or-nodes.* Each node contains a sequent. The root contains the original sequent $P \vdash B$. If a node contains an instance of an axiom of **G1**, the node has 0 children. If a node contains a sequent S, and if there is an instance $S \Longleftrightarrow S'$ of a rule of **G1** which applies to S, then the node has 1 child which contains S'. If a node contains a sequent S, and if there is an instance $S \Longleftrightarrow S'$ or S'' of a rule of **G1** which applies to S, then the node is an or-node and has 2 children which contain respectively S' and S''. If an node contains a sequent S, and if there is an instance $S \Longleftrightarrow S'$ and S'' of a rule of **G1** which applies to S, then the node is an and-node and has 2 children which contain respectively S' and S''. If an node contains a sequent S, and if no rule of **G1** applies to S, then the node has 0 children.

Notice that system **G1** is defined in such a way that at most one rule applies to any given sequent. So for any $P \vdash B$ its full derivation tree is uniquely determined.

Example 2. Consider once again the program P from example 1:
$$p \leftarrow \neg(\neg p \wedge q)$$
$$q \leftarrow \bot$$
The full derivation tree for $P \vdash p$ is infinite:

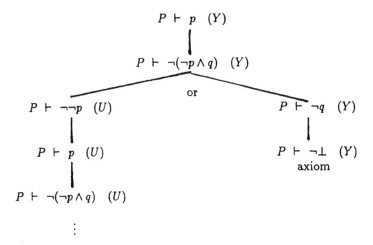

We already labeled the nodes of this tree with labels Y and U. The third possible label is N. The rules of labeling are such that if a node is labeled with Y then its sequent is derivable, and if a node is labeled with N then its sequent is not derivable. This is explained in the following.

Definition 16. Given the full derivation tree for a sequent $P \vdash B$, by a *lazy parallel truth labeling* we understand any assignment of labels Y, N and U to the nodes of the tree, such that the following conditions hold. An and-node is labeled with Y if and only if both its children are labeled with Y; it is labeled with N if and only if at least one of its children is labeled with N. An or-node is labeled with Y if and only if at least one of its children is labeled with Y; it is labeled with N if and only if both its children is labeled with N. Any node which has only 1 child is labeled in the same way as the child. If a node is has 0 children and contains an axiom then it cannot be labeled with N. If a node is has 0 children and does not contain an axiom then it cannot be labeled with Y. Moreover, on any infinite branch, label U occurs infinitely many times.

Theorem 17.

1. *A sequent $P \vdash B$ is G1-derivable if and only if there exists a lazy parallel truth labeling of the full G1-derivation tree for $P \vdash B$ in which the root is labeled with Y.*

2. *A sequent $P \vdash \neg B$ is G1-derivable if and only if there exists a lazy parallel truth labeling of the full G1-derivation tree for $P \vdash B$ in which the root is labeled with N.*

3. *In a full G1-derivation tree with a lazy parallel truth labeling, if a node is labeled with N then its sequent is G1-derivable.*

4. *In a full G1-derivation tree with a lazy parallel truth labeling, if a node is labeled with Y then its sequent is not G1-derivable.*

We should notice that in the definition above the rules of labeling of and-nodes and their children, and or-nodes and their children correspond precisely to the operations ∧ and ∨ of the logic of lazy parallel evaluation.

Definition 18. We define *lazy left-to-right truth labeling* in the same way as in the definition above, except that the rules of labeling of and-nodes, or-nodes and their children correspond to the operations ∧ and ∨ of the logic of lazy left-to-right evaluation.

We say that a sequent S is *G1 lazy left-to-right derivable* or *G1-llr-derivable* if there exists a lazy left to right truth labeling of the full G1-derivation tree for S in which the root is labeled with Y.

With this definition we could formulate a proposition analogous to Theorem 17 characterizing G1-llr-derivability in terms of lazy left-to-right truth labelings.

Given a full derivation tree we could imagine that a lazy parallel truth labeling could be constructed in a nondeterministic way. If we assume this view, the soundness and completeness theorem in the previous section can be interpreted as saying that (nondeterministic) SLDNF-resolution returns answer YES iff a certain nondeterministic procedure constructs a truth labeling of the corresponding full derivation tree in which the root is labeled Y. It is not surprising that nondeterminism is characterized by nondeterminism.

Later we will show that G1-llr-derivations fully characterize PROLOG's resolution. In PROLOG, derivations are constructed by searching the ordered tree of all possible derivations in a deterministic depth-first manner and from the left to the right. So, let us notice that also G1-llr-derivations can be constructed deterministically. We will define a deterministic procedure which attempts construction of a truth labeling of the corresponding full derivation tree.

The procedure is based on a lazy left-to-right post-order traversal: it labels children before it labels their parent, however if visiting the first child gives enough information to label the parent, the second child is not visited. (While this concept is very straightforward, for the sake of the precision of further considerations we need to define the procedure more formally.)

Definition 19. The procedure **traverse** is defined by means of the following pseudo-code. The procedure is supposed to start from the root of a full derivation tree. Initially all the nodes of the tree are labeled with U.

```
PROCEDURE traverse(node);
  BEGIN
    IF type(node)='and' THEN
      BEGIN
        traverse(child1(node));
        IF label(child1(node))='Y' THEN
          BEGIN
            traverse(child2(node));
            label(node):=label(child2(node))
          END;
        IF label(child1(node))='N' THEN label(node):='N'
```

```
      END;
   IF type(node)='or' THEN
      BEGIN
         traverse(child1(node));
         IF label(child1(node))='Y' THEN label(node):='Y';
         IF label(child1(node))='N' THEN
            BEGIN
               traverse(child2(node));
               label(node):=label(child2(node))
            END
      END;
   IF number-of-children(node)=1 THEN
      BEGIN
         traverse(child1(node));
         label(node):=label(child1(node))
      END;
   IF number-of-children(node)=0 AND axiom(node)
      THEN label(node):='Y'
      ELSE label(node):='N'
   END
```

Proposition 20. *Sequent S is G1-llr-derivable if and only if the procedure* traverse *terminates and labels the root of the full G1-derivation tree for S with Y.*

It is interesting to notice that the labeling constructed by traverse is always simultaneously a lazy left-to-right truth labeling and a lazy parallel truth labeling. This is because in the resulting labeling if the first of the two children of a node is labeled with U, so is labeled the second child. The cases with the two children labeled respectively U, Y or U, N never appear, and on the remaining cases there is no difference between the two logics.

The following theorem shows the system **G1** fully characterizes propositional PROLOG's resolution (for the unrestricted class of programs).

Theorem 21 on soundness and completeness. *Let P be a propositional program in Clark's form, and let $\leftarrow B$ be a goal in the language of P. Then:*
1. *$PROLOG(P, \leftarrow B) = Y$ iff $P \vdash B$ is G1-llr-derivable.*
2. *$PROLOG(P, \leftarrow B) = N$ iff $P \vdash \neg B$ is G1-llr-derivable.*
3. *$PROLOG(P, \leftarrow B) = U$ iff neither $P \vdash B$ nor $P \vdash \neg B$ is G1-llr-derivable.*

6   Other Systems

The system **G2**, which will be defined in this section deals with *sequents* such as $P \vdash B$ where P is a propositional program in Clark's form, and $\leftarrow B$ is a propositional goal in the language of P. Below, $\acute{A}$ and A_2 denote any propositional letters.

Definition 22. The system **G2** contains the following rules:
$P, A \leftarrow B_1 \vee B_2 \vdash A \qquad \Longleftrightarrow \qquad P, A \leftarrow B_1 \vdash A \qquad \text{or} \qquad P, A \leftarrow B_2 \vdash A$

$$P, A \leftarrow \neg(B_1 \lor B_2) \vdash A \qquad \Longleftrightarrow \quad P, A \leftarrow \neg B_1 \vdash A \quad \text{and} \quad P, A \leftarrow \neg B_2 \vdash A$$
$$P, A \leftarrow B_1 \land B_2 \vdash A \qquad \Longleftrightarrow \quad P, A \leftarrow B_1 \vdash A \quad \text{and} \quad P, A \leftarrow B_2 \vdash A$$
$$P, A \leftarrow \neg(B_1 \land B_2) \vdash A \quad \Longleftrightarrow \quad P, A \leftarrow \neg B_1 \vdash A \quad \text{or} \quad P, A \leftarrow \neg B_2 \vdash A$$
$$P, A \leftarrow \neg\neg B \vdash A \qquad \Longleftrightarrow \quad P, A \leftarrow B \vdash A$$
$$P, A \leftarrow A_2, \ A_2 \leftarrow B \vdash A \quad \Longleftrightarrow \quad P, A \leftarrow B, \ A_2 \leftarrow B \vdash A$$
$$P, A \leftarrow \neg A_2, \ A_2 \leftarrow B \vdash A \quad \Longleftrightarrow \quad P, A \leftarrow \neg B, \ A_2 \leftarrow B \vdash A$$
$$P \vdash B \quad \Longleftrightarrow \quad P, A \leftarrow B \vdash A \qquad \text{where } B \text{ is not a propositional letter and}$$

A is a propositional letter which does not occur in $P \vdash B$.

G2 contains the following axioms:

$$P, A \leftarrow \top \vdash A$$
$$P, A \leftarrow \neg\bot \vdash A$$

In this system, the second, fourth and fifth rule, and the last axiom, could be replaced by a more general rule $P, A \leftarrow B \vdash A \Longleftrightarrow P, A \leftarrow B^{deM} \vdash A$. Notice that in all the rules of **G2**, except in the last one, the program changes, but the goal stays the same.

The next system, **G3**, incorporates elements of **G1** and **G2**, and is most elaborate of the three. It facilitates easy avoidance of infinite derivations. **G3** deals with *sequents* such as $P \vdash B$ where P is a propositional program in Clark's form, and $\leftarrow B$ is a propositional goal in the language of P.

Below we write L to denote a literal. If L is a negative literal $\neg A$, then $\neg L$ is identified with A.

Definition 23. The system **G3** consists of the following rules:

$$P \vdash B_1 \lor B_2 \qquad \Longleftrightarrow P \vdash B_1 \qquad \text{or} \quad P \vdash B_2$$
$$P \vdash \neg(B_1 \lor B_2) \qquad \Longleftrightarrow P \vdash \neg B_1 \qquad \text{and} \quad P \vdash \neg B_2$$
$$P \vdash B_1 \land B_2 \qquad \Longleftrightarrow P \vdash B_1 \qquad \text{and} \quad P \vdash B_2$$
$$P \vdash \neg(B_1 \land B_2) \qquad \Longleftrightarrow P \vdash \neg B_1 \qquad \text{or} \quad P \vdash \neg B_2$$
$$P \vdash \neg\neg B \qquad \Longleftrightarrow P \vdash B$$
$$P, A \leftarrow B_1 \lor B_2 \vdash A \qquad \Longleftrightarrow P, A \leftarrow B_1 \vdash A \qquad \text{or} \quad P, A \leftarrow B_2 \vdash A$$
$$P, A \leftarrow \neg(B_1 \lor B_2) \vdash A \Longleftrightarrow P, A \leftarrow \neg B_1 \vdash A \qquad \text{and} \quad P, A \leftarrow \neg B_2 \vdash A$$
$$P, A \leftarrow B_1 \lor B_2 \vdash \neg A \qquad \Longleftrightarrow P, A \leftarrow B_1 \vdash \neg A \qquad \text{and} \quad P, A \leftarrow B_2 \vdash \neg A$$
$$P, A \leftarrow \neg(B_1 \lor B_2) \vdash \neg A \Longleftrightarrow P, A \leftarrow \neg B_1 \vdash \neg A \qquad \text{or} \quad P, A \leftarrow \neg B_2 \vdash \neg A$$
$$P, A \leftarrow B_1 \land B_2 \vdash A \qquad \Longleftrightarrow P, A \leftarrow B_1 \vdash A \qquad \text{and} \quad P, A \leftarrow B_2 \vdash A$$
$$P, A \leftarrow \neg(B_1 \land B_2) \vdash A \Longleftrightarrow P, A \leftarrow \neg B_1 \vdash A \qquad \text{or} \quad P, A \leftarrow \neg B_2 \vdash A$$
$$P, A \leftarrow B_1 \land B_2 \vdash \neg A \qquad \Longleftrightarrow P, A \leftarrow B_1 \vdash \neg A \qquad \text{or} \quad P, A \leftarrow B_2 \vdash \neg A$$
$$P, A \leftarrow \neg(B_1 \land B_2) \vdash \neg A \Longleftrightarrow P, A \leftarrow \neg B_1 \vdash \neg A \qquad \text{and} \quad P, A \leftarrow \neg B_2 \vdash \neg A$$
$$P, A \leftarrow \neg\neg B \vdash A \qquad \Longleftrightarrow P, A \leftarrow B \vdash A$$
$$P, A \leftarrow \neg\neg B \vdash \neg A \qquad \Longleftrightarrow P, A \leftarrow B \vdash \neg A$$
$$P, A \leftarrow L \vdash A \qquad \Longleftrightarrow P, A \leftarrow L \vdash L$$
$$P, A \leftarrow L \vdash \neg A \qquad \Longleftrightarrow P, A \leftarrow L \vdash \neg L$$

and of the following axioms:

$$P \vdash \top$$
$$P \vdash \neg\bot$$

In this system some of the rules and the last axiom could be replaced by more general rules: $P \vdash B \Longleftrightarrow P \vdash B^{deM}$ and $P, A \leftarrow B \vdash L \Longleftrightarrow P, A \leftarrow B^{deM} \vdash L$. Notice that in derivations of **G3** both the program and the goal change.

Theorem 24 on soundness and completeness. *Let P be a propositional program in Clark's form, and let $\leftarrow B$ be a goal in the language of P. Then, for $i = 2, 3$:*

1. $SLDNF(P, \leftarrow B) = Y$ iff $P \vdash B$ is Gi-derivable.

2. $SLDNF(P, \leftarrow B) = N$ iff $P \vdash \neg B$ is Gi-derivable.

3. $SLDNF(P, \leftarrow B) = U$ iff   neither $P \vdash B$ nor $P \vdash \neg B$ is Gi-derivable.

In a similar way as for **G1** we can define llr-derivations also for **G2** and **G3**. Then, we have the following characterization of PROLOG's resolution.

Theorem 25 on soundness and completeness. *Let P be a propositional program in Clark's form, and let $\leftarrow B$ be a goal in the language of P. Then, for $i = 2, 3$:*

1. $PROLOG(P, \leftarrow B) = Y$ iff $P \vdash B$ is Gi-llr-derivable.

2. $PROLOG(P, \leftarrow B) = N$ iff $P \vdash \neg B$ is Gi-llr-derivable.

3. $PROLOG(P, \leftarrow B) = U$ iff   neither $P \vdash B$ nor $P \vdash \neg B$ is Gi-llr-derivable.

The following straightforward proposition will be useful for cutting off infinite derivations and showing that certain sequents cannot be derived in **G3**.

Proposition 26. *If for every $i \leq k$, L_i is a literal involving propositional letter A_i, and L is a literal involving one of A_i $(i \leq k)$ then*
$$P, A_0 \leftarrow L_1, A_1 \leftarrow L_2, \ldots, A_{k-1} \leftarrow L_k, A_k \leftarrow L_0 \vdash L \text{ is not G3-derivable.}$$

Example 9. Let P be the following program:
$$q \leftarrow q \vee p$$
$$p \leftarrow \neg p \vee q$$
We have $SLDNF(P, \leftarrow p) \neq Y$ and $P \vdash p$ is not G3-derivable. Let us see how this is this can be demonstrated if we use the system **G3** together with the proposition above.

$$q \leftarrow q \vee p, \; p \leftarrow \neg p \vee q \; \vdash \; p$$
iff
$$q \leftarrow q \vee p, \; p \leftarrow \neg p \vdash p \quad \text{or} \quad q \leftarrow q \vee p, \; p \leftarrow q \vdash p$$
iff
$$\text{false} \quad \text{or} \quad q \leftarrow q \vee p, \; p \leftarrow q \vdash p$$
iff
$$q \leftarrow q \vee p, \; p \leftarrow q \vdash p$$
iff
$$q \leftarrow q \vee p, \; p \leftarrow q \vdash q$$
iff
$$q \leftarrow q, \; p \leftarrow q \vdash q \quad \text{or} \quad q \leftarrow p, \; p \leftarrow q \vdash q$$
iff
$$\text{false or} \quad q \leftarrow p, \; p \leftarrow q \vdash q$$
iff
$$p \leftarrow q, \; q \leftarrow q \vdash q$$
iff
$$\text{false}$$

In this derivation we used Proposition 26 three times to replace certain sequents by "false" by which we avoided infinite branches. The final "false" means that $P \vdash p$ has no G3-derivation.

The idea presented in this example underlies the following decidability result.

Proposition 27.

1. *For propositional programs P and goals $\leftarrow B$ it is decidable whether $P \vdash B$ is Gi-derivable ($i = 1, 2, 3$).*
2. *For propositional programs P and goals $\leftarrow B$ it is decidable whether $P \vdash B$ is Gi-llr-derivable ($i = 1, 2, 3$).*

Algorithms for deciding whether $SLDNF(P, \leftarrow B) = Y$ or N or U, and whether $PROLOG(P, \leftarrow B) = Y$ or N or U are already known. Combination of the decision algorithm for G3 mentioned in the proposition above and soundness and completeness results 24 and 25 gives yet another algorithm for deciding about SLDNF and PROLOG.

7 Conclusion

We formulated three Gentzen-style systems which fully characterize SLDNF-resolution and PROLOG's resolution for the unrestricted class of (pure) propositional programs. We do not assume stratification or any other restrictions on programs. Derivabilities in these Gentzen-style systems can be treated as methods of evaluating propositional programs which are alternative to SLDNF or PROLOG.

We proved that for a propositional program and goal $P, \leftarrow B$, the answer returned by (nondeterministic) SLDNF-resolution is characterized by (nondeterministic) derivability/non-derivability of sequents $P \vdash B$ and $P \vdash \neg B$ in each of the Gentzen-style systems. If the derivability in the Gentzen-style systems is defined by means of a (deterministic) procedure of lazy left-to-right post-order traversal, we obtain an analogous characterization of (deterministic) PROLOG's resolution. In our analysis of these issues we exploited two well known three valued logics: the logic of lazy parallel evaluation (due to Kleene), and the logic of lazy left-to-right evaluation (used in LISP).

This analysis of Negation as Failure can be of interest for rule-based expert systems which represent knowledge in propositional logic. In further research we will consider generalizations of the present results to the situation of rules with attached uncertainty measures. Other work will concern generalizations to the case of first-order programs.

8 Proofs

Proof of **Proposition 4.**
The proposition results from the following two straightforward facts.

1. If Γ_1 is obtained from Γ_2 by selecting one statement and transforming its body by using a single step of the dnf algorithm, then $SLDNF(\Gamma_1^e) = SLDNF(\Gamma_2^e))$

2. If Γ_0 is in dnf then $SLDNF(\Gamma_0^e) = SLDNF(\Gamma_0^n))$

By 1, $SLDNF(\Gamma^e) = SLDNF((\Gamma^{dnf})^e))$, and by 2,
$SLDNF((\Gamma^{dnf})^e)) = SLDNF((\Gamma^{dnf})^n))$.

Proof of **Proposition 8.**
Analogous to the proof of Proposition 4, with "SLDNF" replaced by "PROLOG".

Proof of **Proposition 9.**
Without loosing generality, we can assume that B_1, B_2 are propositional letters. Indeed, if they were not, we could introduce new propositional letters A_1, A_2 and add statements $A_1 \leftarrow B_1$, and $A_2 \leftarrow B_2$ to P, and consider $SLDNF(P', \leftarrow A_i)$ instead of $SLDNF(P', \leftarrow B_i)$, and also $SLDNF(P', \leftarrow A_1 \wedge A_2)$ instead of $SLDNF(P', \leftarrow B_1 \wedge B_2)$, etc. With this simplifying assumption, the proof of the equalities is straightforward.

Proof of **Corollary 11.**
Implications to the left in 1 and 2 follow immediately from Proposition 9. In order to show the implication to the right in 1, we assume that $\leftarrow B_1$, $\leftarrow B_2$ are SLDNF-equivalent, and we need to demonstrate that for every valuation ν : $Prop \longrightarrow \{Y, N, U\}$, $\nu(B_1) = \nu(B_2)$. Given ν, construct program P as follows. If $\nu(A_i) = Y$ put $A_i \leftarrow \top$ into P. If $\nu(A_i) = N$ put $A_i \leftarrow \bot$ into P. If $\nu(A_i) = U$ put $A_i \leftarrow A_i$ into P. We have $\nu(B_i) = SLDNF(P, \leftarrow B_i)$. As we assumed $SLDNF(P, \leftarrow B_i) = SLDNF(P, \leftarrow B_i)$, we have $\nu(B_1) = \nu(B_2)$. The proof of the implication to the right in 2 is analogous.

The next result to prove is Theorem 14. In order to prove it we will need several lemmas.

Lemma 28. *Let* $P, \leftarrow B$ *be a propositional program and goal in Clark's form. Then,*
$G1(((P, \leftarrow B)^e)^C) = G1(P, \leftarrow B)$.

Notice that $((P, \leftarrow B)^e)^C$, mentioned in this lemma is always in Clark's dnf.

Proof. For the proof it is enough to verify that if $P_1, \leftarrow B_1$ is obtained from $P_2, \leftarrow B_2$ by selecting one statement whose body is not a conjunction of literals and replacing it using a single step of the algorithm for transformation $(\cdot)^e$, then $P_1, \leftarrow B_1$ is G1-derivable if and only if $P_2, \leftarrow B_2$ is G1-derivable.

Lemma 29.

1. $P \vdash (B_1 \wedge B_2) \wedge B_3$ *is G1-derivable if and only if* $P \vdash B_1 \wedge (B_2 \wedge B_3)$ *is G1-derivable.*
2. $P \vdash (B_1 \vee B_2) \vee B_3$ *is G1-derivable if and only if* $P \vdash B_1 \vee (B_2 \vee B_3)$ *is G1-derivable.*

Proof. Straightforward.

Due to this lemma we can write $P \vdash B_1 \wedge B_2 \wedge \ldots \wedge B_n$ without specifying the placement of parentheses in $B_1 \wedge B_2 \wedge \ldots \wedge B_n$; and similarly with disjunctions.

Lemma 30. *Let $P, \leftarrow B$ be a propositional program and goal in Clark's form. Then:*

1. *If $SLDNF(P, B) = Y$*   *then* $P \vdash B$ *is G1-derivable.*
2. *If $SLDNF(P, B) = N$*   *then* $P \vdash \neg B$ *is G1-derivable.*

Proof. Notice that $SLDNF(((P, \leftarrow B)^e)^C) = SLDNF(P, \leftarrow B)$. By this fact and by Lemma 28, without loosing generality we can assume that $P, \leftarrow B$ is in Clark's dnf. Notice that

- $SLDNF(P, \leftarrow B_1 \vee \ldots \vee B_n) = Y$ iff there exists $i \leq n$ such that $SLDNF(P, \leftarrow B_i) = Y$
- $P \vdash B_1 \vee \ldots \vee B_n$ is G1-derivable iff there exists $i \leq n$ such that $P \vdash B_i$ is G1-derivable.
- $SLDNF(P, \leftarrow B_1 \vee \ldots \vee B_n) = N$ iff for all $i \leq n$, $SLDNF(P, \leftarrow B_i) = N$
- $P \vdash \neg(B_1 \vee \ldots \vee B_n)$ is G1-derivable iff for all $i \leq n$, $P \vdash \neg B_i$ is G1-derivable.

Because of that, it will be enough to consider a situation in which $\leftarrow B$ is a normal goal. To sum up: we will assume that P is a program in Clark's dnf and $\leftarrow B$ is a normal goal.

We will prove 1 and 2 in parallel by induction on the rank of the SLDNF-refutation or on the height of the finitely failed SLDNF-tree for $P \leftarrow B$.

Case: $P, \leftarrow B$ has SLDNF-refutation of rank 0
We will prove implication 1 by induction on the length l of this refutation.
length=0
    In this case $\leftarrow B$ is $\leftarrow \top \wedge \ldots \wedge \top$. Using the axiom $P \vdash \top$ and the third rule we can construct a G1-derivation of $P \vdash B$.
length=$l+1$
    In this case, B is $\leftarrow A_1 \wedge \ldots \wedge A_k \wedge \ldots \wedge A_m$, and P^n contains normal clause $A_k \leftarrow B'$, and the derived goal $\leftarrow A_1 \wedge \ldots \wedge A_{k-1} \wedge B' \wedge A_{k+1} \wedge \ldots \wedge A_m$ has a refutation of rank 0 and length l. By induction hypothesis (in the induction on l) $P \vdash A_1 \wedge \ldots \wedge A_{k-1} \wedge B' \wedge A_{k+1} \wedge \ldots \wedge A_m$ has a G1-derivation. Thus for each $i \neq k$, $P \vdash A_i$ has G1-derivation, and $P \vdash B'$ has G1-derivation. Using the sixth rule we obtain G1-derivation for $P \vdash A_k$. Using the third rule we obtain a G1-derivation for $P \vdash B$.

Case: $P, \leftarrow B$ has finitely failed SLDNF-tree of rank 0
We will prove implication 2 by induction on the height d of this tree.
height $= 0$
    In this case, $\leftarrow B$ is $\leftarrow L_1 \wedge \ldots \wedge L_k \wedge \ldots \wedge L_m$ where L_k is $\bot$. Using the axiom $P \vdash \neg\bot$ and the fourth rule we obtain a G1-derivation of $P \vdash \neg B$.
height=$d+1$
    In this case, $\leftarrow B$ is $\leftarrow L_1 \wedge \ldots \wedge L_k \wedge \ldots \wedge L_m$ where L_k is a propositional letter, P^n contains normal clauses $L_k \leftarrow B'_1, \ldots, L_k \leftarrow B'_n$, and each of the derived goals $\leftarrow L_1 \wedge \ldots \wedge L_{k-1} \wedge B'_i \wedge L_{k+1} \wedge \ldots \wedge L_m$ has a finitely failed SLDNF-tree of rank 0 and height $\leq d$. By induction hypothesis (in the induction on d) each of $P \vdash \neg(L_1 \wedge \ldots \wedge L_{k-1} \wedge B'_i \wedge L_{k+1} \wedge \ldots \wedge L_m)$ has a G1-derivation. By the fourth rule of **G1**, for each i there exists $j \neq k$

such that either $P \vdash \neg L_j$ is G1-derivable or $P \vdash \neg B'_i$ is G1-derivable. Thus, either there exists $j \neq k$ such that $P \vdash \neg L_j$ is G1-derivable or for each i , $P \vdash \neg B'_i$ is G1-derivable. By the second rule of **G1**, either there exists $j \neq k$ such that $P \vdash \neg L_j$ is G1-derivable or $P \vdash \neg(B'_1 \vee \ldots \vee B'_n)$ is G1-derivable. By the seventh rule of **G1**, either there exists $j \neq k$ such that $P \vdash \neg L_j$ is G1-derivable or $P \vdash \neg L_k$ is G1-derivable. So, there exists j such that $P \vdash \neg L_j$ is G1-derivable, and by the fourth rule $P \vdash \neg B$ is G1-derivable.

Case: $P, \leftarrow B$ has SLDNF-refutation of rank $r + 1$

We will prove implication 1 by induction on the length l of this refutation.

length=0

In this case $\leftarrow B$ is $\leftarrow \top \wedge \ldots \wedge \top$. Using the axiom $P \vdash \top$ and the third rule we can construct a G1-derivation of $P \vdash B$.

length=$l + 1$

We will consider two cases depending on whether a positive or a negative literal is selected in $\leftarrow B$.

1. In this case, B is $\leftarrow L_1 \wedge \ldots \wedge L_k \wedge \ldots \wedge L_m$ where L_k is a propositional letter, P^n contains normal clause $L_k \leftarrow B'$ and the derived goal $\leftarrow L_1 \wedge \ldots \wedge L_{k-1} \wedge B' \wedge L_{k+1} \wedge \ldots \wedge L_m$ has an SLDNF-refutation of rank $r + 1$ and length l. By induction hypothesis (in the induction on l) $P \vdash L_1 \wedge \ldots \wedge L_{k-1} \wedge B' \wedge L_{k+1} \wedge \ldots \wedge L_m$ has a G1-derivation. Thus for each $i \neq k$, $P \vdash L_i$ has G1-derivation and $P \vdash B'$ has G1-derivation. Using the sixth rule we obtain G1-derivation for $P \vdash L_k$. Using the third rule we obtain a G1-derivation for $P \vdash B$.

2. In this case, B is $\leftarrow L_1 \wedge \ldots \wedge L_k \wedge \ldots \wedge L_m$ where L_k is $\neg A$, $\leftarrow L_1 \wedge \ldots \wedge L_{k-1} \wedge L_{k+1} \wedge \ldots \wedge L_m$ has SLDNF-refutation of rank $\leq r$ and length l, and $P, \leftarrow A$ has finitely failed SLDNF-tree of rank $\leq r + 1$. By induction hypothesis (in the induction on l) for each $i \neq k$, $P \vdash L_i$ has a G1-derivation. By induction hypothesis (in the induction on r), $P \vdash \neg A$ has a G1-derivation. So each $P \vdash L_i$ has G1-derivation. Using the third rule we obtain G1-derivation for $P \vdash B$.

Case: $P, \leftarrow B$ has finitely failed SLDNF-tree of rank $r + 1$

We will prove implication 2 by induction on the height d of this tree.

height $= 0$

In this case, $\leftarrow B$ is $\leftarrow L_1 \wedge \ldots \wedge L_k \wedge \ldots \wedge L_m$ where L_k is $\bot$. Using the axiom $P \vdash \neg\bot$ and the fourth rule we obtain a G1-derivation of $P \vdash \neg B$.

height=$d + 1$

We will consider two cases depending on whether a positive or a negative literal is selected in $\leftarrow B$.

1. In this case, B is $\leftarrow L_1 \wedge \ldots \wedge L_k \wedge \ldots \wedge L_m$ where L_k is a propositional letter, P^n contains normal clauses $L_k \leftarrow B'_1, \ldots, L_k \leftarrow B'_n$ and each of the the derived goals $\leftarrow L_1 \wedge \ldots \wedge L_{k-1} \wedge B'_i \wedge L_{k+1} \wedge \ldots \wedge L_m$ has a finitely failed SLDNF-tree of rank $\leq r + 1$ and height $\leq d$. By induction hypothesis (in the induction on d) each of $P \vdash \neg(L_1 \wedge \ldots \wedge L_{k-1} \wedge B'_i \wedge L_{k+1} \wedge \ldots \wedge L_m)$ has a G1-derivation. By the fourth rule of G1, for each i there exists $j \neq k$ such that $P \vdash \neg L_j$ is G1-derivable or $P \vdash \neg B'_i$ is G1-derivable. Thus either there exists $j \neq k$ such that $P \vdash \neg L_j$ is

G1-derivable or for each i $P \vdash \neg B_i'$ is G1-derivable. Thus by the second rule of **G1**, either there exists $j \neq k$ such that $P \vdash \neg L_j$ is G1-derivable or $P \vdash \neg(B_1' \vee \ldots \vee B_l')$ is G1-derivable. By the seventh rule of **G1** there exists $j \neq k$ such that $P \vdash \neg L_j$ is G1-derivable or $P \vdash \neg L_k$ is G1-derivable. So, there exists j such that $P \vdash \neg L_j$ is G1-derivable. By the fourth rule, $P \vdash \neg(L_1 \wedge \ldots \wedge L_m)$ is G1-derivable.

2. In this case, B is $\leftarrow L_1 \wedge \ldots \wedge L_k \wedge \ldots \wedge L_m$ where L_k is $\neg A$, and $P, \leftarrow A$ has a finitely failed SLDNF-tree, and $P, \leftarrow L_1 \wedge \ldots \wedge L_{k-1} \wedge L_{k+1} \wedge \ldots \wedge L_m$ has finitely failed SLDNF-tree of rank $r+1$ and height d. By induction hypothesis (in the induction on l) $P \vdash \neg(L_1 \wedge \ldots \wedge L_{k-1} \wedge L_{k+1} \wedge \ldots \wedge L_m)$ has a G1-derivation. By the fourth rule of **G1**, there exists $i \neq k$ such that $P \vdash \neg L_i$ has G1-derivation. Using the fourth rule we obtain a G1-derivation for $P \vdash \neg(\mathcal{L}_1 \wedge \ldots \wedge L_m)$.

This ends the proof of Lemma 30.

In the following lemma we list without a proof some properties of SLDNF-resolution which are either straightforward or are corollaries to Proposition 9.

Lemma 31.

1. *If* $SLDNF(P, \leftarrow \neg B) = Y$ *then* $SLDNF(P, \leftarrow B) = N$.
2. $SLDNF(P, \leftarrow B_1) = Y$ *and* $SLDNF(P, \leftarrow B_2) = Y$
   then $SLDNF(P, \leftarrow B_1 \wedge B_2) = Y$.
3. *If* $SLDNF(P, \leftarrow \neg B_1) = Y$ *and* $SLDNF(P, \leftarrow \neg B_1) = Y$
   then $SLDNF(P, \leftarrow \neg(B_1 \vee B_2)) = Y$.
4. *If* $SLDNF(P, \leftarrow B_1) = Y$ *or* $SLDNF(P, \leftarrow B_2) = Y$
   then $SLDNF(P, \leftarrow B_1 \vee B_2) = Y$.
5. *If* $SLDNF(P, \leftarrow \neg B_1) = Y$ *or* $SLDNF(P, \leftarrow \neg B_1) = Y$
   then $SLDNF(P, \leftarrow \neg(B_1 \wedge B_2)) = Y$.
6. *If* $SLDNF(P, \leftarrow B) = Y$ *then* $SLDNF(P, \leftarrow \neg\neg B) = Y$.
7. *If* $SLDNF(P, A \leftarrow B, \leftarrow B) = Y$ *then* $SLDNF(P, A \leftarrow B, \leftarrow A) = Y$.
8. *If* $SLDNF(P, A \leftarrow B, \leftarrow \neg B) = Y$ *then* $SLDNF(P, A \leftarrow B, \leftarrow \neg A) = Y$.

Lemma 32. *Let* $P, \leftarrow B$ *be a propositional program and goal in Clark's form. Then:*

1. *If* $P \vdash B$ *is G1-derivable then* $SLDNF(P, \leftarrow B) = Y$.
2. *If* $P \vdash \neg B$ *is G1-derivable then* $SLDNF(P, \leftarrow B) = N$.

Proof. 2 results from 1: if $P \vdash \neg B$ is G1-derivable then by 1, $SLDNF(P, \leftarrow \neg B) = Y$ and by Lemma 31, $SLDNF(P, \leftarrow B) = N$.

We will prove 1 by induction on the height d of the G1-derivation of $P \vdash B$.

height $= 0$

Then $P \vdash B$ is an axiom $P \vdash \top$ or $P \vdash \neg\bot$. We have: $SLDNF(P, \leftarrow \top) = Y$ and $SLDNF(P, \leftarrow \neg\bot) = Y$.

height $= d+1$

Here one should consider cases depending on the structure of B: B is $B_1 \wedge B_2$, B is $\neg(B_1 \wedge B_2)$, B is $B_1 \vee B_2$, B is $\neg(B_1 \vee B_2)$, B is $\neg\neg B_1$, B is a propositional letter A, B is $\neg A$ where A is a propositional letter. All these cases are very similar and we will give details only for the first of them.

If $P \vdash B_1 \wedge B_2$ has G1-derivation of height $d + 1$, by the third rule of G1 $P \vdash B_1$ and $P \vdash B_2$ have G1-derivations of height $\leq d$. By the induction hypothesis $SLDNF(P, \leftarrow B_1) = Y$ and $SLDNF(P, \leftarrow B_2) = Y$. By lemma ... $SLDNF(P, \leftarrow B_1 \wedge B_2) = Y$.

Proof of Theorem 14.

Equivalences 1 and 2 result from Lemma 30 and Lemma 32. Equivalence 3 results from 1 and 2.

Proof of Theorem 17.

1. If $P \vdash B$ is G1-derivable, consider the fragment of the full G!-derivation tree for $P \vdash B$ which corresponds to the G1-derivation of $P \vdash B$, and label all the nodes in that fragment with Y. Label all the remaining nodes of the full derivation tree with U. This gives a valid truth labeling in which the root is labeled with Y.

   If there exists a truth labeling of the full derivation tree for $P \vdash B$ which labels the root with Y, consider the fragment $\mathcal{F}$ of the full derivation tree which consists of nodes x such that x and all the nodes between x and the root are labeled with Y. $\mathcal{F}$ is a finitely branching tree. By the definition of labeling, if $\mathcal{F}$ had an infinite branch it would contain label U, but all the nodes in $\mathcal{F}$ are labeled with Y, so $\mathcal{F}$ does not have any infinite branches. By König's lemma $\mathcal{F}$ is finite. For any or-node in $\mathcal{F}$ delete one of the subtrees of this node. After this pruning, the tree that contains the original root is a G1-derivation of $P \vdash \check{B}$.

2. Consider the full G1-derivation tree T for $P \vdash B$ with a lazy parallel truth labeling which assigns N to the root. Consider the full G1-derivation tree T' for $P \vdash \neg B$.

   Let $\mathcal{D}$ be the set of those nodes of T which contain sequents of the following forms: $P \vdash B_1 \vee B_2$, $P \vdash \neg(B_1 \vee B_2)$, $P \vdash B_1 \wedge B_2$, $P \vdash \neg(B_1 \wedge B_2)$, $P \vdash \neg\neg\neg B_1$, $P \vdash L$, where L is either a literal or $\top$ or $\bot$ or $\neg\top$ or $\neg\bot$. (Roughly speaking, "goals with exactly two negations are excluded.") Let $\mathcal{D}'$ be the set of those nodes of T' which contain sequents of the forms specified above.

   For every node x in T, either $x \in \mathcal{D}$ or x has a unique child, and the child is a member of $\mathcal{D}$. Let

   $$d(x) = \begin{cases} x & \text{if } x \in \mathcal{D} \\ \text{the child of } x & \text{otherwise} \end{cases}$$

   Define a similar function on the nodes of T':

   $$d'(x') = \begin{cases} x' & \text{if } x \in \mathcal{D}' \\ \text{the child of } x' & \text{otherwise} \end{cases}$$

   We will define a one-to-one, onto function $h : \mathcal{D} \longrightarrow \mathcal{D}'$. Denote the root of T by r and the root of T' by r'. Set $h(d(r)) = d'(r')$. If x has children x_1, x_2 and $h(x)$ has children x'_1, x'_2, set $h(d(x_1)) = d'(x'_1)$ and $h(d(x_2)) = d'(x'_2)$. If $x \in \mathcal{D}$ has unique child x_1 and $h(x)$ has unique child x'_1, set $h(d(x_1)) = d'(x'_1)$. (Suggested example: consider G1-derivation trees for $P \vdash B$ and $P \vdash \neg B$ where

$P = \langle p \leftarrow \top, q \leftarrow \bot \rangle$ and B is $\neg\neg\neg(\neg p \wedge \neg q)$.) Using induction we can prove that if $h(x) = x'$ and x contains $P \vdash B_1$ and x' contains $P \vdash B_1'$ then either B_1 is $\neg B_1'$ or B_1' is $\neg B_1$. From that, by induction on the distance from the root it follows that:

(a) if x is an and-node then $h(x)$ is defined and it is an or-node,

(b) if x is an or-node then $h(x)$ is defined and it is an and-node,

(c) if x contains an axiom then $h(x)$ is defined and it is a leaf containing a non-axiom,

(d) if x contains a non-axiom then $h(x)$ is defined and it is a leaf containing an axiom.

(e) if $x \in \mathcal{D}$ has unique child then $h(x)$ is defined and has a unique child.

Using these facts we can show that the domain of h is $\mathcal{D}$. Define a function $h' : \mathcal{D}' \longrightarrow \mathcal{D}$. Set $h'(d(r')) = d(r)$. If x' has children x_1', x_2' and $h'(x')$ has children x_1, x_2, set $h'(d(x_1')) = d(x_1)$ and $h'(d(x_2')) = d(x_2)$. If $x' \in \mathcal{D}'$ has unique child x_1' and $h(x')$ has unique child x_1, set $h'(d(x_1')) = d(x_1)$. By the same reasoning as previously, $\mathcal{D}$ is the domain of h'. By induction on the distance from the root one can prove that for every $x \in \mathcal{D}$, $h'(h(x)) = x$ and that for every $x' \in \mathcal{D}'$, $h(h'(x')) = x'$. Thus $h : \mathcal{D} \longrightarrow \mathcal{D}'$ is one-to-one and onto.

Now we will define a labeling of T'. If $x' \in \mathcal{D}'$ and $h'(x')$ is labeled with Y or N or U, then label x' with (respectively) N or Y or U. If $x' \notin \mathcal{D}$ then label x' in the same way as $d'(x')$. By induction on the distance of x' from the root r', due to properties (a)-(e) one can show that the labeling of x' and its children satisfies the conditions of lazy parallel labeling. As required the root r' is labeled with Y.

3. Follows from 1.

4. Assume that a node x_0 of a full G1-derivation tree T is labeled with N. From the rules of labeling it follows that if a node is labeled with N then either it is a non-axiom leaf or it has a child labeled with F. So, there exists a sequence $x_0, x_1, \ldots$ whose all members are labeled with N and each x_{i+1} is a child of x_i. By the rules of labeling this sequence has to be finite (otherwise it would contain label U). Let x_n be the last element of that sequence. x_n is a leaf of T and as it is labeled with N, it is not an axiom. By induction on the distance from x_n one can prove that no sequent in the sequence is G1-derivable.

Proof of **Proposition 20.**

For the implication to the left it is enough to notice that if **traverse** terminates than it leaves a valid lazy left-to-right labeling.

In order to prove the implication to the right, consider a full G1-derivation tree T with a lazy left-to-right labeling which labels the root with Y. Construct s set $\mathcal{D}$ of nodes of T as follows. Put the root of T into $\mathcal{D}$. If a node is in $\mathcal{D}$ and has a unique child then put the child into $\mathcal{D}$. If an and-node is in $\mathcal{D}$ and its first child is labeled with Y then put both children into $\mathcal{D}$. If an and-node is in $\mathcal{D}$ and its first child is labeled with N then put that child into $\mathcal{D}$. If an or-node is in $\mathcal{D}$ and its first child is labeled with Y then put that child into $\mathcal{D}$. If an and-node is in $\mathcal{D}$ and its first child is labeled with N then put both children into $\mathcal{D}$.

Now modify the original labeling by changing the labels of all nodes which are not in $\mathcal{D}$ to U. From the fact the original labeling was a valid lazy left-to-right labeling, and by the definition of $\mathcal{D}$ it results that no node in $\mathcal{D}$ is labeled with U. One can notice that the new labeling is a valid lazy left-to-right labeling with the root labeled with Y. We will show that the new labeling is the same as the labeling constructed by **traverse**. First notice that $\mathcal{D}$ is a tree whose root is identical with the root of T, and whose leaves are some of the leaves of T. $\mathcal{D}$ does not contain any node labeled with U, so by the rules of lazy left-to-right labeling it has no infinite branch. As $\mathcal{D}$ is also a finitely branching tree, by König's lemma, it is finite. By straightforward induction one can show that the nodes of $\mathcal{D}$ are labeled in the same way as in the labeling constructed by **traverse**.

In order to prove the soundness and completeness theorem for PROLOG's resolution we need to define the rank function for PROLOG refutations and finitely failed trees. While reading the definition, notice that although every PROLOG-refutation is an SLDNF-refutation, its rank as a PROLOG-refutation is in general different from its rank as an SLDNF-refutation; similarly for finitely failed trees. Roughly speaking, rank is the height to which "lemmas" can be nested. In an SLDNF the only "lemmas" are negative subgoals. In a PROLOG-refutation we take as "lemmas" also those subgoals which finitely failed and caused backtracking.

First define an auxiliary operation which removes T's from the beginning of a conjunction of literals.

Definition 33. If L is a literal we define $L^* = L$. If B is a conjunction of literals we define $(\mathsf{T} \wedge B)^* = B^*$.

For instance: $(\mathsf{T} \wedge \mathsf{T} \wedge \neg\mathsf{T} \wedge \mathsf{T} \wedge \mathsf{T} \wedge \perp \wedge p)^* = \neg\mathsf{T} \wedge \mathsf{T} \wedge \mathsf{T} \wedge \perp \wedge p$.

Definition 34. Let $P, \leftarrow B$ be a normal propositional program and goal.

1. By a *PROLOG refutation of rank 0* for $P, \leftarrow B$ we understand a sequence $\leftarrow B_0, \ldots, \leftarrow B_n$ of normal goals such that
   - B_0 is B,
   - B_n^* is T,
   - If B_i^* $(0 \leq i < n)$ is $L_1 \wedge \ldots \wedge L_m$, then L_1 is a propositional letter, and if $L_1 \leftarrow B'$ is the first among those clauses of P whose head is L_1 then B_{i+1} is $B' \wedge L_2 \wedge \ldots \wedge L_m$.

   The number n is called the *length* of that refutation.
2. By a *finitely failed PROLOG-tree of rank 0* for $P, \leftarrow B$ we understand an ordered tree of normal goals such that
   - The tree is finite,
   - The root is $\leftarrow B$,
   - If $\leftarrow B'$ is a leaf then $(B')^*$ is of the form $\perp \wedge \ldots$,
   - If a non-leaf node is $\leftarrow B'$ where $(B')^*$ is $L_1 \wedge \ldots \wedge L_m$, then L_1 is a propositional letter. If $L_1 \leftarrow B_1$, $\ldots$, $L_1 \leftarrow B_n$ are all those clauses of P whose head is L_1, then the node has the following children $\leftarrow B_1 \wedge L_2 \wedge \ldots \wedge L_m$, $\ldots$, $\leftarrow B_n \wedge L_2 \wedge \ldots \wedge L_m$.
3. By a *PROLOG refutation of rank $r+1$* for $P, \leftarrow B$ we understand a sequence $\leftarrow B_0, \ldots, \leftarrow B_n$ of normal goals such that
   - B_0 is B,

- B_n^* is $\top$,
- If B_i^* $(0 \leq i < n)$ is $L_1 \wedge \ldots \wedge L_m$ then L_1 is either a propositional letter or it is of the form $\neg A$.
  - If L_1 is a propositional letter, and if $L_1 \leftarrow B_1'$, ..., $L_1 \leftarrow B_n'$ are all those clauses of P whose head is L_1, then there exists k between 1 and n such that each of the goals $\leftarrow B_1'$, ..., $\leftarrow B_{k-1}'$ has finitely failed PROLOG-tree of rank $\leq r$, and B_{i+1} is $B_k' \wedge L_2 \wedge \ldots \wedge L_m$.
  - If L_1 is $\neg A$, then $\leftarrow A$ has a finitely failed PROLOG-tree of rank $\leq r$, and B_{i+1} is $L_2 \wedge \ldots \wedge L_m$.
4. By a *finitely failed PROLOG-tree of rank* $r + 1$ for $P, \leftarrow B$ we understand an ordered tree of normal goals such that
   - The tree is finite,
   - The root is $\leftarrow B$,
   - If $\leftarrow B'$ is a leaf and $(B')^*$ is $L_1 \wedge \ldots \wedge L_m$ then either L_1 is $\bot$ or it is of the form $\neg A$. If L_1 is $\neg A$, then $\leftarrow A$ has a PROLOG-refutation of rank $\leq r$,
   - If a non-leaf node is $\leftarrow B'$ and $(B')^*$ is $L_1 \wedge \ldots \wedge L_m$, then either L_1 is a propositional letter or it is of the form $\neg A$.
     - If L_1 is a propositional letter, and if $L_1 \leftarrow B_1$, ..., $L_1 \leftarrow B_n$ are all those clauses of P whose head is L_1, then the node has the following children $\leftarrow B_1 \wedge L_2 \wedge \ldots \wedge L_m$, ..., $\leftarrow B_n \wedge L_2 \wedge \ldots \wedge L_m$.
     - If L_1 is $\neg A$, then $\leftarrow A$ has a finitely failed PROLOG-tree of rank $\leq r$, and the unique child of the node is $L_2 \wedge \ldots \wedge L_m$.

If $\leftarrow B_0, \ldots, \leftarrow B_n$ is a PROLOG-refutation then n is called its *length*. The number of edges of the longest branch of a finitely failed PROLOG-tree is called its *height*.

Notice that length and height can be 0. Notice that a PROLOG-refutation of rank r is also a PROLOG-refutation of rank $r + 1$; and similarly for finitely failed PROLOG trees.

Proof of Theorem 21.

The proof is analogous to the proof of Theorem 14. Formulate lemmas analogous to 28, 30, 31 and 32 by replacing "SLDNF" by "PROLOG" and "G1-derivation" by "G1-llr-derivation". Proofs these lemmas cannot be obtained by a simple replacement but they use the same ideas as the original lemmas. (The rank of PROLOG refutations and finitely failed trees has been defined above in such a way that makes induction similar to that of Lemma 30 possible).

Lemma 35.

1. $SLDNF(P, A \leftarrow B_1 \vee B_2, \leftarrow A) = Y$ *if and only if*
   $SLDNF(P, A \leftarrow B_1, \leftarrow A) = Y$ *or* $SLDNF(P, A \leftarrow B_2, \leftarrow A) = Y$.

2. $SLDNF(P, A \leftarrow B_1 \wedge B_2, \leftarrow A) = Y$ *if and only if*
   $SLDNF(P, A \leftarrow B_1, \leftarrow A) = Y$ *and* $SLDNF(P, A \leftarrow B_2, \leftarrow A) = Y$.

3. $SLDNF(P, A_1 \leftarrow A_2, A_2 \leftarrow B, \leftarrow A_1) = Y$ *if and only if*
   $SLDNF(P, A_1 \leftarrow B, A_2 \leftarrow B, \leftarrow A_1) = Y$.

4. $SLDNF(P, A_1 \leftarrow \neg A_2, A_2 \leftarrow B, \leftarrow A_1) = Y$ *if and only if*
   $SLDNF(P, A_1 \leftarrow \neg B, A_2 \leftarrow B, \leftarrow A_1) = Y.$

Proof.

1. **implication to the right** We can assume that B_1 and B_2 are in dnf. If each
   disjunct in $B_1 \vee B_2$ contained a literal involving A, there would be no SLDNF-
   refutation of $P, A \leftarrow B_1 \vee B_2, \leftarrow A$. So, some disjuncts do not involve A. Let
   B_i' ($i = 1, 2$) be obtained from B_i by deleting those disjuncts which contained
   A. We have $SLDNF(P, A \leftarrow B_1' \vee B_2', \leftarrow A) = Y$. Let B_1' be $B_1^1 \vee \ldots \vee B_1^n$ and
   let B_2' be $B_2^1 \vee \ldots \vee B_2^m$. We have: there exists i such that $SLDNF(P, A \leftarrow$
   $B_1' \vee B_2', \leftarrow B_1^i) = Y$ or there exists j such that $SLDNF(P, A \leftarrow B_1' \vee B_2', \leftarrow$
   $B_2^j) = Y$. So, there exists i such that $SLDNF(P, A \leftarrow B_1' \vee B_2', \leftarrow A) =$
   Y or there exists j such that $SLDNF(P, A \leftarrow B_1' \vee B_2', \leftarrow A) = Y$. So,
   $SLDNF(P, A \leftarrow B_1, \leftarrow A) = Y$ or $SLDNF(P, A \leftarrow B_2, \leftarrow A) = Y$.
   implication to the left Obvious.
2. **implication to the right** Take the refutation of $SLDNF(P, A \leftarrow B_1 \wedge B_2, \leftarrow$
   $A) = Y$: $G_0 =\leftarrow A$, $G_1 =\leftarrow B_1 \wedge B_2, \ldots$. Delete goals in which a subgoal
   of B_2 is selected, then from each goal delete subgoals which belong to B_2. In
   this way we obtain a refutation: $SLDNF(P, A \leftarrow B_1, \leftarrow A) = Y$. Similarly
   we can obtain $SLDNF(P, A \leftarrow B_2, \leftarrow A) = Y$.
   implication to the left Assume $SLDNF(P, A \leftarrow B_1, \leftarrow A) = Y$ and $SLDNF(P, A \leftarrow$
   $B_2, \leftarrow A) = Y$. These refutations could not have existed if B_1 contained a
   literal involving A or if B_2 contained a literal involving A. So B_1 and B_2 do
   not involve A. So, $SLDNF(P, A \leftarrow B_1 \wedge B_2, \leftarrow B_1) = Y$ and $SLDNF(P, A \leftarrow$
   $B_1 \wedge B_2, \leftarrow B_2) = Y$. By combining these refutations we obtain a refutation
   $SLDNF(P, A \leftarrow B_1 \wedge B_2, \leftarrow B_1 \wedge B_2) = Y$, so $SLDNF(P, A \leftarrow B_1 \wedge B_2, \leftarrow$
   $A) = Y$
3. Straightforward.
4. Straightforward.

Proof of **Theorem 24.**

For the system G2 the proof is analogous to the proof of Theorem 14 and it can
be obtained in the following way. Formulate lemma analogous to 30 by replacing
"G1-derivable" by "G2-derivable" and prove it using Lemma 28 similarly as we
proved 30. Formulate lemma analogous to 32 by replacing "G1-derivable" by "G2-
derivable" and prove it using Lemma 35 similarly as we proved 32. The proofs cannot
be obtained by a simple replacement, but they use the same ideas. From the two
lemmas obtained in this way follows the equivalence of SLDNF and G2-derivations
(as in the proof of 14).

System G3 is a combination of G1 and G2. The proof G3 combines elements
of the proofs for G1 and G2.

Proof of Theorem 25.

Analogous to that of Theorem 24.

Proof of **Proposition 26.**

Straightforward.

Proof of **Proposition 27.**

From the soundness and completeness theorems 14, 21, 25 and 26 it follows that for $i = 2, 3$, a sequent is derivable in Gi if and only if it is derivable in G3. So, it will be enough to give an algorithm for G3. Consider the full G3-derivation tree for $P \vdash B$. Call a node *cyclic* if it contains a sequent such as that of Proposition 24. Notice that a branch is infinite if and only if it contains a cyclic node.

1. Modify procedure **traverse** so that before it visits any of the children of the current node, it checks if the node is cyclic. If the node is cyclic the procedure labels it with N and does not visit the children. This modified procedure always terminates, and labels the root with Y if and only if $P \vdash B$ is G3-derivable. It labels the root with N if and only if $P \vdash B$ is not G3-derivable

2. Modify procedure **traverse** so that before it visits any of the children of the current node, it checks whether the node is cyclic. If the node is cyclic, the procedure terminates and returns answer "$P \vdash B$ is not G3-llr-derivable". After this modification, notice that if the procedure does not enter a cyclic node it terminates and labels the root with Y or N. In such cases Y means that $P \vdash B$ is G3-llr-derivable, and N that it is not.

References

1. K. R. Apt, Introduction to Logic Programming, in J. Van Leeuwen, editor, *Handbook of Theoretical Computer Science*, North Holland, 1989.
2. K. R. Apt, H. A. Blair and A. Walker, Towards a Theory of Declarative knowledge, in [28], pp. 89-148.
3. K. R. Apt and M. H. van Emden, Contributions to the Theory of Logic Programming, *J. ACM* 29, 3, July 1982, pp. 841-862.
4. L. Cavedon and J. W. Lloyd, A Completeness Theorem for SLDNF Resolution, *Journal of logic programming*, 1989, vol. 7, No 3, pp. 177-192.
5. D. Chan, Constructive Negation Based on Completed Database, in [20], pp. 111-125.
6. K. L. Clark, Negation as Failure, in *Logic and Databases*, H. Gallaire and J. Minker (eds.), Plenum Press, New York, 1978, 193-322.
7. K. L. Clark, Predicate Logic as a Computational Formalism, *Research Report DOC 79/59*, Dept. of Computing, Imperial College, 1979.
8. P. M. Dung and K. Kanchanasut, On the Generalized Predicate Completion of Non-Horn Programs, in [27], pp. 587-603.
9. P. M. Dung and K. Kanchanasut, A Fixpoint Approach to Declarative Semantics of Logic Programs, in [27], pp. 604-625.
10. M. Falaschi, G. Levi, M. Martelli, C. Palamidessi, A New Declarative Semantics for Logic Languages, in [20], pp. 993-1005.
11. M. Fitting, A Kripke-Kleene Semantics for Logic Programs, *Journal of Logic Programming*, 1985, No 4, pp 295-312.
12. M. Fitting, Partial Models and Logic Programming, *Journal of Theoretical Computer Science* 48 (1986), pp. 229-255.
13. M. Fitting and M. Ben Jacob, Stratified and Three-valued Logic Programming Semantics, in [20], pp. 1054-1069.
14. D. Gabbay, Modal Provability Foundations for Negation as Failure I, 4th draft, Feb 1989, unpublished.

15. M. Gelfond and V. Lifschitz, The Stable Model Semantics for Logic Programming, in [20], pp. 1070-1080.

16. J. Harland, A Kripke-like Model for Negation as Failure, in [27], pp 626-644. J. Jaffar, J.-L. Lassez and J. W. Lloyd, Completeness of the negation as failure rule, *IJCAI-83*, Karlsruhe, 1983, pp. 500-506.

17. R. A. Kowalski, Predicate Logic as a Programming Language, *Information Processing '74*, Stockholm, North Holland, 1974, pp. 569-574.

18. R. A. Kowalski, Algorithm = Logic + Control, *Communications of the ACM 22*, 7, July 1979, pp. 424-436.

19. R. A. Kowalski, The Relation Between Logic Progamming and Logic Specification, in C. A. R. Hoare and J. C. Shepherdson (eds.) *Mathematical Logic and Programming Languages*, Prentice Hall, Englewood Cliffs, N.J. 1985, pp. 11-27.

20. R. A. Kowalski and K. A. Bowen (eds.), *Logic Programming, Proceedings of the Fifth International Conference and Symposium*, MIT Press, 1988.

21. K. Kunen, Some Remarks on the Completed Database, in [20], pp. 978-992.

22. K. Kunen, Negation in Logic Programming, *Journal of logic programming* 1987, No 4, pp 289-308.

23. K. Kunen, Signed Data Dependencies in Logic Programs, *Journal of logic programming*, 1989, vol. 7, No 3, pp.231-247.

24. G. Levi and M. Martelli (eds.), *Logic Programming, Proceedings of the Sixth International Conference*, MIT Press, 1989.

25. V. Lifschitz, On the Declarative Semantics of Logic Programs with Negation, in [28], pp. 177-192.

26. J. W. Lloyd, *Foundations of Logic Programming*, Second extended edition, Springer Verlag, 1987.

27. E. L. Lusk and R. A. Overbeek (eds.), *Logic Programming, Proceedings of the North American Conference 1989*, MIT Press, 1989.

28. J. Minker (ed.), *Foundations of Deductive Databases and Logic Programming*, Morgan Kaufmann, 1988.

29. D. Pearce and G. Wagner, Reasoning with Negative Information I: Strong negation in Logic Programs, draft 1989.

30. J. A. Plaza, *Fully Declarative Programming with Logic – Mathematical Foundations*, Ph.D. Dissertation, City University of New York, July 1990.

31. J. A. Plaza, Completeness for Propositional Logic Programs with Negation, in: Z.W. Ras and M. Zemankowa (eds.), *Methodologies for Intelligent Systems – Proceedings of the 6th International Symposium 1991*, Lecture Notes in Artificial Intelligence 542, Springer Verlag, 1991.

32. H. Przymusińska and T. Przymusiński, Weakly Perfect Model Semantics for Logic Programs, in [20], pp. 1106-1123.

33. T. C. Przymusiński, On the Declarative Semantics of Deductive Databases and Logic Programs, in [28], pp. 193-216.

34. T. C. Przymusiński, On Constructive Negation in Logic Programming, in [27], addendum.

35. T. C. Przymusiński, On the Declarative and Procedural Semantics of Logic Programs, to appear in *Journal of Logic Programming*.

36. J. C. Shepherdson, Negation in Logic Programming, in [28], pp. 19-88.

37. J. R. Shoenfield, *Mathematical Logic*, Addison-Wesley, 1967.

38. L. Sterling and E. Shapiro, *The Art of PROLOG*, MIT Press, 1986.

39. M. Wallace, A Computable Semantics for General Logic Programs, *Journal of Logic Programming* 1989, vol. 6, No 3, pp. 269-297.

A New Translation from Deduction into Integer Programming

Reiner Hähnle*

Institut für Logik, Komplexität und Deduktionssysteme, Fakultät für Informatik,
Universität Karlsruhe, Kaiserstr. 12, 7500 Karlsruhe, Germany
e–mail: haehnle@ira.uka.de

Abstract. We generalize propositional analytic tableaux for classical and many–valued logics to *constraint tableaux*. We show that this technique provides a new translation from deduction into integer programming. The main advantages are (i) an efficient satisfiability checking procedure for classical and, for the first time, for a wide range of many–valued, including infinitely–valued propositional logics; (ii) a new point of view on classifying complexity of many–valued logics; (iii) easy NP-containment proofs for many–valued logics.

1 Introduction

We assume the reader is familiar with tableau calculus for classical propositional logic [10]. We assume further some basic knowledge of propositional many–valued logics [11]. Due to space limitations no proofs are included; most of them can be found in [4].

We consider a finitely–valued logic $\mathcal{L}$ consisting of a propositional language **L** and a n–valued matrix **A**. As the set N of truth values we take equidistant rational numbers from the unit interval i.e.

$$N = \{0, \frac{1}{n-1}, \ldots, \frac{n-2}{n-1}, 1\}$$

In a three–valued logic, for example, $N = \{0, \frac{1}{2}, 1\}$. We fix the set of designated truth values (i.e. the truth values that support validity of a statement) as $\{1\}$. The semantic matrix of a logic $\mathcal{L}$ may be given by many–valued truth tables for each logical connective. For example, we might define so–called three–valued Kleene disjunction with the truth table shown on the left in Figure 1.

While in classical signed tableau calculus the signs are either 1 or 0 and thus correspond exactly to a classical truth value, the author has introduced in [2] arbitrary subsets of the truth value set as signs. As an example we give the tableau rule corresponding to the sign $\{\frac{1}{2}, 1\}$ and three–valued Kleene disjunction (see Figure 1).

---

* The research described in this paper has been conducted while the author was sponsored from IBM Germany.

$$
\begin{array}{c|c|c|c}
\vee & 0 & \frac{1}{2} & 1 \\
\hline
0 & 0 & \frac{1}{2} & 1 \\
\frac{1}{2} & \frac{1}{2} & \frac{1}{2} & 1 \\
1 & 1 & 1 & 1
\end{array}
\qquad
\frac{\{\frac{1}{2},1\}\ \phi \vee \psi}{\{\frac{1}{2},1\}\ \phi \,|\, \{\frac{1}{2},1\}\ \psi}
$$

Fig. 1. Truth table and tableau rule for Kleene disjunction.

The truth value of the formula in the premise is *undefined* or *true* if and only if the truth value of one of its direct subformulas is *undefined* or *true*.

In [3] it has been shown that for a non-trivial class of many-valued logics Smullyan style tableau systems[2] result if this approach is being used.

On the other hand, for some important classes of many-valued logics, in particular for Lukasiewicz logics (which are defined below), this approach is not completely satisfying, since the branching factor of rules can become as big as the number of truth values. Moreover, the method does not extend to infinitely-valued logics.

For this reason, we develop in the following section the concept of a constraint tableau rule which is then used to construct a compact tableau system for Lukasiewicz (and many other) logics.

2 Tableau Proofs with Constraints

As mentioned before, the signs used in our tableau systems correspond to subsets of the set of truth values. We will, however, not admit arbitrary truth value sets as signs, but only signs of a certain shape (cf. [3]). We define the following abbreviations:

$$
\boxed{\leq i} := [0, j] \cap N \qquad\qquad \boxed{\geq i} := [j, 1] \cap N
$$

We do not impose any restrictions on the connectives.

Consider a signed formula $\phi = \boxed{\geq i}\, F(\phi_1, \phi_2)$, where F is any 2-place connective. A signed formula of this type is satisfiable iff for some valuation v the value of $v(F(\phi_1, \phi_2))$ is greater than or equal to i. The key idea in the following is to leave i as well as the signs in the rule extensions uninstantiated. For example, we could write down a rule like

$$
\frac{\boxed{\geq i}\, F(\phi_1, \phi_2)}{\boxed{\geq i_1}\, F(\phi_1, \phi_2)}
$$
$$
\boxed{\geq i_2}\, F(\phi_1, \phi_2) \ .
$$

---

[2] That is, the rules may be classified according to Smullyan's uniform notation [10] into type $\alpha, \beta, \gamma, \delta$.

For most instances of i, i_1, i_2, however, such a rule does not properly reflect the semantics of F, hence we must impose some additional constraints. Let us become a little bit more concrete and consider a signed formula $\boxed{\leq i}\,(\phi_1 \supset_L \phi_2)$, where $\supset_L$ denotes n-valued Lukasiewicz implication which can be defined as

$$i_1 \supset_L i_2 = \min\{1, 1 - i_1 + i_2\} \qquad \text{or, for } n = 3:$$

$\supset_L$	0	$\frac{1}{2}$	1
0	1	1	1
$\frac{1}{2}$	$\frac{1}{2}$	1	1
1	0	$\frac{1}{2}$	1

If $i = 1$ the signed formula $\boxed{\leq i}\,(\phi_1 \supset_L \phi_2)$ is trivially satisfied and can be omitted from the further analysis of the current branch, otherwise we have

Proposition 1. *If* $i \leq \frac{n-2}{n-1}$ *then* $\boxed{\leq i}\,(\phi_1 \supset_L \phi_2)$ *is satisfiable iff both,* $\boxed{\geq i_1}\,\phi_1$ *and* $\boxed{\leq i_2}\,\phi_2$ *are satisfied by the same valuation and* $i = 1 - i_1 + i_2$ *holds.*

From this proposition we may derive a tableau rule for $\boxed{\leq i}\,(\phi_1 \supset_L \phi_2)$. It has *provisos* which have to be satisfied in any proof the rule is used in. A similar technique is used in the classical quantifier rules, only that the proviso can be checked immediately in the case of quantifier rules, whereas we delay the check until tableau completion in the present case. Moreover, there may be different constraints associated with each extension. Let us call rules of the new kind **constraint rules**. The rule for $\boxed{\leq i}$ and $\supset_L$ as well as its counterpart for $\boxed{\geq i}$ are given in Table 1. Note that the left extension of the rule for $\boxed{\leq i}$ is empty; only the constraint information $i = 1$ (in which case $\boxed{\leq i}\,(\phi_1 \supset \phi_2)$ is trivially satisfied) counts in the corresponding branch.

Table 1. Constraint Rules for $\boxed{\leq i}\,(\phi_1 \supset_L \phi_2)$ and $\boxed{\geq i}\,(\phi_1 \supset_L \phi_2)$.

$$
\begin{array}{ll}
\boxed{\leq i}\,(\phi_1 \supset_L \phi_2) & \boxed{\geq i}\,(\phi_1 \supset_L \phi_2) \\
\hline
\quad\boxed{\geq i_1}\,\phi_1 \quad i \leq \frac{n-2}{n-1} & \boxed{\leq i_1}\,\phi_1 \\
i = 1 \quad \boxed{\leq i_2}\,\phi_2 \;\; 1 - i_1 + i_2 = i & \boxed{\geq i_2}\,\phi_2 \quad 1 - i_1 + i_2 = i
\end{array}
$$

Recall that the number of extensions for n-valued Lukasiewicz implication rules was up to n with the old rules, while now it is constant for arbitrary n. Different values of n are handled in the constraints. Thus, a proof tree looks the same for the same root formula and different n.

It is instructive to instantiate the premise of a rule and compute all solutions of its constraint system. Consider, for example, the rule for $\boxed{\geq \frac{1}{2}}\,(\phi_1 \supset_L \phi_2)$ in 5-valued Lukasiewicz logic. The constraint system consists of the single equation

$1 - i_1 + i_2 = \frac{1}{2}$ which has to be solved over N. The values of the $(i_1, i_2) \in N^2 = \{0, \frac{1}{4}, \frac{1}{2}, \frac{3}{4}, 1\}^2$ solving this equation are $\{(\frac{1}{2}, 0), (\frac{3}{4}, \frac{1}{4}), (1, \frac{1}{2})\}$.

Each different solution of the constraint system corresponds to a conventional rule extension, so we can backtranslate our example into a conventional rule with three extensions and no constraints:

$$\frac{\boxed{\geq\frac{1}{2}}(\phi_1 \supset_L \phi_2)}{\boxed{\leq\frac{1}{2}}\phi_1 \; \boxed{\leq\frac{3}{4}}\phi_1 \; \boxed{\leq 1}\phi_1 \\ \boxed{\geq 0}\phi_2 \; \boxed{\geq\frac{1}{4}}\phi_2 \; \boxed{\geq\frac{1}{2}}\phi_2}$$

Eliminating the trivially satisfiable formulas $\boxed{\geq 0}\,\phi_2$ and $\boxed{\leq 1}\,\phi_1$ yields exactly the same rule for $\boxed{\geq\frac{1}{2}}(\phi_1 \supset_L \phi_2)$ as the method in [2, 4]. The constraint $1 - i_1 + i_2 = \frac{1}{2}$ is merely an implicit representation of the extensions in the conventional rule.

We postpone the question of branch closures involving signs of the new kind for the moment and observe that with each branch of a completed tableau which is not yet tested for closure a system of linear inequalities whose variables are ranging over N, in other words an *Integer Programming* (IP) problem with solutions in N, is associated.[3]

We modify the usual *tableau construction process* as follows:

When a constraint rule is applied, the constraints from each extension are added to the constraints already present on the current branch and thus form a new constraint system associated with each newly generated branch. It may happen that for one or more of the newly generated branches the new constraint system becomes unsolvable. Such branches are deleted immediately, since they cannot possibly represent a satisfiable extension of the current branch.

In general, this is how a constraint tableau rule does look like:

$$\frac{S\,\phi}{\begin{array}{c|c|c} C_1 & \cdots & C_r \\ (IP_1) & & (IP_r) \end{array}} \quad \text{where} \quad \begin{aligned} \phi &= F(\phi_1, \ldots, \phi_k) \\ S &= \boxed{\lessgtr i} \\ C_j &= \boxed{\lessgtr i_{j1}}\phi_1 \circ \cdots \circ \boxed{\lessgtr i_{jk}}\phi_k \\ IP_j &= A_j I_j \leq B_j \\ I_j &= i, i_{j1}, \ldots, i_{jk} \, . \end{aligned}$$

---

[3] Strictly speaking, this is only the *feasability* part of an IP problem. Nevertheless we speak always of IP problems in the following, although we never need to minimize a cost function. If one wants to use standard IP software to solve such problems it is easy to reformulate as minimization problems, see [5]. Also the elements of N are not integer, but since N is finite it is trivial to transform these problems into true *integer* programming problems.

Together with the specification of a general constraint rule we must prove an analogue to Proposition 1:

Proposition 2 Constraint Rule Adequacy. *Using the notation as above we have that* $S\,\phi$ *is satisfiable iff for some instantiation of the variables in* I_j *in at least one extension (i)* IP_j *is solved and (ii)* C_j *is satisfiable.*

With the following definition one may link conventional tableaux and tableaux constructed with the help of constraint rules:

Definition 3 Partial Tableau. Let **T** be a completed tableau for a set of formulas Φ constructed with the help of constraint rules. Let I_l be the set of variables occurring in the IP problem associated with a branch B_l of **T**. If all variables in all I_l are instantiated with values from N such that all IP problems are solved we call the resulting tree **T*** **partial tableau** for Φ.

The idea is that each solution instance of a constraint tableau corresponds to the partial description of a conventional tableau. The collection of completed partial tableaux for Φ determines some completed conventional tableau for Φ.

Based on these concepts one may derive suitable definitions of Hintikka set and Analytic Consistency Property involving partial tableaux. For these Hintikka's Lemma and a Model Existence Theorem can be proved in rather the same way as in [2].

If we are interested in IP problems the constraints generated by the rules have to be linear inequalities. Later we will investigate the question for which class of connectives such constraints allow substantial simplifications of rules. For the moment, it is sufficient to note that all many–valued connectives principally fit into the new framework, since the old rules are just a special case using the empty constraint.

Intuitively, a formula is valid when every partial tableau of its negation can be closed. Let us see how we can deal with tableau closure in the present setting.

First, we note that in a completed tableau it is sufficient to look for atomic closure. Also it is sufficient to look for *pairs* of contradictory formulas (instead of tuples) when signs are restricted as above. Moreover, closure can only occur between atomic formulas with a different type of sign, that is, it can never occur between formulas like $\boxed{\leq_{i_1}}\,\phi$ and $\boxed{\leq_{i_2}}\,\phi$. Thus, a branch can only be closed when either two atomic formulas $\boxed{\leq_{i_1}}\,\phi,\ \boxed{\geq_{i_2}}\,\phi$ are present or a single formula $\boxed{\leq_j}\,\psi$ (or $\boxed{\geq_j}\,\psi$) such that no rule is defined for some j.[4]

In the first case, the branch is closed iff the signs have an empty intersection iff $i_1 < i_2$. In the second case, consider $\boxed{\leq_j}\,\psi$. There must be a greatest j_0 such that no rule for $\boxed{\leq_{j_0}}\,\psi$ is defined. Then, obviously no rule for $\boxed{\leq_j}\,\psi$ is defined

---

[4] If no truth value in $\boxed{\leq_j}$ does occur in the truth table of a connective F, a signed formula $\boxed{\leq_j}\,F(\phi,\psi)$ is unsatisfiable and no corresponding tableau rule is defined. We call such formulas *self–contradictory.*

iff $j \leq j_0$ iff $j < j_0 + \frac{1}{n-1}$. The value of j_0 is easily obtained from the truth table of the top level connective of ψ. For $\boxed{\geq j}\, \psi$ we proceed similar.

We obtain thus for each tableau branch $\mathbf{B}_l$ a set of strict linear inequalities

$$\{c_{l_1} i_{l_1} < d_{l_1}, \ldots, c_{l_p} i_{l_p} < d_{l_p}\},$$

such that solving *any* of them results in the closure of $\mathbf{B}_l$, in other words, a disjunction of strict linear inequalities. By definition, a branch is open when it cannot be closed. If we negate the disjunction

$$\bigvee_{k=1}^{p} (c_{l_k} i_{l_k} < d_{l_k}),$$

apply DeMorgan's law and observe that $c_{l_k} i_{l_k} \not< d_{l_k}$ iff $c_{l_k} i_{l_k} \geq d_{l_k}$ we get

Definition 4. Let $\mathbf{T}$ be a completed tableau for Φ built up using constraint rules and let $\mathbf{B}_1, \ldots, \mathbf{B}_m$ be the branches of $\mathbf{T}$. Moreover, let $C_l I_l \leq D_l$ be the IP problem (that is, $\bigwedge_{k=1}^{p} c_{l_k} i_{l_k} \geq d_{l_k}$) corresponding to the closure of $\mathbf{B}_l$ as above. Then $\mathbf{B}_l$ is called **open** iff $C_l I_l \leq D_l$ has a solution that solves also the IP problem $A_l I_l \leq B_l$ associated with the provisos on $\mathbf{B}_l$. $\mathbf{T}$ is a **constraint tableau proof** of Φ iff it has no open branch.

Actually, there is another, in some sense simpler way to represent branch closure. If we view atomic formulas (that is, propositional variables) as object variables ranging over the set of truth values we can take advantage from the fact that the (meta) variables in the signs and (object) variables are of the same type and mix them together in a single constraint. If p is atomic and $\boxed{\geq i}\, p$ is present on the branch $\mathbf{B}_l$ we simply add the constraint $p \geq i$ and we do similar for $\boxed{\leq i}\, p$. The resulting constraint system on a branch has then to be solved over $I_l \cup P_l$, where P_l are the propositional variables occurring on $\mathbf{B}_l$. This representation has the advantage of being shorter than the one without object variables, but it has the disadvantage of involving a greater number of variables.[5]

Theorem 5. *ϕ is a tautology iff there is a completed tableau for $\boxed{\leq \frac{n-2}{n-1}}\, \phi$ built up using constraint rules which represents a constraint tableau proof.*

3 Example

Before we proceed, let us give an example of a tableau proof using constraint rules. We will give a proof of the formula $p \supset_L (q \supset_L p)$ in three–valued Lukasiewicz logic. In the upper part of Figure 2 the conventional proof is shown, in the lower part the proof tree using constraint rules. Note that rule application to formula (3) in the lower tree does yield only one branch, since adding the condition $i_2 = 1$ would make the constraint system unsolvable.

The following IP problem corresponds to the only branch of the tree on the bottom (note that an equality is represented by two inequalities):

---

[5] This phenomenon is characteristic for MIP representations, see [6, p. 8].

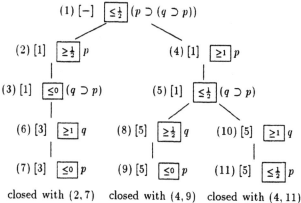

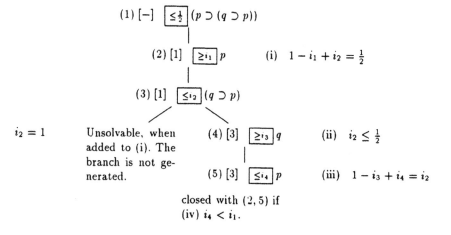

Fig. 2. A simple derivation in three-valued Lukasiewicz logic with and without constraints.

(i-a)	$-i_1 +i_2$	$\geq -\frac{1}{2}$
(i-b)	$+i_1 -i_2$	$\geq \frac{1}{2}$
(ii)	$-i_2$	$\geq -\frac{1}{2}$
(iii-a)	$-i_2 -i_3 +i_4 \geq -1$	
(iii-b)	$+i_2 +i_3 -i_4 \geq 1$	
(iv)	$-i_1$ $+i_4 \geq 0$	

To complete the proof we must show the infeasibility of this problem over $N = \{0, \frac{1}{2}, 1\}$.

This may seem not a great achievement if compared to constructing the conventional proof tree, but we must take into account that there exist very efficient algorithms for solving IP problems and, more important, while the IP problem *does not become substantially more complex when n grows*, the conventional proof tree becomes bigger and bigger (in the example it grows with $\mathcal{O}(n^2)$, but in general it grows with $\mathcal{O}(n^k)$, where k is the depth of the formula to be proved).

Another important point is that the system of inequalities for each branch can be computed incrementally, while the tree is constructed, thus using information that was computed only once in more than one branch.

4 Infinitely Valued Logic

As the proof trees for all finite Lukasiewicz logics look the same modulo some constants, it is a natural question to ask, whether the method can be generalized to *infinitely valued* logic.[6] The answer is in the affirmative and all we must do is to handle strict inequalities as signs such as in $\boxed{<i}$. As a consequence, also in the constraints strict inequalities will occur. In Table 2 we have summarized the rules for infinitely valued Lukasiewicz logic L_ω, while in Figure 3 we give the infinitely valued version of the example from above. An analogue to Proposition 1 can be used to show soundness and completeness of that system.

Table 2. Constraint rules for infinitely valued Lukasiewicz logic.

$$\frac{\boxed{\leq i}\,(\phi_1 \supset_L \phi_2)}{\boxed{\geq i_1}\,\phi_1 \quad i < 1}$$
$$i = 1 \qquad \boxed{\leq i_2}\,\phi_2 \;\; 1 - i_1 + i_2 = i$$

$$\frac{\boxed{\geq i}\,(\phi_1 \supset_L \phi_2)}{\boxed{\leq i_1}\,\phi_1}$$
$$\boxed{\geq i_2}\,\phi_2 \quad 1 - i_1 + i_2 = i$$

$$\frac{\boxed{< i}\,(\phi_1 \supset_L \phi_2)}{\boxed{\geq i_1}\,\phi_1}$$
$$\boxed{\leq i_2}\,\phi_2 \quad 1 - i_1 + i_2 < i$$

$$\frac{\boxed{> i}\,(\phi_1 \supset_L \phi_2)}{\boxed{\leq i_1}\,\phi_1}$$
$$\boxed{\geq i_2}\,\phi_2 \quad 1 - i_1 + i_2 < i$$

Note that if we work in infinitely valued logic we must try to solve the constraint systems over the unit interval of the rational line and thus have *Linear Programming* (LP) problems instead of IP problems. Since LP is known to be in P we have polynomial satisfiability in infinitely-valued Lukasiewicz logic for all formulas whose tableau proof trees branch at most polynomially often.

The signs $\boxed{<1}$ and $\boxed{>1}$ occur only in the initial tableau.

---

[6] The definition of the operators in infinitely valued Lukasiewicz logic is the same as in the finite case, but the set of truth values becomes the rational interval $[0, 1]$.

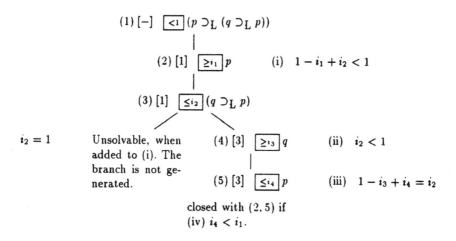

Fig. 3. A derivation in infinitely valued Lukasiewicz logic using constraints.

5 Complexity of Many–Valued Logics

At this point we can ask ourselves which classes of many-valued logics can be axiomatized naturally using constraint tableau rules. Consider the $k + 1$-dimensional region R that is defined by each combination of sign variable and k-ary connective f. Let $S(i)$ be a member of $\{\boxed{\leq i}, \boxed{\geq i}, \{i\}\}$. Define

$$R_f(S(i)) = \{(i, i_1, \ldots, i_k)| f(i_1, \ldots, i_k) \in S(i); i, i_1, \ldots, i_k \in N\}$$

R can be represented as a union or *disjunction* of bIP problems over the $S(i_j)$. The number of different bIP problems that are necessary gives us the number of extensions of the rule for $S(i)$ and f.

We arrive at a new complexity classification of many-valued logics. Given a logic with many-valued connectives and a certain set of signs S taken from $\{\boxed{\leq i}, \boxed{\geq i}, \{i\}\}$, our complexity measure will be the maximal number of disjuncts needed in the representation of $R_f(S(i))$ for any connective f and sign $S(i)$ over literals from S. Note that not for all logics such a representation is necessarily possible. In two-valued logic, for instance, it is not possible to represent $\boxed{\geq i}\,\phi \rightarrow \psi$ with only $\boxed{\geq i}$ and $\boxed{\leq i}$.

From this point of view the complexity of classical logic, Post logic and strong Kleene logic is 1, that of classical logic with the equivalence connective and Lukasiewicz logic is 2. This complexity measure reflects *topological* properties of the truth tables, since the minimal number of extensions of a rule corresponds to the minimal number of convex regions needed to cover the relevant entries in the truth table. All we can say of the mentioned logics in terms of traditional complexity classes is that they are NP-complete; at the moment we do not know how to relate both notions of complexity. We do not want to pursue this topic further here—we are not even sure whether it has an interesting theoretical perspective, but from a practical point of view it seemed worthwhile mentioning.

6   A Reduction from Many–Valued Deduction to MIP

We will now see that constraint tableaux can in fact be linearized, in other words, a tableau can be translated into a single constraint system. In Operations Research merging of disjunctively connected IP/LP problems has been extensively investigated. Given a set of IP/LP problems over the same set of variables the task is to find a single constraint system, possibly involving both integer and rational variables, such that it comprises exactly the disjunction of the solutions of the input systems. The discipline is called *Disjunctive Programming* and we will refer to a constraint system involving linear inequalities over real *and* integer variables with the term *Mixed Integer Programming* (MIP) problem. Fairly recent overviews on MIP research can be found in [6, 9].[7]

Our first example will be a formulation of tableau rules for classical propositional logic which consists only of *linear rules*. It is given in Table 3. Double negations are always eliminated before rule application. Note that only two rules introduce new sign variables at all and thus have associated constraints. The MIP problems resulting from tableaux constructed with these rules are in fact pure IP problems as is the case for all finitely–valued logics.

Table 3. Classical tableau rules in disjunctive constraint formulation.

$\boxed{\leq_i}\,\alpha$		$\boxed{\geq_i}\,\alpha$	$\boxed{\geq_i}\,\beta$		$\boxed{\leq_i}\,\beta$
$\boxed{\leq_{i_1}}\,\alpha_1\;\;i_1+i_2\leq i+1$		$\boxed{\geq_i}\,\alpha_1$	$\boxed{\geq_{i_1}}\,\beta_1\;\;i_1+i_2\geq i$		$\boxed{\leq_i}\,\beta_1$
$\boxed{\leq_{i_2}}\,\alpha_2$		$\boxed{\geq_i}\,\alpha_2$	$\boxed{\geq_{i_2}}\,\beta_2$		$\boxed{\leq_i}\,\beta_2$

A prototype implementation of a model checker for classical propositional logic has been implemented with the constraint mechanism of PrologIII, a logic programming language which is able to handle linear inequality constraints. First results are encouraging, for instance, the pigeon hole problem for $n = 8$ is solved in few minutes. The logic capabilities of PrologIII are hardly needed, after the

---

[7] In the former work some connections to automated deduction in classical logics are explained. Recently, some very fast satisfiability checking algorithms for classical propositional logic have been designed on the basis of LP, see [5]. The difference between these approaches and the present one, besides that the former are only suitable for classical logic, is that they are motivated by resolution or the Davis–Putnam procedure while our approach is motivated by the tableau method. One consequence is that we are not confined to use conjunctive normal form for the input.

input is transformed into an IP problem; we took PrologIII only for the benefit of having a prototype without having to spend much time on coding. We expect that the performance can be increased considerably by experimenting with the IP representation and using a tailor–made IP solving algorithm implemented in C.

Many other representations of tableau rules than the one in Table 3 are possible and some of them pay in increased efficiency in an implementation. A variant of the two more complicated of the classical constraint rules where one sign variable is saved is shown in Table 4.[8] The prize is to admit linear expressions as signs, but this is a mere technical difficulty which does not cause any problems. A similar optimization is possible for the rules of Lukasiewicz logic. Moreover, it can be shown that the standard translation of propositional CNF formulas into IP problems [5] can be obtained as a special case of our approach.[9]

Table 4. Improved constraint rules for classical logic.

$$
\frac{\boxed{\leq i}\,\alpha}{\boxed{\leq i-j+1}\,\alpha_1 \;\; i \leq j} \qquad \frac{\boxed{\geq i}\,\beta}{\boxed{\geq i-j}\,\beta_1 \;\; i \geq j}
$$
$$
\boxed{\leq j}\,\alpha_2 \qquad\qquad\quad \boxed{\geq j}\,\beta_2
$$

7 Negation

As in conventional tableaux there are two possible ways of dealing with negation. The first one was chosen in the rules above; negation operators in front of complex connectives are treated together with these connectives as α– or β– formulas, double negations are eliminated and negated atoms are transformed into inequalities. This works only for some many–valued logics. The second method consists of viewing negations as proper connectives with their own rules and this works for all kinds of many-valued logics. Possible rules for negation (whenever it is defined as $\neg i = 1 - i$) are

$$
\frac{\boxed{\leq i}\,\neg\phi}{\boxed{\geq 1-i}\,\phi} \qquad \frac{\boxed{\geq i}\,\neg\phi}{\boxed{\leq 1-i}\,\phi}
$$

---

[8] The rules are even sound and complete when the constraints are omitted completely. However, although redundant, the information reduces time for solving the IP considerably.

[9] We will present the details in a forthcoming paper.

8 MIP Formulation of Infinitely Valued Logic

In this section we demonstrate that constraint tableaux provide us with a decision procedure even for some infinitely valued logics. The only approaches to theorem proving in infinitely valued logics are, to our best knowledge, [1, 8]. Both are restricted to infinitely valued Lukasiewicz logic. Our approach can handle *any* infinitely valued logic whose connectives can be characterized by linear inequalities; moreover, we claim that it can be implemented much more efficiently. To ease comparison with other approaches we use L_ω as an example.

Let us first note that the left rule in Table 1 can be reformulated a little bit more redundant, but still sound:

$$\boxed{\leq i}\,(\phi_1 \supset_L \phi_2)$$

$$i = 1 \qquad \begin{array}{ll} \boxed{\geq i_1}\,\phi_1 & i \leq 1 \\ \boxed{\leq i_2}\,\phi_2 & 1 - i_1 + i_2 = i \end{array}$$

The only difference is $i \leq 1$ instead of $i < \frac{n-2}{n-1}$ in the right hand side constraint. A moment's thought reveals that this rule is still sound; when $i = 1$ some of the truth table entries in $\boxed{\leq 1}$ are now covered in both branches. This is not dramatic, however, since it happens also in the usual classical β–rules.

Next we note that this rule is still sound and complete for L_ω, since N does not occur in it explicitly.

Applications of disjunctive representation methods [6] and simplification then yields a linear formulation of the same rule

$$\boxed{\leq i}\,(\phi_1 \supset_L \phi_2)$$

$$\begin{array}{ll} \boxed{\geq i_1}\,\phi_1 & y \leq i \leq 1, \quad i_1 \leq 1 - y \\ \boxed{\leq i_2}\,\phi_2 & 1 - i_1 + i_2 = y + i, \quad y \leq i_2 \end{array}$$

where y is binary and i, i_1, i_2 range over N. If $y = 0$ the right extension of the rule above is selected, the left extension if $y = 1$. For this reason y is called **control variable**. The rule on the right in Table 1 as well as rules for negation stay unchanged.

In this way we can translate every many–valued deduction problem from L_ω into a single MIP problem whose integer part has not more variables than the input formula has connectives.

Corollary 6. $SAT_{L_\omega} \in NP$.

This result was obtained in [7] in a rather more complicated way using McNaughton's Theorem. Our completely different method is not only much simpler, but renders itself also to many other logics for which such results as McNaughton's Theorem do not exist.[10]

It is not trivial to compute MIP representations of many–valued tableau rules as can be seen in the example above nor is it solving MIP problems, but the work done by Jeroslow and other researchers in the field of Operations Research, where a considerable amount of knowledge about MIP methods has been accumulated, fits in here exactly.

9 Further Research

Our presentation remains on a somewhat sketchy level and many details have to be filled in yet. Nevertheless we hope to have convinced the reader that there are natural and promising connections between classical tableaux and IP on the one side and between many–valued tableaux and MIP on the other. We summarize the directions for further research which are in our eyes the most promising:

1. Apply fast satisfiablity checkers for many–valued logics to fuzzy reasoning and verification of integrated circuits.
2. The translation of tableau speed–up methods such as additional inference rules, lemma generation, indexing etc. into the MIP representation should be investigated.
3. While going from search trees to MIP representations structural information is lost. On the other hand, we might propagate such information from the tableau to the MIP representation, for instance, by specifying a partial order on variables which is then used to determine the order in which they are fixed when solving the MIP.
4. IP is hard to do while LP is not. It has been shown in other contexts that with certain inference rules (for example unit resolution) it is safe to substitute LP for IP. It would be interesting to identify such situations in the present context.
5. (Sub–)matrices of the resulting MIP problems are usually structured and sparse. Specialized MIP solving methods could improve performance considerably.
6. Perhaps the most interesting and challenging task is the extension to first–order logic. Several approaches are possible and are currently investigated.
7. The technique may well be applicable to other non–classical logics, such as modal or temporal logics.

---

[10] The other direction, NP–hardness, was shown, too, in [7]. The idea is to define for each set of propositional variables $p_1, \ldots p_k$ a L_ω–formula $\mathtt{two}(p_1, \ldots p_k)$ such that $\mathtt{two}(p_1, \ldots p_k)$ is satisfiable in L_ω iff $p_1, \ldots p_k$ are assigned binary truth values. Then for any formula ϕ which contains the propositional variables $p_1, \ldots p_k$ it is true that ϕ is satisfiable in classical logic iff $\mathtt{two}(p_1, \ldots p_k) \supset \phi$ is satisfiable in L_ω. If $\mathtt{two}$ has polynomial size in k this property reduces SAT to SAT_{L_ω}. In L_ω the function $\mathtt{two}$ is a bit awkward to define, since there is no connective corresponding to truth value set complement. The technique works for many other non–standard logics.

Acknowledgements

I am grateful to Daniele Mundici and Klaus Ries for valuable discussions and suggestions.

References

1. M. G. Beavers. Automated theorem proving for Lukasiewicz logics. Manuscript of talk given at 1991 Meeting of Society for Exact Philosophy, Victoria, Canada, may 1991.
2. Reiner Hähnle. Towards an efficient tableau proof procedure for multiple–valued logics. In *Proceedings Workshop on Computer Science Logic, Heidelberg*, pages 248 – 260. Springer, LNCS 533, 1990.
3. Reiner Hähnle. Uniform notation of tableaux rules for multiple–valued logics. In *Proc. International Symposium on Multiple-Valued Logic, Victoria*, pages 238 – 245. IEEE Press, 1991.
4. Reiner Hähnle. *Tableaux-Based Theorem Proving in Multiple-Valued Logics*. PhD thesis, University of Karlsruhe, Dept. of Computer Science, may 1992.
5. John N. Hooker. A quantitative approach to logical inference. *Decision Support Systems*, 4:45 – 69, 1988.
6. Robert S. Jeroslow. *Logic-Based Decision Support. Mixed Integer Model Formulation*. Elsevier, Amsterdam, 1988.
7. Daniele Mundici. Satisfiability in many-valued sentential logic is NP–complete. *Theoretical Computer Science*, 52:145 – 153, 1987.
8. Daniele Mundici. Normal forms in infinite–valued logic: The case of one variable. In *Proceedings Workshop Computer Science Logic 91, Berne*. Springer, LNCS, 1991.
9. George L. Nemhauser and Laurence A. Wolsey. Integer programing. In G. L. Nemhauser, A. H. G. Rinnooy Kan, and M .J. Todd, editors, *Handbooks in Operations Research and Management Science, Vol. II: Optimization*, chapter VI, pages 447 – 527. North-Holland, Amsterdam, 1989.
10. Raymond Smullyan. *First-Order Logic*. Springer, New York, 1968.
11. Alasdair Urquhart. Many-valued logic. In D. Gabbay and F. Guenthner, editors, *Handbook of Philosophical Logic, Vol. III: Alternatives in Classical Logic*, chapter 2, pages 71–116. Reidel, Dordrecht, 1986.

Reasoning About Time

Martin Charles Golumbic

IBM Israel Scientific Center, Technion City, Haifa, Israel
and Bar-Ilan University, Ramat Gan, Israel
golumbic@israearn.bitnet

Abstract

Reasoning about time is essential for applications in artificial intelligence and in many other disciplines. Given certain explicit relationships between a set of events, we would like to have the ability to infer additional relationships which are implicit in those given. For example, the transitivity of "before" and "contains" may allow us to infer information regarding the sequence of events. Such inferences are essential in story understanding, planning and causal reasoning. There are a great number of practical problems in which one is interested in constructing a time line where each particular event or phenomenon corresponds to an interval representing its duration. These include seriation in archeology, behavioral psychology, temporal reasoning, scheduling, and combinatorics. Other applications arise in non-temporal context, for example, in molecular biology, arrangement of DNA segments along a linear DNA chain involves similar problems.

Interval consistency problems deal with events, each of which is assumed to be an interval on the real line or on any other linearly ordered set, and reasoning about such intervals when the precise topological relationships between them is unknown or only partially specified. In Golumbic and Shamir (1991), we relate the two notions of interval algebra from the temporal reasoning community and interval graphs from the combinatorics community, obtaining new algorithmic complexity results of interest to both disciplines. Several versions of the *satisfiability, minimum labeling* and *all consistent solutions* problems for temporal (interval) data are investigated. The satisfiability question is shown to be NP-Complete even when restricting the possible interval relationships to subsets of the relations intersection and precedence only. On the other hand, we give efficient algorithm for several other restrictions of the problem.

References

J.F. Allen, Maintaining knowledge about temporal intervals, *Comm. ACM 26*, 832-843, 1983.

A. Belfer and M.C. Golumbic, The role of combinatorial structures in temporal reasoning, *Proc. AAAI Workshop on Constraint Satisfaction*, Boston, Mass., July 1990.

M.C. Golumbic, *Algorithmic Graph Theory and Perfect Graphs*, Academic Press, New York, 1980.

M.C. Golumbic and R. Shamir, Complexity and algorithms for reasoning about time: A graph-theoretic approach, DIMACS Tech. Rep. 91-54, Rutgers Univ., 1991. To appear in *J. ACM*.

Learning Qualitative Physics Reasoning from Regime Analysis

Waldir L. Roque*

Departamento de Ciência da Computação
Universidade Federal de Santa Catarina
88040.970 Florianópolis, SC - Brazil

Abstract. One of the very basic *rules* we learn in an introductory course of physics is that two physical quantities, say Q_1 and Q_2, can only be compared when they have the same dimensional representation and, in addition, should be in the same system of units. This rule is also frequently applied to discard non-matching physcial formulae. In other words, the dimensional consistence of physical formulae must be fulfilled.

The rule we are talking about is in fact known as the *Principle of Dimensional Homogeneity* (PDH), which states that any physical law has to be dimensionally consistent to be meaningful.

In this paper we show that using the PDH and results from the *Regime Analysis*, it is possible to reason qualitatively about a physical system. In addition, as the whole process is algorithmic, this allows automating the reasoning through a symbolic computer system. The system QDR – **Q**ualitative **D**imensional **R**easoner has been developed for this task.

1 Introduction

In the learning process of basic classical physics concepts, it has become very often procedure to emphasize strongly towards the quantitative and formula-driven description of a system, giving a weaker, an sometimes poor, attention to the qualitative and common sense behaviour of the system. It is clear that the actual formula that rules the system's behaviour and the quantitative results obtained through it are quite important. Nevertheless, in many simple everyday physical problems the qualitative analysis of a system can suffice and respond to many questions about the system's behaviour as well.

As a matter of fact, in many engineering applications in industry and computer science, qualitative analysis is sometimes enough to describe the behaviour

---

*Bitnet: CEC1WLR@BRUFSC.

of a process or device. Therefore, it is growing the necessity for a better comprehension of physical processes and devices via qualitative reasoning techniques. This lead us to the necessity of focusing a bit more on the qualitative behaviour and on the study of qualitative methodologies that can be applied without too much appeal to the actual formulae and quantitative data describing a physical system.

The need of better understanding of the qualitative behaviour of physical systems has been realized by computer scientists in their trends for new computer systems with better *intelligence* power. This new approach to solve physical problems has been named *Qualitative Physics* [1, 2] and many styles of qualitative reasoning have already appeared in the literature of Artificial Intelligence.

Qualitative Physics is a new field with strong relationship to Artificial Intelligence (AI) and other areas. The basic scope of this subject has been nicely stated by Forbus in [2], which we quote here:

> *Qualitative physics is concerned with representing and reasoning about the physical world. The goal of qualitative physics is to capture both the common sense knowledge of the person on the street and the tacit knowledge underlying the quantitative knowledge used by engineers and scientists.*

Although a new subject, it has shown to be very important and promising in many applications in engineering processes. In particular in diagnosing faulty devices [3, 4], control and monitoring of industrial plants [5, 6], qualitative modeling and simulation [7, 8, 9], machine learning [10], robotics [11], medicine [12] etc.

Several approaches for reasoning about physics has been proposed in the literature. Among them, we can cite some of the first and basic papers like *Naive Physics* [13], *Qualitative Process Theory* [1], *Qualitative Physics based on Confluences* [14] and *Qualitative Simulation* [15], to name a few. However, as the field has grown and many other relevant papers were scattered in several journals, a book collecting many of them together by their concerning subject has been published recently [16].

In a paper on qualitative reasoning about physics [17] the authors make use of the Theory of Dimensional Analysis - (TDA) as a supporting technique to investigate the qualitative behaviour of physical systems. The idea behind is to use the dimensional representation of the variables describing the physical system to obtain some dimensionless functionals, which provide relationships among the systems' variables. Through these dimensionless functionals, it is possible to obtain *qualitative partial derivatives* and then analyse the behaviour of some systems' variables with respect to changes in the others.

There the notions of *regimes*, which are particular dimensionless functionals, and *partials* were introduced as the primary tools to qualitatively analyse a physical system. Nevertheless, several additional analyses were not covered and some points needed a better understanding. Therefore, an extension of these ideas have been forwarded in [18] and the results will appear elsewhere. This approcah as a whole constitutes the *Regime Analysis*.

The goal of this paper is twofold. Firstly, to give an account of how the regime analysis can provide a qualitative understanding of simple physical systems, and secondly, to report the development of a symbolic computer system capable to automate the whole qualitative reasoning process.

In section 2 a short and simple view of the quantitative (formula-driven) and qualitative (via regime analysis) approaches is given illustrated by a physical example. Section 3 gives an introduction to the regime analysis with focus directed to qualitative reasoning. Section 4 reports the system QDR and its qualitative reasoning capabilities. Section 5 concludes the paper.

2 Quantitative and Qualitative Physics

In this section a simple physical system will be looked at according to the quantitative and qualitative analyses approaches. Quantitative shall be understood as formula-driven approach to the problem.

2.1 The Mass×Spring System

Let us assume a very simple physical system, namely, the mass-spring harmonic oscillator. For that, let us consider a spring with elasticity constant k, attached to the ceiling by one of its extremity and having attached in the other extremity a mass m. It will be taking for granted that there is no air resistance or any other dissipative effect. The only force acting in the system is due to the gravitational acceleration given by g and due to the spring.

Quantitative Analysis. From Newton's law it is known that the system is in equilibrium when the force due to gravity equals the (opposite) force due to the spring. This equilibrium is broken when the mass is pulled down and freed. Assuming that the motion is only allowed in the vertical direction and that at the equilibrium point the spring displacement is set to $x = x_0$, the laws ruling the system are such that at the equilibrium point x_0 the sum of the forces due to gravity and due to the spring must vanish. In other words, the force due to gravity given by $F_g = m\,g$ and, from Hooke's law, the force due to the spring, given by $F_s = -k\,x$, must add to zero. Thus,

$$F_g + F_s = 0 \quad \Longrightarrow \quad m\,g = k\,x_0\,. \tag{1}$$

When the mass is pulled down a distance x from the equilibrium point, then at the point x the sum of the forces acting on the system must add to zero. Thus,

$$F + F_g + F_s + F_k = 0\,, \tag{2}$$

where $F = m\,a$ and $F_k = -k\,x$.

It is convenient to reset the displacement $x_0 = 0$ and consider the origin of the motion at the point where the mass is released. That is, at time $t = 0$ the mass is at the point $x = 0$ and to take x positive downwards. We will adopt these here.

The equation 2 leads to the differential equation

$$\frac{d^2 x}{dt^2} + \omega^2 x = 0 ,$$ (3)

where the acceleration was written as $a = \frac{dv}{dt} = \frac{d^2 x}{dt^2}$ and $\omega^2 = \frac{k}{m}$, as usual.

The solution of 3 taking into account the initial condition, is given by:

$$x(t) = A \sin(\omega t) ,$$ (4)

where A is a constant.

It is easy to see from the equation for the displacement x, that the system has an oscillatory behaviour due to the functional dependence on $\sin(\omega t)$.

Now by simple trigonometric manipulations the well known period P of oscillation can be obtained. It is given by,

$$P = \frac{2\pi}{\omega} = 2\pi \sqrt{\frac{m}{k}} .$$ (5)

From the equation for the period above we can immediately see that:

i) The period P of oscillation increases when the mass m increases.

ii) The period P of oscillation decreases when the elasticity constant k of the spring increases.

The amplitude of the oscillation is found when the displacement of the spring attains its maximum (with respect to the equilibrium point $x = 0$) which is reached at $\sin(\omega t) = \pm 1$. At the point where $\sin(\omega t) = -1$, the acceleration is only due to gravity. Thus, from the acceleration equation obtained from $x(t)$, we get that the absolute value of the amplitude of oscillation is given by

$$A = \frac{g}{\omega^2} = \frac{mg}{k} .$$ (6)

The equation above tell us that[1]:

iii) The amplitude A of the oscillation increases when the mass m increases.

iv) The amplitude A of the oscillation increases when the gravitational acceleration g increases.

v) The amplitude A of the oscillation decreases when the elasticity constant k of the spring increases.

---

[1] In principle A is a fixed constant (from the differential equation solution), but our common sense allows the reasoning that follows.

From the relations for the amplitude A and the period P, we obtain the relationship among them as

$$P = 2\pi \sqrt{A/g}.$$ (7)

Thus, we can get the following informations out of it:

vi) The period P of the oscillation increases when the amplitude A increases.

vii) The period P of the oscillation decreases when the gravitational acceleration g increases.

The behaviour of the mass-spring system given in (**i**,...,**vii**) where reached just looking at the equations. It is *not* very common to reason about how the period of oscillation varies with respect to the gravitational pull, because it has been *a priori* assumed, when teaching this problem, that the value of g is a (fixed) constant. Therefore, in this regard, it does not make sense to think of it. In addition, there is a tendency to reason wrongly at a first glance, just saying that if the gravitational acceleration increases, than the period should also increases. This reasoning is not correct because there might be other variables involved as we have shown in **vi**. If the amplitude is mantained constant, the right solution is that the period should decrease (this can be easily seen in a plot of the function $x(t)$).

Qualitative Analysis. Let us now look at the same problem through a different approach. The goal here is to be able to get qualitative informations about the period of the system and its amplitude without going through Newton's and Hooke's laws, solutions of differential equations, setting of reference's frame or initial conditions.

The vertical mass-spring system can be described by the following variables: the mass m, the elasticity constant k, the gravitational acceleration g and as we are interested in reasoning about the period of oscillation, it is necessary to include the variable P for period. Let us call this set of variables of *process variables* as they are the ones involved in our process.

Let us call the dimensional representation for the mass m by [MASS], for the elasticity constant k by [MASS TIME^{-2}], for the gravitational acceleration g by [LENGTH TIME^{-2}] and for the period P by [TIME].

An argument that can be raised by someone is that, the dimensional representation of the elasticity constant k or the gravitational acceleration g inherit the historical physical knowledge, and so, in some respect the dimensional representation has underneath the knowledge of Newton's and Hooke's laws.

This argument cannot be taking for granted as it can be put in the other way around. The definition of secondary quantities obey the *Product Theorem* (see section 3) and the PDH imposes further restrictions on the dimensional representation of the physical variables. Therefore, the formula discovery could well departure from the dimensional analysis of the physical variables involved in a system and, with additional informations provided by other means (experiments, observations, logic, etc), be finally reached.

Another feasible possibility where there is no need to know whether there is or not a specific law describing the systems' behaviour would be the *layman approach*. He could just have looked at in a table what are the dimensional representations of k and g.

It can be seen from the dimensional representation of the process variables for the mass-spring system that only three dimensions are independent, namely: [MASS, TIME, LENGTH]. Thus we have $n = 4$ process variables and $d = 3$ independent dimensions.

As we are interested in reasoning about the period of oscillation, let us try to find out a simple relation such that,

$$\Pi = P\, m^{\alpha}\, k^{\beta}\, g^{\lambda}\,, \tag{8}$$

where the solution of (α, β, λ) generates a dimensionless functional Π. In other words, the combination of the process variables are such that their dimensional representation vanishes.

The equation 8 is simple to be solved. It is necessary to substitute each of the process variable by its corresponding dimensional representation and then the exponents of the independent dimensions must add up to zero. It becomes,

$$\Pi = \text{TIME} \, \text{MASS}^{\alpha} \, \text{MASS}^{\beta} \, \text{TIME}^{-2\beta} \, \text{LENGTH}^{\lambda} \, \text{TIME}^{-2\lambda}\,, \tag{9}$$

which has the solution $\alpha = -1/2$, $\beta = 1/2$ and $\lambda = 0$. Thus, 8 can be written as

$$\Pi = P\sqrt{\frac{k}{m}}\,. \tag{10}$$

Notice that the process variable g does not appear in the expression of the dimensionless functional. In this case, the gravitational acceleration can be seen as a *superfluos variable* for this consideration.

From the expression of the dimensionless functional Π the variable P, which will be called of *performance variable*, can be written as

$$P = \Pi\sqrt{\frac{m}{k}}\,. \tag{11}$$

Taking now the partial derivative of the performance variable P with respect to the m and k, we get

$$\frac{\partial P}{\partial m} = \frac{P}{2\,m}\,, \quad \frac{\partial P}{\partial k} = -\frac{P}{2\,k}\,, \tag{12}$$

where the expression of the dimensionless functional Π has been substituted afterwards.

By inspection of the *partials*, it is easy to see that their signs are, respectively, given by,

$$\frac{\partial P}{\partial m} > 0, \quad \frac{\partial P}{\partial k} < 0. \tag{13}$$

Therefore, the *product form* of the period (eq. 11) and the signs of the partials tell us that:

i') The period P of the system increases when the mass m increases.

ii') The period P of the system decreases when the elasticity constant k of the spring increases.

These results are in full (qualitative) agreement with the ones obtained throughout the quantitative analysis (**i** and **ii**) done before.

The product form of the period becomes exactly the physical law for the period of motion of the mass-spring harmonic oscillator, eq.(5), if the value of $\Pi = 2\pi$. In fact, dimensionless functional can be seen as defining a family of hypersurfaces in the *process variables space* [19], which is defined by the process variables.

Nevertheless, more precisely, the actual physical law is an *orbit* on the hyper-surface defined by $\Pi = 2\pi$, as the values of the process variables are constrained by physics (for instance, $m > 0$).

Some hypersurfaces are very important as they divide space in regions with distinct qualitative process behaviours. Unfortunately, to find out the actual value which defines a hypersuface corresponding to a physical law is not possible just through dimensional analysis. There is a need of external knowledge, which can come from experimental or observational counterparts of the physical phenomena.

Enriching the process. Suppose that now someone wants to reason about the amplitude of the oscillation. However, the only thing that is actually known is the period of oscillation. The approach is to include in the list of process variables a new variable, say A, regarding as the amplitude. The dimensional representation of the amplitude is [LENGTH]. Thus, the number of process variables now is $n = 5$, nevertheless the number of independent dimensions is still the same, $d = 3$.

By a similar calculation as done for the period, if the amplitude A is chosen as a performace variable, then a new dimensionless functional can be easily obtained. It is given by,

$$\Pi' = \frac{A k}{g m}. \tag{14}$$

Notice that in the dimensionless functional Π' the gravitational acceleration appears. In other words, it is not superfluous in this new process. The inclusion of the amplitude in the list of process variables has shown that it is not a superfluous variable. In fact, including it a more useful information was added to the system. Thus, we say that the physical process has been *enriched* by the inclusion of the amplitude.

By writing the amplitude in the product form, and by computing the partials, the signs obtained are:

$$\frac{\partial A}{\partial m} > 0, \quad \frac{\partial A}{\partial k} < 0, \quad \frac{\partial A}{\partial g} > 0. \tag{15}$$

Therefore, from the above partial signs and the product form, it can be concluded that:

iii') The amplitude A of the oscillation increases when the mass m increases.

iv') The amplitude A of the oscillation increases when the gravitational acceleration g increases.

v') The amplitude A of the oscillation decreases when the elasticity constant k of the spring increases.

These results are in full agreement with the ones obtained by the formula-driven quantitative approach (**iii, iv** and **v**) obtained previously. Inasmuch, the question, that is **not** normally thought of, is easely answered: Would the amplitude of oscillation be changed if we were in another place with a different gravitational acceleration? The answer is yes. Qualitative analysis provided a straightforward reply to that question as it comes out explicitly, while it is somehow hidden in the quantitative analysis, as it is normally taken for granted that the value of the gravitational acceleration is constant and fixed. This can be seen a *predicting* feature of the quantitative analysis in extending our empirical knowledge upon the physical process.

In this case the actual physical law given in eq.(6) is described by the hyper-surface $\Pi' = 1$.

How can the qualitative analysis provides a similar information that has been achieved by the quantitative analysis in statements **vi** and **vii**? The answer can be reached as follows: Combining the two dimensionless functionals Π and Π' through the common process variable m and looking the signs of the partials, we obtain

$$\frac{\partial P}{\partial A} > 0, \quad \frac{\partial P}{\partial g} < 0. \tag{16}$$

Therefore, it can be concluded that:

vi') The period P of the oscillation increases when the amplitude A increases.

vii') The period P of the oscillation decreases when the gravitational acceleration g increases.

These, as well, are exactly the contents stated in **vi** and **vii** of the quantitative analysis. The reasoning in this case was performed by an *interchange* of informations contained in different dimensionless functionals.

To conclude this section, it is worthwhile to stress that at least as far as the qualitative behaviour of the mass-spring system is concerned, the qualitative analysis performed here, through the dimensional analysis and the signs of the partials, is completely in agreement with the results provided by the quantitative formula-driven approach. In addition, the behaviour of the system shown in **vii** was rather easily found in the qualitative analysis.

The questions that remain are essentially, a) How far can we go with this approach? and b) How can we obtain all the important dimensionless functionals of a physical system?

The former question may not be answered accurately. For instance, a drawback of the dimensional analysis is that it is not fully capable to deal with variables that their dimensional representations are not well defined or enter in the system as arguments of functions like exponentials, logarithms, trigonometric, etc., despite of their vanishing dimensional representation. In this regard, the formula for the displacement $x(t)$ has no hope to be achieved by dimensional analysis due to its functional dependence in $\sin(\omega t)$.

However, for many physical systems the qualitative reasoning is just adequate. The latter will be precisely stated in the following section.

3 Regime Analysis

The TDA provides the framework for the regime analysis developed in [17, 18]. In this section a short and formal discussion to this qualitative reasoning approach will be addressed.

3.1 Theory of Dimensional Analysis

Theory of Dimensional Analysis (TDA) is quite old when compared with Qualitative Physics. It goes far back to Newton [20] and Fourier [21] who seem to be the first ones to call attention to the importance that the concept *dimension* plays to physics. At the end of the last century and the first two decades of this century, most of the eminent physicists have in one way or another contributed to the understanding and establishment of dimensions as an essential content of a physical quantity.

In more recent time we can cite several considerations of dimensional analysis in the field of physics and related subjects [22, 23, 24, 25], but only very recently is that attention has been driven to its *qualitative richness* applied to AI [10, 17].

The main results of the TDA follows from the Principle of Dimensional Homogeneity of physical laws, which states that all physical laws must be dimensionally consistent. These results are expressed in the *Product Theorem* and the Π-*Theorem*, which we give below. However, as our intention here is to give an account of TDA in adequacy to our need, the rigorous mathematical proofs will be skipped off, but they can be found in the references [26, 22, 27, 28, 29] as well as other details.

The Product Theorem. Let a secondary quantity be derived from measurements of primary quantities, say α, β, γ,.... Assuming *absolute significance of relative magnitudes*[2], the value of the secondary quantity is derived as:

$$C\,\alpha^a\,\beta^b\,\gamma^c\,\ldots,$$

where C, a, b, c, ... are constants (numbers).

This theorem establishes that dimensional representations must be product of powers of the fundamental dimensions defined.

The Π-Theorem. Given measurements of physical quantities α, β, γ,..., such that $\phi(\alpha, \beta, \gamma, \ldots) = 0$ is a homogeneous equation, then its solution can be written in the form $F(\Pi_1, \Pi_2, \ldots, \Pi_{n-r}) = 0$, where n is the number of arguments of ϕ and r is the minimal number of dimensions needed to express the variables α, β, γ,.... For all i, the Π_i are dimensionless functionals.

The Π-*Theorem*, due to Buckingham [30], provides a simple way to identify the number of dimensionless functionals that characterizes a physical process and also gives information on how to construct them.

In [17], the authors have shown that *Hall's Theorem*, from combinatorial theory, is useful to regime analysis and so, we include it here too.

Hall's Theorem. Let S be a finite set of indices, $S = \{1, 2, \ldots, n\}$. For each $i \in S$, let S_i be a subset of S. A necessary and sufficient condition for the existence of distinct representatives x_i, $i = 1, 2, \ldots, n$, $x_i \in S_i$, $x_i \neq x_j$, when $i \neq j$, is the condition: For every $k = 1, 2, \ldots, n$ and choice of k distinct indices $i_1, \ldots, i_j$, the subsets $S_{i_1}, \ldots, S_{i_k}$ contain among them at least k distinct elements.

According to Buckingham's theorem, one can construct $p = n - r$ independent dimensionless functionals in a physical system. Let us call these p independent functionals of *regimes*. The regimes retain all the physical content of a system and so, they are the important ones, as all other dimensionless functionals that can be construct with the physical process variables are combinations of products of power of the regimes.

Now, Hall's theorem guarantees that each regime represents exactly one process variable, that will be called *performance variable*, which is **not** in the *process basis* [17]. The basis variables are r variables among the n process variables whose dimensional representations i) are linearly independent (LI) and ii) all fundamental dimensions involved in the process are covered by them. But, how to compute the regimes associated to these variables?

---

[2] In [26] this means that the ratio between two measurements of quantities is independent of the system of units used. Mathematically, this corresponds to the function that forms the secondary quantities be homogeneous.

3.2 Regime Calculus

Let us denote the set of process variables describing a physical system by $\{v_1, v_2, \ldots, v_n\}$. According to Buckinham's and Hall's theorems, we may select out of the set of process variables $p = n - r$ performance variables, where r is the rank of the dimensional matrix[3] M_D. Let us denote the p performance variables by $\{y_1, y_2, \ldots, y_p\}$ and the remaining r variables, which form the basis variables, by $\{x_1, x_2, \ldots, x_r\}$.

The regime Π_i for the performance variable, say $\{y_i\}$, is given by,

$$\Pi_i = y_i \times \left(x_1^{\alpha_{i1}} x_2^{\alpha_{i2}} \ldots x_r^{\alpha_{ir}}\right),$$

where the coefficients α_{ij}, $i = 1, \ldots, p$, $j = 1, \ldots, r$ are such that Π_i is a dimensionless functional. The α's are the solution of the system of algebraic homogeneity equations for the dimensional representation of the corresponding performance variable and basis variables.

From the above expression of a regime, we can write the performance variable in the form,

$$y_i = \Pi_i \times \left(x_1^{-\alpha_{i1}} x_2^{-\alpha_{i2}} \ldots x_r^{-\alpha_{ir}}\right).$$

When a performance variable is written in the form above we say it is written in its *product form*.

It is possible that a physical system be composed of subsets of process variables. Thus, we say that each set forms an *ensemble*. When the choice of performance variables is changed and still provide a consistent basis, then we have a new *ensemble representation*, but also with p regimes in this new *ensemble representation*. Any ensemble representation is equally good in the physical sense as the p regimes have the same physical content. Thus, we say that the p regimes are the *regime generators*.

3.3 Qualitative Partials

The qualitative reasoning is obtained through the various *regimes* and *qualitative (partial) derivatives* analyses. The latter has been introduced in [1, 14] and is essentially the analysis of the sign of a partial derivative. They are defined in the following way:

$$\partial s[Q]_x = -1, \quad \partial s[Q]_x = 0 \quad \partial s[Q]_x = +1,$$

where the $\partial s[Q]_x$ (should be read the "qualitative partial of the quantity Q with respect to variable x") denotes the qualitative partial derivative of the quantity Q with respect to changes in the variable x. When $\partial s[Q]_x = -1$ or $\partial s[Q]_x = 0$, or $\partial s[Q]_x = +1$, one says that the quantity Q decreases, is constant

---

[3]This matrix is formed by the exponents of the dimensional representation of the process variables.

or increases according to variations in the variable x, while any other variable is kept constant.

The *intra-regime analysis* is done by computing the qualitative partial of a performance variable with respect to a basis variable, keeping the remaining basis variables fixed. This is given by,

$$\frac{\partial\, y_i}{\partial\, x_j} = -\alpha_{ij}\,\frac{y_i}{x_j}\,,$$

which in qualitative notation is,

$$\partial\, s\,[y_i]_{x_j} = -\frac{\alpha_{ij}}{|\alpha_{ij}|} = \pm 1\,.$$

Therefore, from the qualitative partial above, we can conclude that the performance variable y_i increases (or decreases) with respect to variations in x_j, according to the sign of $\partial\, s[y_i]_{x_j}$.

It is clear from the product form that one can have the qualitative partial derivative of the performance variable with respect to the basis variables present in the regime. Thus, qualitative informations about the physical system can be obtained from the intra-regime analysis.

It is also possible to have, *inter-regime*, *intra-regime-ensemble* and *inter-regime-ensemble* analyses. These allow qualitative partial analyses among the performance variables and across variables from distinct regimes and ensembles. However, we will leave these out of this report. Additional technical details can be found in [18].

Regimes have shown to be very important in many situations carring special meaning in several applications in different fields. Some of them are the *Reynold's number*, which indicates when a fluid flow is laminar ($R < 2000$) or turbulent ($R > 3000$); the *Prandtl numbers* in the study of forced and free convection of heat transfer; the *Thrust coefficients* used in propeller studies; the *heat pump performace parameter* [31] used to compare different performance of working fluids for solar energy heat pumps and many others as can be seen in the reference [32].

4 The Symbolic System QDR

The process of computing the regimes and the partials is fully algorithmic. Thus, it is suitable for a computer implementation. As the manipulations required are symbolic, they were implemented in a symbolic system called QDR – Qualitative Dimensional Reasoner [18] – through the symbolic language REDUCE [33, 34], which comprises both symbolic and algebraic programming facilities.

In [18] the reader can find the algorithm that has been implemented in REDUCE to create the system QDR , the details of several new concepts used this paper and several features of the QDR system. In particular, it is shown how it identifies i) an *incomplete specification problem* [10], ii) the *superfluous variables* of the system and iii) which variables are *global contact variables*.

An outstanding feature of QDR's algorithm is that it gives a constructive procedure to determine the *contact variables* and the *coupling regimes*, which were **not** presented in [17] and, in addition, the *inter-ensemble contact variables*, which is an extension of the contact variables' concept to variables from different ensembles.

Presently QDR is able to compute all regimes for all ensemble representations, write the extended dimensional matrix M_D and process matrix M_P, find the contact variables, the coupling regimes and the inter-ensemble contact variables, compute the various regimes and their qualitative partials, write the qualitative partials matrix M_S and finally perform the qualitative reasoning within and across regimes and ensembles providing a qualitative analysis of the process according to these informations.

The qualitative partials matrix M_S is an important piece of information, as it provides an easy knowledge retrieval about the qualitative behaviour of the process under changes of the process variables. This is relevant for control and monitoring of systems by table look-up.

QDR can run in two different modes: i) the *interactive mode*, where the user chooses the performance variables and ii) in a *batch-like* mode, where the process is fully automated. In the former the system search all possible choices of performance variables and then computes the regimes for all ensemble representations, etc. (see [18]).

As a sign of QDR's capability, all the examples (more than 20) from the classical book of Bridgman [26], for every ensemble representation of each example, have been fully worked out. In addition, in [18] the vertical motion of a projectile, the heat exchanger, the pressure regulator, the RLC circuit, the gravitational attraction, a material engineering problem, the heat pump problem in solar energy, and the cardiac output in medicine, were treated by QDR.

5 Conclusions

The evolution of physics in demand to respond to highly complex systems has somehow imposed stronger and sophisticated mathematical techniques to be developed and used in order to obtain the correct behaviour of the system and, not less important, predict new facts. Nevertheless, in many physical problems found in everyday life and engineering applications, this highly sophisticated and *heavy* mathematics is not necessary. Particularly, when the only interest is to perform a qualitative analysis of the physical process involved and learn about its behaviour under various circumstances.

In this sense, qualitative reasoning about the behaviour of physical systems is losing ground and little attention is devoted to it as a methodology for learning about the system. As a consequence it has been weakening the teaching of qualitative physics methods to the new generations. However, with the present demand for computers and systems simulating the reasoning capabilities of humankind, this tendency seems to have already started changing its direction.

In fact, many applications of Artificial Intelligence techniques in engineering have appealed to a better understanding of the qualitative behaviour of the physical process involved. This led to the birth of the *Qualitative Physics* [2, 16] field.

In general, Qualitative Physics techniques seem appropriate to analyse physical systems and devices where,

- the actual qualitative analysis of the physical process (or of a device) suffices the requirements,

- the computational costs of a full numerical (quantitative) analysis is high to be performed just to obtain at the end qualitative informations about the process,

- the process is defined by a large number of variables and parameters and the response of changing in one of these variables has to be analysed against the others,

- the actual physical law that rules the process or device is not known,

- in situations where rapid and rough estimates are enough to be considered.

The system QDR is applicable in the above circumstances and in addition it may also be used for pointing out *incomplete specification problem* [10], in simulation of new *landmark points* [15, 35] and formulae checking.

It is clear that qualitative physics by its own nature does not allow a **fully** accurate analysis of a physical process. In particular, the methodology that has been presented here, as it was pointed out before. Therefore, the merging of Qualitative Physics with other means of analysis cannot be discarded. Perhaps the better outcome is achieved when several means of analysis are simultaneously applied.

In this paper, a style of qualitative reasoning has been discussed based on Regime Analysis. Several other qualitative reasoning approaches have appeared in the literature of Artificial Intelligence. This suggests that there is a need for development of new qualitative physics methodologies, which is a challenge to the physicists, computer scientists, applied mathematicians and engineers.

As the regime analysis of a physical system is algorithmic, a symbolic system – QDR – Qualitative Dimensional Reasoner – has been developed to automate this qualitative reasoning methodology. QDR was implemented using the symbolic mathematical manipulation software REDUCE. The algebraic manipulation power of REDUCE was very important in the overall development of QDR. Although QDR runs on top of a REDUCE session, just very few and simple knowledge of it is needed to be able to profit from QDR.

From the computer science point of view, the idea of intelligent computer systems with reasoning capabilities has shown to be possible, viable and necessary. QDR gives support, among several other systems already developed, that the implementation of qualitative reasoning methods is feasible and works.

In engineering, several systems where the number of process variables is fairly large and/or no formal law exist to describe the system as a whole, can be qualitatively analysed with the aid of QDR. As possible candidates we can mention engineering plants. It might be useful also for practical applications in control and monitoring of systems, process engineering, simulations, decision making and other technological tasks. In addition, for those fields where the notion of intransmutable dimensions can be set up (meteorology, economy, ecology, etc.).

On the physics side, QDR serves in the first place as an example ponting out the importance of this new trend that is Qualitative Physics in its own, and on the other hand, the system can be used for education purposes. Particularly, in teaching the notions of dimensions (primary and secondary quantities), physical relationship discovering and prediction through process enrichment as well as physical laws consistency. For students, it can help in developing their learning skills and in their homework.

Ultimately, we hope that the ideas forwarded here serve to emphasize and stimulate educators, engineers and other researchers to the necessity of discovering and understanding the qualitatively rich behaviour of physical systems encoutered in everyday's real life problems.

6 Acknowledgments

I would like to thank Prof. B. Buchberger for the warm hospitality I received personally from him during my visit to the RISC-Linz and the Conselho Nacional de Desenvolvimento Científico e Tecnológico - CNPq, Brazil, for the financial support.

References

[1] K. D. Forbus: Qualitative process theory. Artificial Intelligence, **24**, 85-168, (1984).

[2] K. D. Forbus: Qualitative physics: past, present and future. In: D. S. Weld and J. de Kleer (eds.): Readings in qualitative reasoning about physical systems. Morgan Kaufmann Publishers, Inc. 1990, pp. 11-39.

[3] J. de Kleer and B. C. Williams: Diagnosis with behavioral modes. In: Proceedings of International Joint Conference in Artificial Intelligence, 1989, pp. 1324-1330.

[4] J. Douglass and J. W. Roach: Hoist: A second-generation expert system based on qualitative physics. AI Magazine, 108-114, 1990.

[5] D. Dvorak and B. J. Kuipers: Model-Based monitoring of dynamic systems. In: Proceedings of International Joint Conference in Artificial Intelligence, 1989, pp. 1238-1243.

[6] M. M. Kokar: Qualitative monitoring of time-varying physical systems. In: Proceedings of the 29th. IEEE Conference on Decision and Control, vol. 3, 1990, pp. 1504-1508.

[7] B. J. Kuipers: Qualitative simulation using time-scale abstraction. International Journal of AI in Engineering, 3, 185-191, 1988.

[8] B. J. Kuipers: Artificial intelligence: a new approach to modeling and control. In: Proceedings of the 1st IFAC Symposium on Modeling and Control in Biomedical Systems, Venice, Italy, 1988.

[9] D. T. Molle, B. J. Kuipers and T. F. Edgar: Qualitative modeling and simulation of dynamic systems. Computers and Chemical Engineering, 12, 853-866, 1988.

[10] M. M. Kokar: Determining arguments of invariant functional descriptions. Machine Learning, 1, 403-422, 1986.

[11] B. J. Kuipers and Y. T. Byun: A robust qualitative method for spatial learning in unknown environments. In: Proceedings of the American Association of Artificial Intelligence. Morgan Kaufman, Los Altos, CA, 1988.

[12] B. J. Kuipers and J. P. Kassirer: Causal reasoning in medicine: analysis of a protocol. Cognitive Science, 8, 363-385, 1984.

[13] P. J. Hayes: Naive physics manifesto. In: D. Michie (ed.): Expert Systems in the Micro Eletronic Age. Edinburg University Press, 1979, pp. 242-270.

[14] J. de Kleer and J. S. Brown: A qualitative physics based on confluences. Artificial Intelligence, 24, 7-83, 1984.

[15] B. J. Kuipers: Qualitative simulation. Artificial Intelligence, 29, 289-338, 1986.

[16] D. S.Weld and J. de Kleer (eds.): Readings in Qualitative Reasoning about Physical Systems. Morgan Kaufmann Publishers, Inc. 1990.

[17] R. Bhaskar and A. Nigam: qualitative physics using dimensional analysis. Artificial Intelligence, 45, 73-111, 1990.

[18] W. L. Roque: Automated qualitative reasoning with dimensional analysis. Technical report # 255, Departamento de Matemática, Universidade de Brasília, Brazil, 1991.

[19] M. M. Kokar: Critical hypersurfaces and the quantity space. In: Proceedings of American Association of Artificial Intelligence, 1987, pp. 616-620.

[20] I. Newton: Philosophiae Naturalis, Principia Mathematica II, prop. 32 (1713), ansl. A. Motte. University of California Press, Berkeley, 1946.

[21] J-B. Fourier: Théorie Analytique de la Chaleur. Gauthier-Villars, Paris, 1888.

[22] L. I. Sedov: Similarity and Dimensional Methods in Mechanics. Translated from Russian by M. Holt. Academic Press, New York, 1959.

[23] R. Kurth: Dimensional Analysis and Group Theory in Astrophysics. Pergamon Press, Oxford, 1972.

[24] J. P. Catchpole and G. Fulford: Dimensionless Groups. Ind. Eng. Chem, **58**, 46-60, 1966. G. Fulford and J. P. Catchpole: Dimemsionless groups. Ind. Eng. Chem, **60**, 71-78, 1968.

[25] J. D. Barrow and F. J. Tipler: The Anthropic Cosmological Principle. Oxford University Press, 1986.

[26] P. W. Bridgman: Dimensional Analysis. Yale University Press, 1922.

[27] S. Dobrot: On the Foundations of Dimensional Analysis. Studia Mathematica, **14**, 84-99, 1953.

[28] H. Whitney: The mathematical of physical quantities, Part I: Mathematical models for measurements. American Mathematical Monthly, **75**, 115-138, 1968. Part II: Quantity structures and dimensional analysis. American Mathematical Monthly, **75**, 227-256, 1968.

[29] E. de St Q. Issacson and M. de St Q., Issacson: Dimensional Methods in Engineering and Physics. John Wiley & Sons, New York, 1975.

[30] E. Buckingham: On physically similar systems: Illustrations of the use of dimensional equations. Physical Review, **IV**, 345-376, 1914.

[31] K. Srinivasan: Choice of vapour-compression heat pump working fluids. Int. Jour. Energy Research, **15**, 41-47, 1990.

[32] Encyclopedia of Science and Technology, 5th. edition, McGrow-Hill, 1982.

[33] A. C. Hearn: REDUCE 3.3 User's Manual. The Rand Corporation, Santa Monica, CA, 1987.

[34] M. A. H. MacCallum and F. J. Wright: Algebraic Computing with REDUCE. In: M. J. Rebouças and W. L. Roque (eds.): Proceedings of the First Brazilian School on Computer Algebra, vol. 1, Oxford University Press, 1991.

[35] M. M. Kokar: Physical similarity generalizaion rule: learning and qualitative reasoning. Pre-print Northeastern University, Boston, MA, 1991.

Qualitative Mathematical Modelling of Genetic Algorithms

R. Garigliano, D. J. Nettleton

Artificial Intelligence Systems Research Group,
School of Engineering and Computer Science,
University of Durham, UK. DH1 3LE

Abstract. Genetic algorithms are adaptive search algorithms which generate and test a population of individuals where each individual corresponds to a solution. They have been successfully applied to a range of problems in both artificial intelligence research and industry. The selection of the optimal parameters for a genetic algorithm is often a problem. This is especially true if the genetic algorithm has a protracted run-time in which case the setting of the parameters by trial and error is often unrealistic. This paper proposes the use of probability distribution functions and random walks to model various operators used in genetic algorithms. In this way it is hoped that a qualitatively accurate model with a very short run-time can be produced.

1 Introduction

The subject of computer vision is a well established area of artificial intelligence. Research is currently in progress [1] on shape representation and recognition, the aim of which is to automatically generate an encoding for a shape and successfully match it with a library of known shape encodings. The shapes are encoded using an iterated function system (IFS) representation which involves representing the shape by a collage of smaller copies of the original shape. A further explanation of IFS encoding is beyond the scope of this paper, but details may be found in [2] [3] [4]. Once encoded a shape is represented by eighteen numbers (in the current implementation) each of which have various constraints placed upon them.

One of the main problems of a shape recognition algorithm that uses an IFS representation for the shape is the sheer size and complexity of the search space in which the solutions exist. For any search algorithm in such a space there exists a fundamental trade-off between exploration and exploitation. An example of a totally exploratory algorithm would involve evaluating all possible solutions and selecting the best. This, although eventually giving the optimal solution, is far too inefficient to be of any practical use in all but the smallest of search spaces. A hill climbing algorithm on the other hand works around the best solution found so far. Although this method finds a solution in the search space, it is highly likely that in all but the simplest of search spaces the solution found will be a sub-optimal one.

It has been shown by Holland [5] that a genetic algorithm (GA) achieves a near optimal trade-off between exploration and exploitation. This was therefore chosen as the search algorithm to be used for the exploration of IFS space. As shall be seen later in this paper there are many different operators and parameters that need to be considered in the setting up and running of a GA. If (as in the case of IFSs) the run-time for the genetic algorithm is excessive (more than one hour, say), then to optimise the parameter settings by examining many different settings is not feasible. What is required is some model of a GA that simulates the convergence of the GA both quickly and accurately. This can then be run many times for various parameter combinations in a fraction of the time that it would take for all of these combinations to be ran through the full GA. The results of the model can then be used to guide the choice of an optimal set of parameters for the original GA.

Other problems that have been tackled using GAs include the travelling salesman problem, (TSP), in which the shortest route between a list of cities is required; the prisoners dilemma and the game of "Go".

2 Genetic Algorithms

Genetic algorithms are loosely based upon the Darwinian principles of biological evolution and in fact many of the strategies and operators used in their application bear similar names to their biological counterparts, e.g. survival of the fittest, crossover and mutation. A GA takes a set of possible (usually randomly generated) solutions to the problem under consideration and calls this the initial generation. Succesive generations are produced via a series of operators that act on the previous generation. This process continues until the desired number of generations has been completed. The main advantage of GAs is their ability to consider many solutions 'at the same time' and from these solutions produce other solutions that converge to the optimal.

The main operators used in GAs are those of crossover and mutation. Crossover is a way of producing new offspring via the recombination of two 'parent' solutions. Mutation prevents the irretrievable loss of any solutions by introducing random changes in some other solution. A genetic algorithm also needs a way of evaluating the fitness of a solution. In some cases of GA application the evaluation function to be used is fairly clear. For example in the travelling salesman problem, the evaluation function would be the total distance travelled between some ordering of the cities. However, in the case of many other applications of GAs the best evaluation function is not always obvious. In the case of shape representation where a numerical measure of the similarity between two shapes is needed there are many possible evaluation functions, including percentage shape coverage and the Hausdorff distance.

It would appear, at first, that a particular generation of a GA possesses only a small selection of possible solutions. However each of these solutions is made up of a string of data and within each string there are many smaller strings of data, usually called schemata. Throughout a single generation the solutions contain many schemata which can be combined using a reproductive plan to represent other individuals that are not present in the population at that time. An approach using a GA allows for the highly fit, short schemata to be propagated quickly through a population whilst at the same time considering other less fit possibilities. This is, in effect, what is responsible for the intrinsic parallelism and success of genetic algorithms.

Before discussing the details of the proposed model a brief revision of some mathematics that will be used later is given. A fuller description of which may be found in [6] and [7].

3 Probability Distribution Functions

A probability distribution function (pdf) is a way of assigning probabilities to the possible outcomes of an event. In the case of an event such as the tossing of a die, the space of possible outcomes is a discrete space, i.e. the die lands with a 1, 2, 3, 4, 5 or 6 facing upwards. Discrete cases can be handled quite easily and in the case of n equally likely outcomes the probability of any particular outcome occurring is $\frac{1}{n}$. The pdf, $f(x)$, for this would be:

$$f(A) = \begin{cases} \frac{1}{n} & \text{if } A \text{ is a possible outcome} \\ 0 & \text{otherwise} \end{cases}$$

Many random experiments, however, can be considered to have as outcomes all possible values on an interval of the line of real numbers (the real line), for example, the exact time it takes in seconds for an individual to drive to work. The outcome spaces of such measurements are said to be continuous. More formally:

Definition: An outcome space S of a random experiment is called continuous if it is composed of an interval, or a union of a countable number of intervals, of the real line.

Often, the interval is taken to have infinite length, for example, $(-\infty, \infty)$ or $(0, \infty)$, even though in reality there is a limit to the size of measurements that can be realistically taken. A continuous pdf assigns probabilities to a continuous outcome space S. In the continuous case, probabilities cannot be assigned to each possible event since any interval of the real line contains an infinite number of points and so an infinite amount of probability would be assigned which is not allowed. Instead probabilities are assigned to intervals of the real line by means of a non-negative function $f(x)$. The area under $f(x)$ between two points on the real line represents the probability associated with that interval.

Definition: A probability distribution function for a continuous outcome space S is defined by a real-valued function $f(x)$ satisfying,

1. $f(x) > 0$, for all real x.
2. The function $f(x)$ has, at most, a finite number of discontinuities on any finite interval of the real line.
3.

$$\int_{-\infty}^{\infty} f(x)dx = \int_{S} f(x)dx = 1.$$

An example of a pdf assigning probability to the interval [1,2] would be:

$$f(x) = \begin{cases} 1 & \text{if } 1 < x < 2 \\ 0 & \text{otherwise} \end{cases}$$

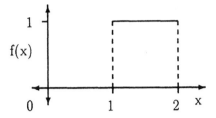

For the above pdf the probability of choosing a number in the range $[0.5, 1.5]$ is calculated as follows:

$$\int_{0.5}^{1.5} f(x)dx = \int_{0.5}^{1.0} 0dx + \int_{1.0}^{1.5} 1dx = 0 + 0.5 = 0.5.$$

Another important function associated with the pdfs, in a continuous space S, is the cumulative distribution function (cdf),

$$F(x) = \int_{-\infty}^{x} f(x)dx$$

where $f(x)$ is a pdf. $F(x)$ represents the probability of the event $\{\omega | \omega \leq x\}$. It can be seen that at all but the discontinuous points of $f(x)$, the following relationship (where the prime denotes differentiation) holds:

$$F'(x) = f(x).$$

The cdf for the example above is:

$$F(x) = \begin{cases} 0 & \text{if } x < 1, \\ x - 1 & \text{if } 1 \leq x \leq 2, \\ 1 & \text{if } x > 2. \end{cases}$$

Hence, $\int_{0.5}^{1.5} f(x)dx = F(1.5) - F(0.5) = 0.5 - 0 = 0.5$

4   Random Walks

A *stochastic process* is a system that develops in time (or space) in accordance with probabilistic laws. A random walk is an example of such a process.

A simple example of a random walk would be the path followed by a drunk who takes steps forward or staggers backwards with equal probabilities. The basic requirements for a random walk are a starting state, a range of other possible states that can be visited and a list of possible movements together with the probability of each occurring. The walk starts in the initial state, follows one of the possible movements, 'forgets' where it started (a Markov process) and takes as its new starting point, the point where the last movement left it.

In the above example, the possible movements allowed form a discrete set (a step forward or a step back); this is not always realistic and in fact the set of movements can be an infinite set. In fact, any random walk on the real line that allows movement by an amount $x \in [a, b]$ where $a, b \in \mathcal{R}$, i.e. any section of the real line, has an infinite set of possible movements. To handle the assignment of probabilities to an infinite set of possible movements, probability distribution functions are used.

An example of a random walk on the real line would be a point starting at 0 and moving in either direction by an amount less than or equal to 1, to any other point on the real line. This has two of the required elements of a random walk: a starting point and an (infinite) set of possible movements. However, the allowed movements have not yet been assigned probabilities. If starting at a point x to allow movement to any other point x', as long as $x' \in [x-1, x+1]$, a probability distribution function must be used to assign probabilities to this infinite set of points. Examples of pdfs for the random walk described above would be:

 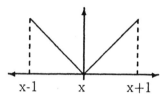

The first pdf allows movement to any other point in the range $[x - 1, x + 1]$ with equal probability. The second assigns a greater probability to larger movements than to smaller ones. To move from one point to another, a random number in the range [0,1] is generated. The new starting point of the random walk is then taken to be the point x where the cdf $F(x)$ is equal to the random number.

5   Modelling

In this section, models for various operators that can act on the solutions in any given generation shall be examined. Before moving on to the actual modelling of a genetic algorithm some assumptions shall be made about the genetic algorithm and the search space that it is to be used in. These assumptions act as a starting point for the model.

1. **Representation** – The search space can be represented as an n dimensional space, each dimension of which has some constraints placed upon it. A point in the space can then be represented by the set of coordinates $\{x_n\}$. The evaluation function takes a point in this space and returns a value for it.

2. **Optimal solutions** – The target solutions (both optimal and sub-optimal) are represented as points in n dimensional space and are weighted such that the better solutions are given a higher weight. For example, if there are two possible target solutions m and n, with weights 1 and 10 respectively, m would be a sub-optimal solution and n an optimal solution.

3. **Crossover convergence** – The overall effect of crossover on one generation is that it is more likely to reduce than increase the average distance between the solutions in that generation and the target solutions. This is a reasonable enough assumption to make since otherwise a genetic algorithm would never converge to any of the target solutions and this clearly does not happen in practice (if it did, GAs would not be used as search algorithms!).

Holland [5] identifies four components of a genetic algorithm: an environment, a set of structures, an adaptive plan and an evaluation function. The interpretation of these for the purposes of modelling is:

1. The environment under consideration is not specific to any particular problem. An aim of the model is that it can be applied to many different environments.
2. The set of structures used are points in n dimensional space. A particular generation consists of a fixed number of such points — one for each individual.
3. The adaptive plan is the process by which operators act on one generation to produce the next. There are many operators that can be used in this process and a means of modelling them shall be discussed in this document.
4. The evaluation function, when applied to one of the structures (solutions) gives a measure of how 'good' that solution is. Since the structures in the model are points in n dimensional space, the evaluation function shall be taken as some form of distance measure between a point and the target solutions.

5.1   Hill climbing model

Hill climbing is not traditionally an operator that is used in genetic algorithms. However, certain applications of GAs are using a form of it in conjunction with the other better known genetic operators of mutation and crossover.

Consider the case of a smooth function that takes one parameter and returns a number. If the function has a single maximum then a hill climbing algorithm will produce solutions that are progressively closer to that maximum. It is then possible to take two parameters, one of which is slightly smaller than the original value the other slightly greater, and evaluate them using the function. If one of the evaluated values is greater than the evaluated value of the original, then this value is taken as the new starting value. If both evaluated values are less than that of the original then the original value is retained. In the diagram below, it can be seen that a hill climbing algorithm would move from the point X to the point a since $f(a) > f(X) > f(b)$.

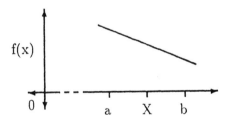

The model under development starts with a random set of allowed solutions and a set of target solutions (both sub-optimal and optimal), all represented as points in n dimensional space. A hill climbing algorithm acting on this space can be thought of as an algorithm that attracts all of the points in the space to a selection of sub-optimal and optimal solutions. Once a solution starts to 'climb' toward a particular value it will, if not interfered with, continue to do so until it reaches that value. A simple way of modelling this is to calculate the Euclidean distance between each of the randomly generated points and each of the target solutions. Then let the target solution, the one which a particular solution 'climbs' toward, be the closest target to that solution. The actual amount of climb is calculated using a random walk with a probability distribution function that allows only a reduction in the distance between the solution and its target solution and does not allow the solution to 'overshoot' the target. So, if a solution is represented by the point $\{x_n\}$ in n dimensional space, the hill climbing operator model produces a new solution at $\{x'_n\}$ as follows:

1. Calculate the Euclidean distance d between the solution under consideration and the closest target solution to it.
2. Generate a random number $p \in [0, 1]$.
3. The new distance of the solution from the target solution is the value d' such that,

$$\int_0^{d'} f(x)dx = p \quad \text{or} \quad F(d') = p$$

where $f(x)$ is the pdf representing the hill climbing operator and $F(x)$ is the corresponding cdf.
4. Calculate the new coordinates $\{x'_n\}$ of the solution by assuming that it moves on a straight line towards the target solution until a distance d' from it, i.e.

$$x'_n = x_n \left(\frac{d'}{d}\right) + \left(1 - \frac{d'}{d}\right)t^n$$

where t^n is the coordinates of the target solution.

The modelling of the hill climbing operator is now reduced to the problem of selecting a suitable pdf. In the present implementation, a pdf that assigns equal probability to a range of movements is used. Each movement results in the solution moving closer to its target by no more than a fixed amount and does not allow the solution to 'overshoot' the target.

In more complex search spaces (such as IFS space), there may be many local maxima that correspond to sub-optimal solutions. Indeed there may also be more than one optimal solution. In this case a pure hill climbing approach would lead to a final generation containing many sub-optimal solutions, with no guarantee that one of the optimal solutions will be among them. For this reason it is not sufficient to consider hill climbing alone.

5.2 Mutation model

The purpose of mutation is to ensure that no schemata (partial solutions) are permanently lost from the pool of schemata that a generation can contain. Mutation is usually kept as a background process and so the probability of it occurring is kept small. In the implementation used, if the mutation operator is invoked on the point $\{x_n\}$ then x_i where $i \in [1, 2, .., n]$ is replaced by a random number that is in valid range for dimension i.

5.3 Abstract Crossover model

As the name suggests, this is a model of the crossover operator at a very abstract level. Consider the case of a very simple search space, with just one target solution. From the assumption on crossover, as progress is made from one generation to the next, the average distance of the solutions from the target solutions will decrease. However, it is also quite possible that some solutions will be further away from the target solutions than any of those in the previous generation. To model this, an overall standard probability of improvement (SPI) is assigned to the crossover operator. The larger the value of the SPI, the quicker the crossover operator converges. If the SPI is less than 0.5, the crossover operator will not converge to any solution since this means that it is more likely for solutions to move further away from the target solution than toward it. The standard probability of improvement for the pdf $f(x)$ for a point a distance d from a target solution is calculated using:

$$\text{SPI} = \frac{\int_0^d (d-x)f(x)dx}{\int_0^\infty |x-d|f(x)dx}$$

The abstract crossover model in an n dimensional space with one target solution produces a new generation from a previous one as follows:

1. Select a parent solution from the last generation.
2. Calculate its distance d from the target solution.
3. Generate a random number $p \in [0,1]$.
4. The distance of the child solution from the target solution is the value d' such that,

$$\int_0^{d'} f(x)dx = p \quad \text{or} \quad F(d') = p$$

where $f(x)$ is the pdf representing the crossover operator and $F(x)$ is the corresponding cdf.

5. The coordinates of the child solution are then chosen at random with the proviso that they must be at a distance d' from the target solution.
6. The child solution is then placed in the next generation.
7. Repeat for every element of the last generation.

For ease of explanation, only one target solution has been considered. However, any smooth search space with just one optimal solution can be searched more efficiently using a pure hill climbing algorithm than by using a GA involving crossover. For a model of a GA to be of use in solution spaces, with many sub-optimal and possibly multiple optimal solutions, a more complex abstraction of the crossover operator is needed. One way of doing this is to weight the solutions in such a way that the optimal solutions are more likely to be picked as the target that a particular solution in a population converges toward. This can be done at the stage at which the model is deciding which of the target solutions is closest. By dividing the distance of the solution from each target solution by the weight of that target solution, the solution is more likely to favour convergence to a target solution with a higher weight.

As in the modelling of hill climbing, the problem now reduces to the selection of a suitable pdf for the crossover operator. One simple example of a pdf for a solution at a distance d from a target solution is:

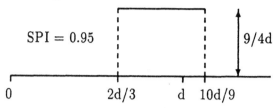

The implementation of these operators is now considered.

6 Implementation

The model described in the previous section has been implemented in the programming language C. The program reads in population data together with data on the optimal and sub-optimal solutions. The population data may either be randomly produced (within dimensional constraints) or can be the population produced at the end of some previous program run. The main program parameters are:

POP – population size
GEN – generation size.
OPT – number of optimal and sub-optimal solutions.
DIM – dimension of the search space that the solutions are in.

There are other parameters that can be set, depending on which operators are under consideration in that program run – these are mentioned later.

As stated earlier, it is necessary for the the model to be qualitatively correct as well as have short run-times. When the program was run on a Sun-4, the following cpu times (in seconds) for various parameter combinations were obtained.

POP	GEN	OPT	DIM	time
100	100	4	2	4
100	100	20	2	15
100	100	20	10	47
100	100	50	18	165

These run-times are considerably less than the run-times of the GA (with similar parameter settings) which is used in searching IFS space. However, these fast run-times would be of no use if the results of the model did not agree with those of real GAs. In the next section the results of the model are studied.

7 Results

7.1 Hill climbing

An approach using just the hill climbing model was considered. The pdf used was rectangular and allowed the solutions to move closer by no more than a small fixed amount. Initially, a population of 100 was randomly generated in two dimensional space with each coordinate constrained to be in the range $[-50, 50]$. Four optimal solutions A, B, C, D were chosen with positions $(10, 0), (0, 10), (-10, 0)$ and $(0, -10)$ respectively. As expected the solutions converged (if the Euclidean distance between the solution and A was less than 1.0) to the optimal they started closest to. The final result was that roughly equivalent numbers of solutions converged to each optimal. The table below shows the numbers of solutions that converged to one of the target solutions after a fixed number of generations:

GEN	0	10	20	30	40	50	60	70	80	90	100
A	0	0	0	1	4	12	21	26	26	26	26
B	0	0	0	2	2	7	24	27	27	27	27
C	0	0	1	1	4	10	23	27	27	27	27
D	1	1	1	1	1	9	17	20	20	20	20

This was then extended to eighteen dimensions with more sub-optimal and optimal solutions (fifty in all). The final result was again that roughly equal numbers of solutions converged to each of the target solutions regardless of whether they were optimal or sub-optimal. This is exactly what would be expected of a pure hill climbing approach.

7.2 Abstract Crossover

A genetic algorithm using only the abstract crossover model to produce succesive generations was considered next. A rectangular pdf, similar to that described previously was used, with a standard probability of improvement (SPI) of 0.65. A search space identical to that in the previous section was considered (i.e. POP=100, OPT=4, and DIM=2). In this case the optimal solution A was given a weight of 10 whilst the others were left with weight 1, this models B, C, D as sub-optimal solutions. The results achieved were as follows:

GEN	0	10	20	30	40	50	60	70	80	90	100
A	0	0	0	0	1	1	4	11	30	40	52
B	0	0	3	5	6	7	8	8	8	8	8
C	0	2	4	7	10	12	12	13	13	13	13
D	1	2	2	3	3	4	4	4	4	4	4

From this data it can be seen that more solutions converged to the optimal solution A than to the other sub-optimal solutions. This is the behaviour that would be expected with a crossover operator acting in a space containing both optimal and sub-optimal solutions. Another interesting point to note is that not all of the solutions have converged (only 77 of them). This is not surprising since the model used was given an SPI of only 0.65. For a greater value more solutions converge to the optimal and for smaller values fewer converge.

A search space with four different optimal solutions and no sub-optimal solutions was examined. The final generation contained solutions that had converged in equal numbers (± 5) to each of the optimals. This indicates that, for identical optimal solutions, the crossover operator used cannot tell between equally good solutions. This is not really a problem provided convergence to optimal solutions still occurs.

7.3 Hill-Climbing and Crossover

The effect of combining the hill climbing and abstract crossover models is now considered. This requires another parameter which shall be called RAT (short for ratio) and is the ratio of the number of times the hill climbing operator was applied before applying crossover. For example if RAT was chosen to be two, two successive generations would be produced using hill climbing and the next by crossover, the whole process then being repeated. A search space identical to that used in the previous section was used with the additional parameter RAT set at 3. The following results were then achieved:

GEN	0	10	20	30	40	50	60	70	80	90	100
A	0	0	4	4	24	49	68	82	82	83	83
B	0	0	4	5	5	5	5	5	5	5	5
C	0	1	2	4	5	5	5	5	5	5	5
D	1	1	3	7	7	7	7	7	7	7	7

Using a combination of hill climbing and crossover led to the best results so far obtained. For very large values of RAT the solution convergence tends to that of pure hill climbing. Similarily when RAT is 1 results similar to those for pure crossover were achieved. For values of RAT in between these two extremities the amount of convergence to the unique optimal solution (as opposed to a sub-optimal one) varied.

A search space containing 19 sub-optimal solutions and one optimal solution in a 10 dimensional space was then considered. All the other parameters were left unchanged i.e. POP=100, GEN=100, RAT=3 and SPI=0.65. In this case similar behaviour occurred and by the end of 100 generations 90% of the solutions had converged to the optimal.

This section has briefly examined some ways in which the model can compare different combinations of operators that can be used in GAs. In the next section some other possible uses for the model are briefly discussed.

8 Further work

Selection of the probability distribution function — So far only simple rectangular probability distribution functions have been considered. There are clearly many other discontinuous probability distribution functions that could be used as well as continuous pdfs such as modified normal distributions. Once suitable pdfs have been chosen for the operators in the GA under consideration, the model can be used. It is hoped that a list of suitable pdfs can be produced, from which the appropriate one may be chosen.

Distance measure — In the model described the Euclidean distance between two points in n dimensional space has been used. In some search spaces other distance measures may be more appropriate. Fortunately there are many other distance measures that can be used, some of which may be more suited to some problems than to others.

Reducing search space — In some applications of GAs the search space can be reduced beforehand by the imposition of constraints (IFS space being an example).

The model described here can be used to examine the effect of reducing the search space on the speed of convergence.

Crossover operators — The crossover operator used here is very abstract in nature. In practice crossover operators involving one-point and two-point crossover [8] are often used. The modelling of such operators is possible within the framework provided here. The effect of using different crossover operators and different reproductive plans (the method of selecting parent solutions) on a problem can then be considered.

9 Conclusion

This paper has examined the possibility of modelling genetic algorithms using probability distribution functions and random walks. The model proposed has a short run-time when compared with that of a full genetic algorithm and the results produced so far have been qualitatively correct. The main difficulty remaining is in the selection of suitable probability distribution functions to model the various operators. However, the model described provides a good foundation on which to base a qualitatively accurate mathematical model of genetic algorithms.

References

1. R. Garigliano, D.J. Nettleton, Shape Representation and Recognition using Iterated Function Systems and Genetic Algorithms. Technical Report 7/92, Computer Science, University of Durham, England (1992)
2. M.F. Barnsley: Fractals Everywhere. Academic Press (1988)
3. P.A. Giles: Iterated Function Systems and Shape Representation. PhD Thesis, University of Durham (1990)
4. R. Garigliano, A. Purvis, P.A. Giles, D.J. Nettleton: Genetic Algorithms and Shape Representation. to be published in Proceedings of Second Annual Conference on Evolutionary Programming, San Diego, USA (1993)
5. J.H. Holland: Adaption in Natural and Artificial Systems. University of Michigan Press (1975)
6. G.R. Grimmett, D.R. Stirzaker: Probability and Random Processes. Clarendon Press (1982)
7. K.L. Chung: Elementary Probability Theory with Stochastic Processes: Springer-Verlag (1982)
8. L. Booker: Improving Search in Genetic Algorithms. in Genetic Algorithms and Simulated Annealing, ed Davis L., Pitman (1987)
9. L. Davis M. Steenstrup: Genetic Algorithms and Simulated Annealing: An Overview. in Genetic Algorithms and Simulated Annealing, ed Davis L., Pitman (1987)

Springer-Verlag
and the Environment

We at Springer-Verlag firmly believe that an international science publisher has a special obligation to the environment, and our corporate policies consistently reflect this conviction.

We also expect our business partners – paper mills, printers, packaging manufacturers, etc. – to commit themselves to using environmentally friendly materials and production processes.

The paper in this book is made from low- or no-chlorine pulp and is acid free, in conformance with international standards for paper permanency.

Lecture Notes in Computer Science

For information about Vols. 1–665
please contact your bookseller or Springer-Verlag